T0141662

DENDROCHILUM
OF BORNEO

Dedicated to the memory of

Gunnar Seidenfaden
(1908–2001)

whose example has been
a constant source of inspiration.

Mount Kinabalu is the richest locality for *Dendrochilum* diversity in Borneo, from where 32 species are recorded, representing 39.5% of the Bornean total. Of these, twelve species and two varieties are endemic to the mountain. (Photo: Cede Prudente)

DENDROCHILUM OF BORNEO

Jeffrey J. Wood

Line illustrations by

Susanna Stuart-Smith

Eleanor Catherine, Maureen Church, Linda Gurr,
Judi Stone, and Oliver Whalley

Photographs by

Todd J. Barkman, Kath Barrett, C.L. Chan, James B. Comber,
Phillip Cribb, Gerald Cubitt, Paul Davies, Hans P. Hazebroek,
Peter Jongejan, Anthony Lamb, Gwilym Lewis, Robert New,
Cede Prudente, Takashi Sato, and André Schuiteman

Natural History Publications (Borneo)
Kota Kinabalu

in association with

2001

"It would seem that Borneo will rival the Philippines as a home of *Dendrochilum* species and furnish material that will help to crystalize our knowledge of one of the most fascinating orchid genera of the eastern tropics."

Oakes Ames (1920).

Published by

Natural History Publications (Borneo) Sdn. Bhd. (216807-X)
A913, 9th Floor, Wisma Merdeka
P.O. Box 15566
88864 Kota Kinabalu, Sabah, Malaysia
Tel: 088-233098 Fax: 088-240768
e-mail: chewlun@tm.net.my

in association with

The Royal Botanic Gardens
Kew, Richmond
Surrey TW9 3AB, England

Dendrochilum of Borneo
by Jeffrey J. Wood

General Editor: P.J. Cribb

ISBN 983-812-047-2

Design and layout by C.L. Chan

Half title page: *Dendrochilum pterogyne* (Photo: T. Sato)

Printed in Malaysia.

CONTENTS

LIST OF FIGURES

(including plates depicting herbarium material of *D. pallidiflavens* var. *pallidiflavens*)

Taxa listed alphabetically (Line drawings only)

viii

LIST OF TABLES

LIST OF GEOGRAPHICAL DISTRIBUTION MAPS

NEW TAXA

D. gracile (Hook.f.) J.J. Sm. var. **bicornutum** J.J. Wood
D. haslamii Ames var. **quadrilobum** J.J. Wood
D. jiewhoei J.J. Wood
D. kingii (Hook.f.) J.J. Sm. var. **tenuichilum** J.J. Wood
D. pallidiflavens Blume var. **oblongum** J.J. Wood

NEW COMBINATIONS AND TAXA GIVEN NEW STATUS

D. murudense (J.J. Wood) J.J. Wood
D. pallidiflavens Blume var. **brevilabratum** (Rendle) J.J. Wood
D. tenompokense Carr var. **papillilabium** (J.J. Wood) J.J. Wood

NEW SYNONYMS

D. brevilabratum (Rendle) Pfitzer var. *petiolatum* = **D. pallidiflavens** Blume var.
pallidiflavens
D. bulbophylloides Schltr. = **D. pallidiflavens** Blume var. **pallidiflavens**
D. conopseum Ridl. = **D. pallidiflavens** Blume var. **pallidiflavens**
D. dewindtianum W.W. Sm. var. *sarawakense* Carr = **D. dewindtianum** W.W. Sm.
D. ellipticum Ridl. = **D. pallidiflavens** Blume var. **pallidiflavens**
D. intermedium Ridl. = **D. pallidiflavens** Blume var. **pallidiflavens**
D. meijeri J.J. Wood = **D. dolichobrachium** (Schltr.) Merr.
D. micranthum Schltr. = **D. pallidiflavens** Blume var. **pallidiflavens**
D. spathulatum J.J. Sm. = **D. pallidiflavens** Blume var. **pallidiflavens**
D. tardum J.J. Sm. = **D. gracile** (Hook.f.) J.J. Sm. var. **gracile**

Fig. 1. Mount Batu Lawi, an isolated sandstone pillar in the Kelabit Highlands, Sarawak, is the type locality of *D. longipes*; several other species occur in the vicinity, including *D. geesinkii* and *D. simplex*. (Photo: A. Lamb)

INTRODUCTION

*D*endrochilum (subtribe Coelogyninae) was established by Carl Ludwig von Blume (1796–1862) in 1825 who subdivided it into two unnamed sections. Two monographs have appeared subsequently, *viz.* J.J. Smith (1904) and Pfitzer & Kraenzlin (1907), both of which are now inadequate since over 65% of the currently recognised species had yet to be described at that time. Consequently, the generic subdivisions they proposed do not reflect the full spectrum of variation known today. The total number of species recognised today is 263. Their distribution conforms to the Indo-Malesian pattern as defined by van Steenis (1979: 112). The subgenera and sections occupy variously delimited areas within the total geographical range. *Dendrochilum* is distributed from Myanmar (Burma) in the west, south-eastwards across peninsular Thailand, Peninsular Malaysia and Singapore to Borneo, south through the islands of the Indonesian Archipelago, north to Taiwan and the Philippines, and east as far as New Britain in the Bismarck Archipelago of Papua New Guinea. Centres of diversity are located in Sumatra, Borneo and the Philippines. A revised subdivision proposed by Pedersen, Wood & Comber (1997) recognises four subgenera and thirteen sections within *Dendrochilum*. Of these, all of the subgenera and six sections are represented in Borneo. This system is followed here.

Most species of *Dendrochilum* inhabit cool, humid and often exposed habitats in lower and upper montane forest. This is mirrored by the high diversity encountered in the mountains of Sumatra, Borneo and the Philippines. Areas of more modest elevation, e.g., Peninsular Malaysia, or those with a distinct dry season, e.g., East Java and the Lesser Sunda Islands, are relatively poor in species. The geographical isolation of populations, resulting from mostly montane preferences may help to explain the very high percentage of endemicity within the genus. For example, 91.9% of the Bornean taxa are endemic. Of the 87 taxa recorded, 33 (37.9%) occur on Mount Kinabalu (34 if Mount Tembuyuken is included). Most of these endemic high-elevation species are probably of recent and local origin (see section on orographical and climatic correlations).

Only two (probably conspecific) species have been recorded from New Guinea, a mountainous island which would appear to offer conditions conducive to diversification. *Dendrochilum*, however, is almost exclusively restricted to areas belonging to the Sunda Shelf, including Borneo, whereas areas of very low species diversity, such as eastern Sulawesi and New Guinea, form part of the Sahul Shelf. Such paucity may possibly be explained, providing that the genus is sufficiently old, by the biogeographical theory first postulated by Alfred Russell Wallace in 1863. Effective long-range dispersal of dust-like orchid seed by wind has not resulted in the subsequent successful colonisation of montane New Guinea. This is perhaps not surprising given the vulnerability of orchid seed which lacks endosperm and is very susceptible to dessication. In addition, mycorrhizal relationships in many *Dendrochilum* may be very host specific and certain symbiotic fungal partners are absent in New Guinea.

As one would expect, several lower elevation taxa have a wide distribution outside Borneo, including *D. kingii* var. *kingii*, *D. longifolium* and *D. pallidiflavens* var. *pallidiflavens*. Curiously, other low elevation taxa such as *D. pallidiflavens* var. *brevilabratum, D. havilandii, D. oxylobum, D. pubescens* and *D. simplex*, appear to be endemic. A similar pattern is found in *Pholidota*.

MORPHOLOGICAL CHARACTERS

A mix of vegetative and reproductive characters have been used for this study. Particular attention, however, has been paid to characters of the column which are generally less variable than most others.

Habit

Two predominant sympodial growth-forms can be recognised in *Dendrochilum*: the first, pendent or creeping plants with elongate rhizomes and widely spaced pseudobulbs, and the second, densely to loosely caespitose plants with short to slightly elongate rhizomes and more or less clustered, and often approximate, pseudobulbs. The first type of habit is universal in subgenus *Dendrochilum*, e.g., *D. crassum* and *D. gravenhorstii*, and scattered within subgenus *Platyclinis* sections *Eurybrachium* (non-Bornean species only) and *Platyclinis*, e.g., *D. longipes* and *D. simplex*. Pseudobulbs may be spaced as much as 12 cm apart in the Bornean *D. subintegrum* (section *Platyclinis*). Species with this habit generally prefer rather sheltered habitats and are often pendent epiphytes, tangled scramblers over rocks and other vegetation or, more rarely, forest floor terrestrials. The second type of habit, however, predominates throughout section *Platyclinis*. The dense tufted habit is clearly advantageous to the numerous species inhabiting more exposed sites such as canopy branches and twigs, or windswept rock faces. The narrow leaved, lithophytic *D. stachyodes* is a perfect example.

Rhizome

The rhizome is usually more or less branched and is initially covered with imbricate, scarious cataphylls which, after some time, disintegrate into fibres and finally disappear. The internodes below each leaf-and/or inflorescence-producing shoot are very short, the cataphylls produced from the nodes between being accordingly abbreviated and covering the developing, leaf-producing or potentially flowering, shoot.

Roots

The roots appear from the rhizome and sometimes from the bases of the pseudobulbs. They may be short or elongate, simple or sparsely to densely branching. They vary in diameter, are usually filiform or sometimes thick and fleshy. The variation in these characters occurs within and between species. The velamen may be either smooth and glabrous, minutely papillose or, very rarely, hirsute. Root hairs are generally restricted to points of contact with the substrate. Root type appears to be species specific.

Cataphylls

The young developing pseudobulbs are subtended by one or more imbricate, tubular cataphylls, the number varying between species. These soon disintegrate into fibres that either decay immediately or are persistent. The mature cataphylls, particularly the lowermost, may often have a long cleft along one side. In most species this character is too variable, even on the same specimen, to be useful. In some species, the cataphyll surface may be mottled or speckled with darker pigment, e.g., *D. imitator* and *D. lacinilobum*, or, as in *D. subintegrum*, have deeper pigmented apical margins.

The condition of only one cataphyll subtending each pseudobulb was applied by Pfitzer & Kraenzlin (1907: 87) as a diagnostic character to define a new subgenus *Monochlamys*. Subsequent observations have shown that the number of subtending cataphylls is not constant and has little significance in a natural subdivision.

Pseudobulbs

The pseudobulbs, which are well-developed and consist of only one internode, vary a great deal in size and are always unifoliate. Although a considerable amount of intraspecific variation occurs, the shape of the pseudobulbs remains a useful character in distinguishing some species. For example, very narrowly cylindrical, pencil-like pseudobulbs are typical of *D. gracilipes* and *D. hologyne*. The shape in most other taxa, however, is predominantly globose, elliptic, fusiform or narrowly conical. In *D. pubescens* they are often somewhat compressed and adpressed to the rhizome. The surface of the young pseudobulbs in the fresh condition is generally smooth, although it may sometimes be wrinkled, notably in *D. corrugatum*. When dessicated the surface becomes furrowed or wrinkled, the pattern usually being constant in a particular species but variable within each section. The pigmentation of the pseudobulbs is sometimes of diagnostic value. Many high elevation species in section *Eurybrachium*, e.g., *D. alatum* and *D. corrugatum*, have orange or red pseudobulbs.

Vernation

The folding or rolling of the individual leaves in the bud stage is significant at subgeneric and sectional level. The majority of species in subgenera *Acoridium* and *Platyclinis* have a convolute vernation, while in subgenus *Dendrochilum* it is conduplicate. Both vernation types are found in subgenus *Platyclinis* section *Eurybrachium*. In subgenus *Acoridium* section *Falsiloba*, which is endemic to Borneo, the vernation type is conduplicate. The peduncle is entirely free from the subtending leaf at the time of flowering in section *Falsiloba* and the two species of section *Acoridium* represented in Borneo. The remaining species of section *Acoridium*, all of which are native to the Philippines, have the peduncle tightly enclosed, to a greater or lesser extent, by the unexpanded leaf, or borne through a median slit of the fully expanded leaf. Conduplicate vernation facilitates this more specialised condition and it would seem reasonable to regard it as an apomorphic character state. Convolute vernation, on the other hand, appears, at least in subgenus *Acoridium*, to represent the plesiomorphic condition.

Leaves

The leaves of *Dendrochilum*, each borne at the apex of a pseudobulb, are articulate, petiolate, rarely almost sessile, and always more or less xeromorphic, as demonstrated by Rosinski (1992: 97). An abscission layer is present between the leaf and the apex of the pseudobulb. The majority of species have a dorsiventrally complanate leaf blade. In Borneo the semi-terete state found in many Philippine species of section *Acoridium*, a result of extreme xeromorphic adaptation, is absent. Among Bornean species, however, *D. pachyphyllum* (section *Platyclinis*) is unique in having thick, rigid, fleshy leaves that are shallowly v-shaped in cross-section. The outline of the leaf blade in Bornean taxa is most commonly narrowly elliptic or linear-ligulate. Leaf width is often variable within a single species, e.g., *D. gibbsiae*, and is of little use as a diagnostic character in Bornean taxa. An exception is *D. johannis-winkleri* where the leaf blade is often broadest distally. The leaf apex varies from shallowly retuse, e.g., *D. johannis-winkleri*, or obtuse and mucronate, e.g., *D. pandurichilum*, to narrowly acute, e.g., *D. imitator*, but is never unequally bilobed (see Fig. 2). In many species the distal leaf margin is distinctly constricted at varying distances below the apex (see Fig. 2D & E). This may possibly function as a drip tip, a feature of many plants of high rainfall areas in the tropics.

Leaf texture varies from thin, e.g., *D. angustipetalum*, to tough and leathery, e.g., *D. crassifolium*. Texture is usually relatively constant per species. Most species have smooth, glabrous leaf blades, although there are exceptions. The leaves of *D. ochrolabium*, for example, are very minutely black punctate-ramentaceous, particularly on the lower surface. Those of *D. pubescens* and *D. vestitum* are unique among Bornean species in having a sparse to dense, short, black or brown, furfuraceous-pubescence composed of simple or branching trichomes (see Fig. 132). This, however, may gradually wear off, especially in *D. vestitum*, as the leaf matures. Petioles vary from very short to elongate and are always sulcate or caniculate.

Each leaf contains a prominent, raised mid nerve in addition to from two to ten distinct and from several to many indistinct, parallel to curved, lateral nerves. Horizontal secondary nerves may also be present, e.g., *D. grandiflorum*. *D. minimiflorum* has very narrow, one-nerved leaves.

A characteristic feature of certain members of subgenus *Platyclinis* section *Cruciformia*, e.g., *D. gibbsiae* and *D. hastilobum*, are the numerous crystalline calcium oxalate bodies present in the leaf blades. These take the form of small crystals, styloids and druses deposited in the epidermal cells. Their shape may be simple, or complicated, sometimes resembling an

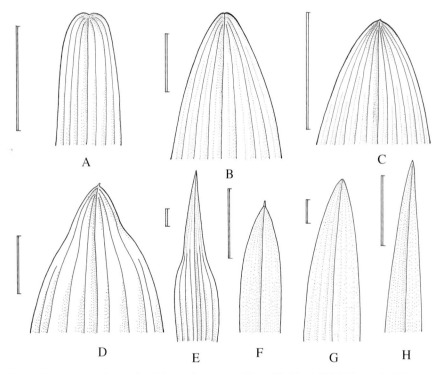

Fig. 2. Leaf apex shape. **A** *Dendrochilum johannis-winkleri, Winkler* 1495 (holotype); **B** *D. gracile* var. *gracile*, cult. Bogor, ex *Hallier* (holotype of *D. tardum*); **C** *D. angustilobum, Carr* C. 3233, SFN 26874 (holotype); **D** *D. murudense, Nooteboom & Chai* 01995 (holotype of *D. crassifolium* var. *murudense*); **E** *D. kingii* var. *kingii, Hallier* 1312 (syntype of *D. bigibbosum*); **F** *D. cupulatum, Hansen* 498; **G** *D. havilandii, Yii* S.39541; **H** *D. imitator, de Vogel* 8451 (holotype).Scale: double bar = 1 cm. Drawn by Susanna Stuart-Smith.

asymmetrical snowflake. Sometimes few in number, they more commonly occur densely distributed over the entire surface and are clearly visible in dried and spirit material.

Inflorescence

In *Dendrochilum* inflorescence type is of prime importance at subgeneric and, to a lesser extent, sectional level. In the sections of subgenus *Acoridium* represented in Borneo, and in subgenus *Platyclinis*, the inflorescence is synanthous. Here the inflorescence is produced at the apex of the pseudobulb and simultaneously with its growth. In subgenus *Dendrochilum* the inflorescence is heteranthous with the production of an independent specialised flowering shoot which develops rudimentary pseudobulbs and leaves only. The non-flowering leafy pseudobulbs are produced separately at the base of which are borne the flowering shoots. The heteranthous inflorescence is a development from the proteranthous condition found, for example, in some species of *Chelonistele*, in which the growth of the pseudobulbs and leaves on the flowering shoots is deferred until after flowering.

All species of *Dendrochilum* have a racemose inflorescence which is differentiated into a peduncle and a rachis. The length of the inflorescence expressed as the rachis to peduncle ratio is often of diagnostic value. The rachis is usually separated from the peduncle by one or more imbricate, adpressed, non-floriferous bracts. These are lacking in subgenus *Dendrochilum*.

Peduncle

The peduncle is terete (laterally compressed in *D. planiscapum*), naked along its length, usually very slender and often filiform, or rarely a little stouter, e.g., *D. hosei*. In the majority of species the weight of the flowers on the rachis causes the peduncle to curve or arch. Although generally glabrous, there may sometimes be a few scattered trichomes or, more rarely, the surface is pubescent (in *D. pubescens* and *D. vestitum* only). The peduncle is generally very short in subgenus *Dendrochilum*.

Non-floriferous bracts

The presence or absence of adpressed non-floriferous bracts borne at the apex of the peduncle at the junction with the rachis is sometimes useful in distinguishing certain species, although this may not always be constant. In some species they are solitary, in others there may be several in number which are imbricate. In a few species the number of bracts varies from specimen to specimen. This is particularly true of the widespread and variable *D. gracile*.

Rachis

The rachis is few- to many-flowered, lax to dense, always either obscurely to strongly quadrangular in transverse section, and often concave to channelled along two opposite sides. The flowers normally alternate distichously, but the rachis axis may sometimes be spirally twisted, the flowers therefore facing all sides (quaquaversal). The surface is usually provided with minute, scattered, brown or blackish trichomes, while in *D. pubescens* and *D. vestitum* it is densely pubescent.

The weight of the flowers combined with an often slender peduncle causes the rachis to become pendent, although erect rachises are not uncommon. Some species, such as *D. dulitense*, have a very short rachis, while others like *D. longirachis*, may measure 45 cm or more in length.

Bornean species, in common with most other *Dendrochilum*, normally commence flowering from the basal part of the rachis. Exceptions include *D. hosei* where the central flowers appear to open first, and *D. joclemensii, D. pandurichilum* and *D. trusmadiense* which commence flowering from the top portion of the rachis.

Floral bracts

The floral bracts are persistent, glumaceous, and variable in size, but usually broad and shallowly concave to cymbiform. They are often many-nerved, the nerves usually prominent and raised on the lower surface. The margin varies from entire to distinctly erose. A scarious texture is predominant, although some are membranous (especially in subgenus *Dendrochilum*).

The bracts are fully developed prior to the flowers they subtend thereby providing effective protection to the buds. The morphology is very variable within most species, often in a single inflorescence, and consequently of little taxonomic importance. In a few species though, the floral bracts are highly distinctive, e.g., *D. glumaceum* and *D. imbricatum*. Those of *D. imbricatum* are large enough to protect the fully expanded flowers.

Indumentum

In *Dendrochilum* a scattered ramentaceous or 'finely setose' indumentum composed of one- to three-celled trichomes is found on the cataphylls, leaves, inflorescences, bracts, and sometimes the sepals, petals and ovary. In addition, patches of sparse, short hairs occur toward the base of the sepals and petals in many species. *D. pubescens* and *D. vestitum* are notable in having a dense furfuraceous indumentum. In *D. anomalum* the ovary is somewhat papillose-hairy.

General floral characters

The flowers of *Dendrochilum* are long-lasting and always open simultaneously on the inflorescence, despite differences in flowering sequence. Although generally non-resupinate, many of them often assume a position with the labellum pointing downwards as the inflorescence elongates, curves forward and the peduncle and rachis become pendent. A few species with erect inflorescences, e.g., *D. cupulatum*, have resupinate flowers.

Although the flowers are modest in size, their number per inflorescence is often large. Many of the clump-forming species, such as *D. graminoides, D. kamborangense* and *D. trusmadiense*, produce a showy floral display of graceful appearance. The smallest-flowered Bornean species, *D. microscopicum*, has flowers barely 2 mm across, while the recently discovered *D. flos-susannae* is the largest with flowers measuring up to *c.* 1.8 cm long. Many species have a distinct scent, often reminiscent of cucumber, while others have no discernible fragrance. Flower colour is variable between and within species. In Bornean species colouring varies from mostly pale green, yellowish green and lemon-yellow to white, with a labellum of a contrasting darker, usually brown or ochre, colour. Flesh-coloured to salmon-pink or brownish ochre flowers are also widespread, while pure yellow or red are very rare.

Sepals and petals

The sepals and petals often provide useful characters at specific level. These may be gently recurved or incurved and, unlike certain species of *Pholidota*, are never bent at right angles. In a few high elevation species of section *Eurybrachium*, e.g., *D. alpinum* and *D. pseudoscriptum*, the petals are twisted to become aligned at 90° from the vertical. The reason for this is unclear. The sepals are free in all Bornean species, while the petals are always free throughout the genus. The outline of both may vary from linear to almost orbicular, the sepals, in particular, often being concave to a varying degree. The apex of both may be rounded or acute to acuminate, although the petals are more often less acute than the sepals. The sepals are always entire, whereas the petals are frequently erose to fimbriate or irregularly serrate, e.g., *D. corrugatum*, *D. crassifolium* and *D. kamborangense*. The sepals and petals are 1- or 3-nerved from the base. The nerves are often sparsely branched.

D. papillitepalum is unique among Bornean species in having a sparse covering of short papillae on the upper surface of the sepals and petals.

Labellum

The morphology of the labellum is very diverse and provides important characters at subgeneric, and especially sectional and specific level. The nature of the point of attachment of the labellum, ignored by earlier authors, is of great importance. Sessile or shortly clawed, it is either firmly (in sections *Acoridium*, *Cruciformia*, *Eurybrachium* and *Mammosa*) or elastically attached (in subgenus *Dendrochilum* and section *Platyclinis*) to the column. Either way, the point of attachment is a column-foot or the base of a footless column. The role of an elastically attached labellum (or epichile only) in specialised pollination mechanisms is well documented (e.g. Darwin 1877, van der Pijl & Dodson 1966, Nilsson 1978). The elastically attached labellum represents the apomorphic condition in *Dendrochilum*. The presence or absence of a column-foot is of significance at subgeneric and sectional level.

The outline of the labellum of Bornean species is less varied than those of the Philippines where section *Acoridium* has diversified and additional subgenera and sections are also present. Although intermediate forms do occur, six main types of labellum outline can be distinguished among Bornean taxa:

1. Entire to subpandurate: side lobes are absent, although the blade may be slightly auriculate at the base. The margin may vary from entire to erose or fimbriate: subgenus *Dendrochilum*; subgenus *Platyclinis* sections *Eurybrachium* and *Platyclinis*.
2. Cruciform: subgenus *Platyclinis* section *Cruciformia* (e.g. *D. cruciforme*).
3. Pandurate: subgenus *Platyclinis* sections *Cruciformia* (e.g. *D. hosei*) and *Falsiloba*.
4. Three-lobed with small, sometimes inconspicuous, often auriculate side lobes and a large, prominent, sometimes transversely-oblong mid-lobe: common in subgenus *Platyclinis* section *Platyclinis*; also subgenus *Platyclinis* sections *Cruciformia* (e.g. *D. dolichobrachium*), *Eurybrachium* (e.g. *D. acuiferum*) and *Mammosa* (e.g. *D. kingii*).
5. Three-lobed with side lobes having a narrow, caudate, free apical portion, and a large prominent mid-lobe: subgenus *Platyclinis* section *Platyclinis* (e.g. *D. longirachis*).
6. Three-lobed with side lobes and mid-lobe of more or less equal size: subgenus *Platyclinis* sections *Mammosa* (*D. rufum*) and *Platyclinis* (e.g. *D. imitator*, *D. lacteum*).

The blade of the labellum may be either straight or somewhat curved, and is often decurved just above the base. The surface of the blade is always flat in subgenus *Dendrochilum* and most often so in subgenus *Platyclinis*. A usually shallow longitudinal furrow is often present in both subgenera which, in a few species (e.g. *D. flos-susannae*, *D. oxylobum*), may be more pronounced on the disc. In several species, particularly in subgenus *Acoridium* (*D. auriculilobum* and *D. hologyne*) and subgenus *Platyclinis* section *Eurybrachium* (e.g. *D. cupulatum* and *D. microscopicum*), the entire labellum is distinctly concave. The disc of the labellum in Bornean taxa is usually somewhat concave, but rarely truly saccate. A saccate disc is, however, found in several Philippine taxa, especially of section *Acoridium*. In Borneo only *D. saccatum* in section *Mammosa* has a truly saccate labellum. The evolution of a saccate labellum in *Dendrochilum* may be a recent adaptation to an increasingly specialised pollination biology.

7

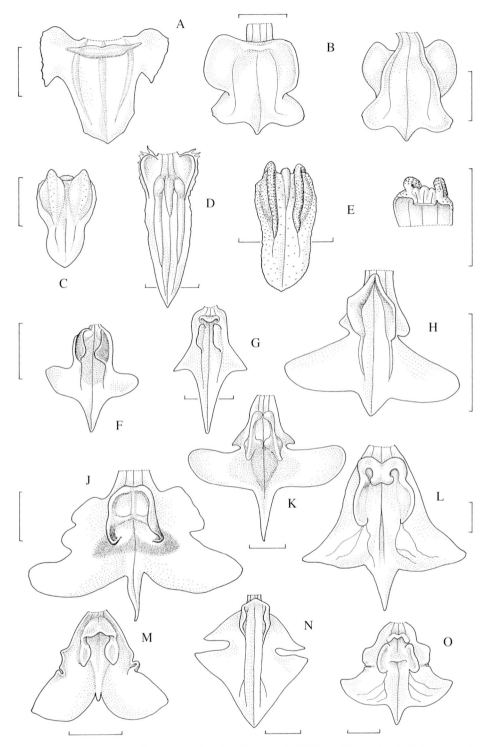

Fig. 3. Labellum keel and callus type. **A** *Dendrochilum auriculilobum, Vermeulen & Duistermaat* 1057 (holotype): **B** *D. pandurichilum,* front view (left), back view (right), *J.B. Comber* 102 (holotype); **C** *D. crassum, Wood* 733; **D** *D. gravenhorstii, Gravenhorst* s.n. (holotype); **E** *D. pallidiflavens* var. *pallidiflavens,* front view (left), basal portion, back view (right), *Hewitt* s.n. (holotype of *D. intermedium*); **F** *D. cruciforme* var. *cruciforme, J. & M.S. Clemens* s.n. (holotype); **G** *D. cruciforme* var. *longiscuspe, Carr* C. 3675, SFN 28004 (holotype); **H** *D. dolichobrachium, Meijer* 599; **J** *D. gibbsiae, Gibbs* 4087 (holotype); **K** *D. exasperatum, Vermeulen & Duistermaat* 668; **L** *D. grandiflorum, Wood* 608; **M** *D. haslamii* var. *haslamii, Collenette* 21535; **N** *D. hastilobum, Awa & Lee* S. 50754 (holotype); **O** *D. hosei, de Vogel* 1244 (holotype). Scale: single bar = 1 mm. Drawn by Susanna Stuart-Smith.

The surface of the disc of the labellum is usually variously ornamented with glabrous to papillose keels and calli, the number, distribution and form of which are useful characters at specific level (see Figs. 3–5). Ornamentation may also sometimes be of significance at sectional level. In section *Mammosa*, for example, two free, prominent calli divide the labellum into a subsaccate or saccate hypochile and a flat epichile (see Fig. 4N). The monospecific subgenus *Acoridium* section *Falsiloba* has an entire labellum provided with two prominent, wing-like proximal keels reminiscent of rounded side lobes (see Fig. 3B). In Borneo the most diverse ornamentation is to be found in section *Platyclinis*. Here the labellum is provided with two longitudinal keels of varying length which are often united at the base to form a u-shaped structure. Each keel may sometimes have a small tooth-like projection at the base, (e.g., *D. crassifolium, D. gracile* var. *bicornutum*). Sometimes a third, usually shorter, median keel is produced forming a letter M. A horseshoe-like or u-shaped structure is prevalent in Bornean species of section *Eurybrachium* (see Fig. 4A–M). In other species, e.g., *D. cupulatum* (see Fig. 4E), the ornamentation is reduced to two small swellings. Ornamentation is generally simplest in subgenus *Dendrochilum* where the labellum is provided with two free proximal longitudinal keels and sometimes an additional shorter, median keel or swelling (see Fig. 3C & E). Rarely, a more complicated structure is found as, for example, in *D. gravenhorstii* (see Fig. 3D).

Many species of *Dendrochilum* have minute to distinct papillae on the surface of the labellum. This is particularly so in subgenus *Dendrochilum*, and subgenus *Platyclinis* sections *Eurybrachium* and *Platyclinis*. Dense papillae are found in some individuals of *D. tenompokense* var. *papillilabium* (section *Platyclinis*).

Column

In *Dendrochilum* the most critical characters at subgeneric, sectional and specific level are found in the morphology of the column. The column is generally sub- or semi-terete to somewhat clavate, stout or slender, and either straight or incurved. Column shape is less variable than other characters and more or less uniform in subgenus *Dendrochilum* and within each section of the remaining subgenera. The column of *D. anomalum* (section *Platyclinis*) is notable for having a papillose ventral surface (see Fig. 14A).

Column-foot

A short, flat column-foot is found in subgenus *Acoridium* section *Falsiloba*, subgenus *Dendrochilum* and subgenus *Platyclinis* sections *Mammosa* (rudimentary in *D. saccatum*) and *Platyclinis*. A foot is absent in subgenus *Platyclinis* sections *Cruciformia* and *Eurybrachium*. No Bornean taxa have a deeply concave column-foot such as is found in some species from the Philippines.

Stelidia

Lateral stelidia (column arms) are absent in subgenus *Acoridium* and vestigial to very well developed in all species of subgenus *Dendrochilum* and in most species of subgenus *Platyclinis*. Their shape varies, sometimes short and cuneate or auriculate, often narrowly elliptic to linear-ligulate, or acicular. In section *Mammosa* they are reduced to tiny teeth (see Fig. 7J & K). The apex may be obtuse to acuminate, or bifurcate. Generally porrect to erect, rarely decurved, the stelidia are borne at the base, from the middle or at the top portion of the column proper. These characters are often highly diagnostic at specific level (see Figs. 6 & 7).

The interpretation of stelidia as probably staminodial in origin by Pfitzer & Kränzlin (1907) and Butzin (1984–1986) is questionable. Non-staminodial origin is supported by the

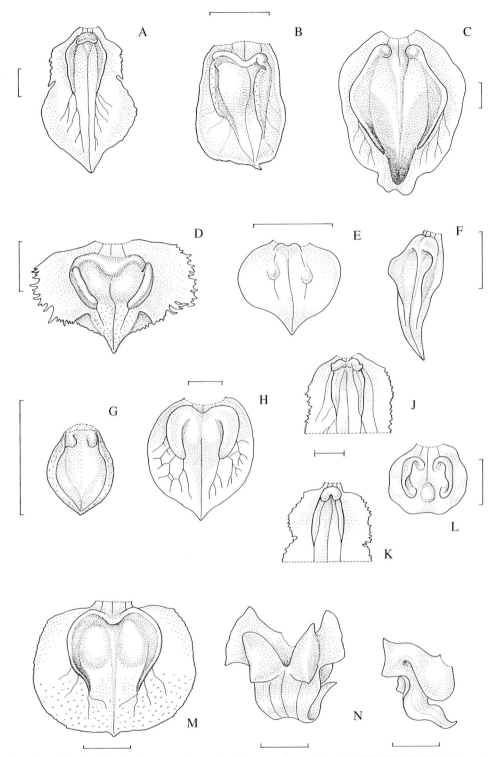

Fig. 4. Labellum keel and callus type. **A** *Dendrochilum acuiferum, Carr* C. 3550, SFN 27645 (holotype); **B** *D. alatum, Lamb* AL 735/87; **C** *D. alpinum, Carr* C. 3545, SFN 27624 (holotype); **D** *D. corrugatum, Bailes & Cribb* 841; **E** *D. cupulatum, de Vogel* 1013; **F** *D. joclemensii, Gunsalam* 10; **G** *D. microscopicum, Endert* 4103 (holotype); **H** *D. pseudoscriptum, Barkman* 198 (holotype); **J** *D. stachyodes,* basal portion, *J. & M.S. Clemens* 33177; **K** *D. stachyodes,* basal portion, *Wood* 605; **L** *D. scriptum, Carr* C. 3597 (holotype); **M** *D. transversum, Carr* C. 3477, SFN 27431 (holotype); **N** *D. saccatum,* front and side views, *Moulton* 6762 (holotype). Scale: single bar = 1 mm. Drawn by Susanna Stuart-Smith.

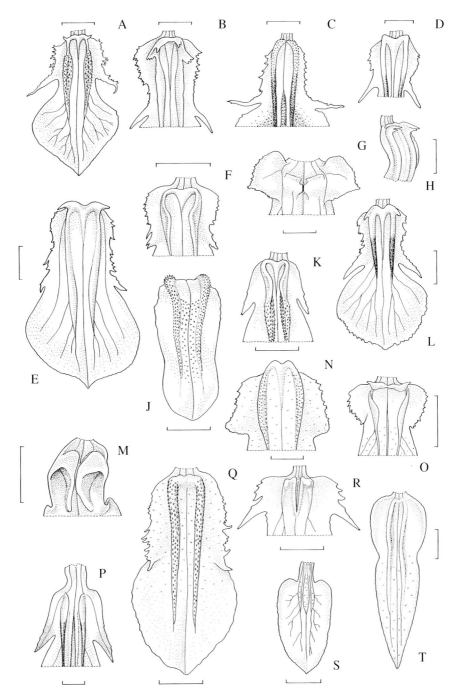

Fig. 5. Labellum keel and callus type. **A** *Dendrochilum crassifolium, Haslam* s.n. (holotype); **B** *D. crassifolium,* basal portion, *Vermeulen & Duistermaat* 1043; **C** *D. crassifolium,* basal portion, *Burtt & Martin* B. 5241; **D** *D. gracile* var. *gracile,* from Peninsular Malaysia, *Lewis* 158; **E** *D. dewindtianum, Shackleton* S. 186 (lectotype of *D. dewindtianum* var. *sarawakense*); **F** *D. graminoides,* basal portion, *Leiden* cult. (*de Vogel*) 911236; **G** *D. kingii* var. *kingii,* basal portion, *Cribb* 89/65; **H** *D. gracile* var. *bicornutum,* basal portion, oblique view, *Sands* 5327 (holotype); **J** *D. lumakuense, J.B. Comber* 108 (holotype); **K** *D. mucronatum,* basal portion, *J. & M.S. Clemens* 22587 (holotype); **L** *D. gracile* var. *bicornutum, Sands* 5327 (holotype); **M** *D. longipes,* basal portion, *Endert* 3991 (holotype of *D. mantis*); **N** *D. lacteum,* basal portion, *Bailes & Cribb* s.n.; **O** *D. ochrolabium,* basal portion, *de Vogel* 8351 (holotype); **P** *D. oxylobum,* basal portion, *Lamb* AL 506/85; **Q** *D. tenompokense* var. *papillilabium, Vermeulen & Duistermaat* 666 (holotype); **R** *D. papillitepalum,* basal portion, *Awa & Lee* S. 47676 (holotype); **S** *D. vestitum, van Steenis* 1435 (holotype); **T** *D. planiscapum, Giles & Woolliams* s.n. Scale: single bar = 1 mm. Drawn by Susanna Stuart-Smith.

fact that stelidia are absent in the primitive, plesiomorphic subgenus *Acoridium*. In addition, these organs also appear to act as 'steering arms' facilitating the longitudinal alignment of a potential pollinator on the labellum. It is interesting that some species of *Dendrochilum* are provided with well-developed stelidia but no labellum side lobes (e.g. *D. dulitense, D. globigerum, D. lancilabium, D. longipes*), while others may lack stelidia but are provided with prominent, erect side lobes (e.g. *D. auriculilobum*), a cupulate labellum with erect margins (e.g. *D. hologyne*), or prominent, suberect keels reminiscent of side lobes (e.g. *D. pandurichilum*). Does this reflect analogous evolution in response to an identical pollination requirement?

Anatomical examination by Pedersen (1997) failed to find any vascular traces of staminodia in the stelidia of the Philippine *D. cobbianum* and *D. filiforme*. The functional and, admittedly limited, anatomical evidence points to the stelidia being derived organs. Stelidia probably evolved only once in a common ancestor of subgenera *Dendrochilum* and *Platyclinis*, although reversals have probably taken place several times within subgenus *Platyclinis*.

Column apex

In subgenus *Acoridium* the top portion of the column is more or less subequal in length to the anther and never prolonged. In subgenera *Dendrochilum* and *Platyclinis* the column apex is elongated into a hood or wing-like structure which exceeds the anther. The presence of an elongated hood or wing is usually associated with the presence of well-developed stelidia. This, together with the stelidia, may act as a guidance or 'steering system' for pollinators. The outline of the hood or wing may be entire, erose or variously toothed, and is often characteristic of certain species, although the extent of toothing in many species may vary among flowers on the same inflorescence (see Fig. 8). The presence or absence of a prolonged, apical hood or wing indicates the existence of two main evolutionary trends within *Dendrochilum*. One trend gave rise to subgenera *Acoridium* and *Pseudacoridium*, the other to subgenera *Dendrochilum* and *Platyclinis*.

Stigma and Rostellum

The stigmatic cavity is commonly narrowly oblong to more or less circular (see Fig. 9). Its margin may often be thickened and elevated, especially so in the ventral region. In *D. tenompokense* var. *tenompokense* and *D. tenuitepalum* the ventral margin is swollen and bilobed (see Fig. 9D–F), while in *D. tenompokense* var. *papillilabium* it is prolonged into a distinctive bilobed flange (see Fig. 9G & H). All Bornean species of *Dendrochilum* have an entire, more or less flat rostellum. The size of the rostellum is significant in a few species, e.g., *D. corrugatum* (section *Eurybrachium*) and *D. saccatum* (section *Mammosa*). In most species it is ovate to cordate, obtuse or acute, but in *D. kingii* (section *Mammosa*) it is commonly rectangular or quadrate, and truncate.

Anther-cap

In *Dendrochilum* the anther-cap is terminal, or somewhat ventral in species with an apical column hood or wing. It is strongly incumbent, flexibly attached by a short connective and often provided with a dorsal wart. The biloculate or incompletely quadriloculate chamber dehisces ventrally by slits. The apex varies from rounded (e.g. *D. grandiflorum*) or retuse (e.g. *D. havilandii*) to acuminate (e.g. *D. tenompokense*) but never deeply emarginate as in several Philippine species. The surface may be smooth (e.g. *D. longifolium*) or, more rarely, minutely papillose (e.g. *D. cupulatum*).

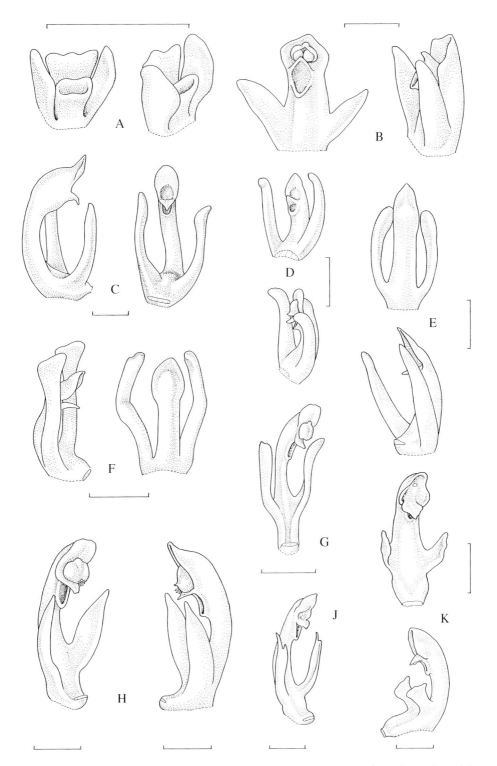

Fig. 6. Column stelidia variation. **A** *Dendrochilum cupulatum*, front view (left), oblique view (right), *Hansen* 498; **B** *D. pterogyne*, front view (left), side view (right), *Carr* C. 3541, SFN 27597 (holotype); **C** *D. murudense*, side view (left), front view (right), *Nooteboom & Chai* 01995 (holotype); **D** *D. gibbsiae*, oblique view (top), side view (bottom), *Wood* 578; **E** *D. lancilabium*, back view (top), oblique view (bottom), *Lewis* 343; **F** *D. longipes*, side view (left), back view (right), *Lewis* 345; **G** *D. angustilobum*, front view, *Carr* C. 3233, SFN 26874 (holotype); **H** *D. lacteum*, oblique view (left), side view (right), *Bailes & Cribb* s.n.; **J** *D. angustilobum*, oblique view, *Beaman* 7309; **K** *D. dulitense*, front view (top), side view (bottom), *Burtt & Martin* B. 4894. Scale: single bar = 1 mm. Drawn by Susanna

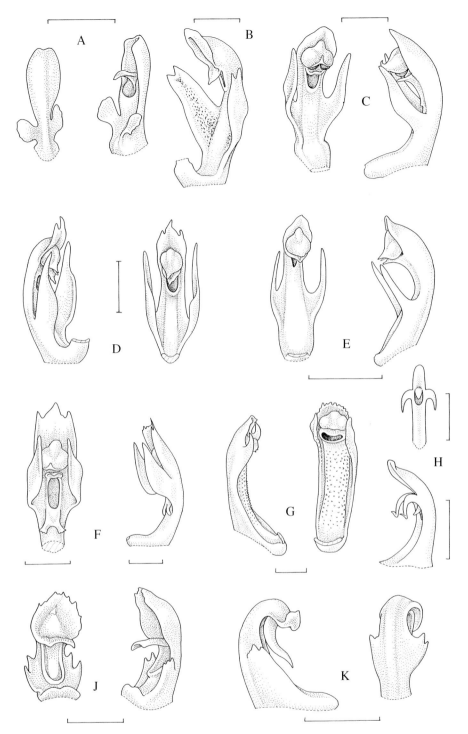

Fig. 7. Column stelidia variation. **A** *Dendrochilum papillitepalum*, back view (left), oblique view (right), *Awa & Lee* S. 47676 (holotype); **B** *D. imitator*, oblique view, *de Vogel* 8451 (holotype); **C** *D. crassifolium*, oblique view (left), side view (right), *Vermeulen & Duistermaat* 686; **D** *D. crassifolium*, side view (left), front view (right), *Burtt & Martin* B. 5241; **E** *D. graminoides*, front view (left), side view (right), *Leiden* cult. (*de Vogel*) 911236; **F** *D. tenompokense* var. *papillilabium*, front view (left), side view (right), *Vermeulen & Duistermaat* 666 (holotype of *D. papillilabium*); **G** *D. anomalum*, side view (left), front view (right), *Giles* 964A; **H** *D. hamatum*, front view (top), side view (bottom), *Beccari* 476 (isotype); **J** *D. kingii* var. *kingii*, front view (left), oblique view (right), *Cribb* 89/65; **K** *D. saccatum*, side view (left), back view (right), *Moulton* 6762 (holotype of *Pholidota gracilis*). Scale: single bar = 1 mm Drawn by Susanna Stuart-Smith.

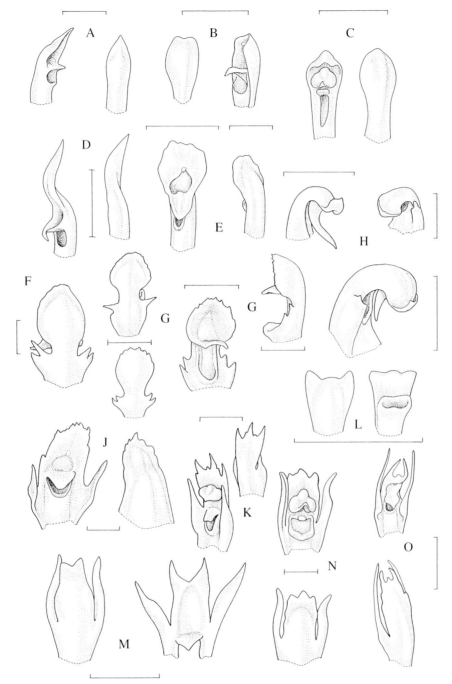

Fig. 8. Column apical hood. **A** *Dendrochilum lancilabium*, side view (left), back view (right), *Lewis* 343; **B** *D. papillitepalum*, back view (left), oblique view (right), *Awa & Lee* S. 47676 (holotype); **C** *D. subintegrum*, front view (left), back view (right), *Moulton* SFN 6670; **D** *D. hastilobum*, oblique view (left), back view (right), *Awa & Lee* S. 50754 (holotype); **E** *D. imitator*, front view (left), back view (right), *de Vogel* 8451 (holotype); **F** *D. kingii* var. *kingii*, back view, *de Vogel* 1584; **G** *D. kingii* var. *kingii*, back views (left), oblique view (middle), side view (right), *Cribb* 89/65; **H** *D. saccatum*, side view (top left), oblique views (top right and bottom), *Moulton* 6762 (holotype of *Pholidota gracilis*); **J** *D. oxylobum*, oblique view (left), back view (right), *Lamb* AL 506/85; **K** *D. glumaceum*, oblique view (below), back view (above), *cult. Bogor*, no. 928.11.46; **L** *D. cupulatum*, back view (left), front view (right), *de Vogel* 1013; **M** *D. pallidiflavens* var. *brevilabratum*, back view (left), front view (right), *Hose* 52 (holotype); **N** *D. crassilabium*, front view (top), back view (bottom), *Leiden* cult. (*de Vogel & Cribb*) 913205A; **O** *D. vestitum*, oblique view (top), back view (bottom), *van Steenis* 1435 (holotype). Scale: single bar = 1 mm. Drawn by Susanna Stuart-Smith.

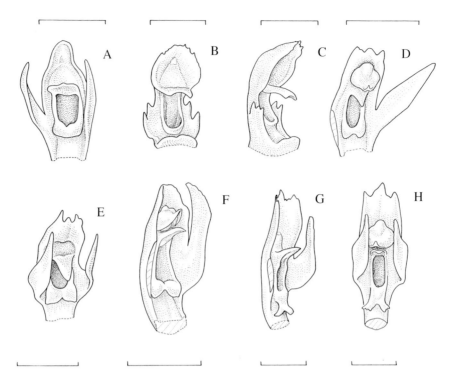

Fig. 9. Stigmatic cavity. **A** *Dendrochilum globigerum, I. Sinclair & Argent* 126; **B** *D. kingii* var. *kingii, Cribb* 89/65; **C** *D. kingii* var. *kingii, de Vogel* 919; **D** *tenompokense* var. *tenompokense, Carr* C. 3623, SFN 27891 (holotype); **E** *D. tenompokense* var. *tenompokense, Jermy* 13200; **F** *D. tenuitepalum, Vermeulen & Duistermaat* 1008 (holotype); **G & H** *D. tenompokense* var. *papillilabium, Vermeulen & Duistermaat* 666 (holotype). Scale: single bar = 1 mm. Drawn by Susanna Stuart-Smith.

Pollinia

The four, rather hard pollinia are narrowly clavate to almost spherical, always entire and attached to the viscidium by usually short, often rudimentary caudiculae. The minute fragile viscidium is orbicular to quadrate and stipites are absent.

Fruit

The fruit is a unilocular, ellipsoid to spherical capsule with three or six longitudinal keels (usually three jugae and three valvae). The rostrate shrivelled column is persistent in many species and, when combined with vegetative characters, is often useful in identification. The labellum is sometimes persistent, particularly in subgenus *Acoridium* and subgenus *Platyclinis* section *Eurybrachium*.

The manner of capsule dehiscence is uncertain but is probably by means of slits leaving three broad, seed-bearing valvae and three narrow valvae constituting a replum. The valvae remain attached at both ends (Pfitzer & Kraenzlin 1907). Hygroscopic hairs are absent.

Seeds

An ellipsoid or fusiform embryo is enclosed in a papery testa. The seed of subgenera *Acoridium* and *Platyclinis* is more or less fusiform in shape, shorter than in subgenus *Dendrochilum* (c. 0.9–2.4 mm long), and with an embryo exceeding half the total length. The central cells of the testa are prosenchymatic, i.e. lengthened and with tapering ends which overlap. The junctions of their anticlinal walls do not protrude (see Figs. 10 C–F, 11–13). In subgenus *Dendrochilum* the seed is longer (4.4–6.7 mm long) and filiform, with an embryo much shorter than the total length. The central cells of the testa are almost isodiametric, i.e. of equal dimensions, and the junctions of their anticlinal walls protrude from the testa giving the seed a rather spiny appearance (see Figs. 10 A & B). A swollen distal oil-secreting body or elaiosome occurs in *D. pallidiflavens* and possibly other members of subgenus *Dendrochilum* (see Fig. 14D).

ANATOMY

Anatomical details of the roots, pseudobulbs, leaves and flowers of mostly Philippine taxa are scattered throughout the literature. A summary of anatomical research pertaining to Bornean taxa is provided below.

Roots

The roots of *Dendrochilum* are polyarch with an actinostele containing six or twelve xylem ridges and were found by Porembski & Barthlott (1988) to be surrounded by a '*Coelogyne* type' of velamen which is unique to the Coelogyninae. This is a usually four- to six-layered velamen lacking a clear differentiation between epi- and endo-velamen. The walls have relatively massive thickenings arranged as helical bands, while non-thickened areas show small pores. Outgrowths of the inner tangential and radial walls of the central velamen cells known as tilosomes ('Stabkörper') are also present. The walls of the exodermis are either partially or totally thickened and tracheoidal idioblasts, i.e. distinct partially thickened, often porate cells, occur in the cortex. *D. gibbsiae* (section *Cruciformia*) and *D. kamborangense* (section *Platyclinis*) were examined by Porembski & Barthlott and found to have a four-layered velamen containing tracheoidal idioblasts and tilosomes. The cortex parenchyma is approximately three times as wide as the velamen in *D. glumaceum* due to a zone of enlarged cells.

Pseudobulbs

The anatomical structure of three species of section *Platyclinis*, including *D. glumaceum*, was investigated by Zörnig (1904). *D. glumaceum* was found to have a relatively strong cuticle which is wavy due to arched periclinal walls of the epidermal cells below. These epidermal cells are one to two times as long as wide, usually hexagonal in periclinal section, and arranged in almost regular longitudinal rows. A chlorenchyma layer containing many mucilage cells and starchy parenchyma cells was found below water-storing tissue. The chlorenchyma cells were also found to contain variously shaped crystals of calcium oxalate. Flat, tabular stegmata cells containing silica were also noted.

Leaves

Zörnig (1904) also examined the leaves of several species of *Dendrochilum*. Again, the only Bornean species examined was *D. glumaceum*. A weak cuticle was found on both sides of the leaf, although generally stronger on the petiole than the leaf blade. The cells of the upper (adaxial) epidermis are usually polygonal in *Dendrochilum*, though often variable in shape and size. In *D. glumaceum* they are arranged in neat longitudinal rows over the entire surface. The lower (abaxial) epidermis in *Dendrochilum* is composed of alternating longitudinal bands of cells which often differ depending upon their position in proportion to the nerves. Cell wall

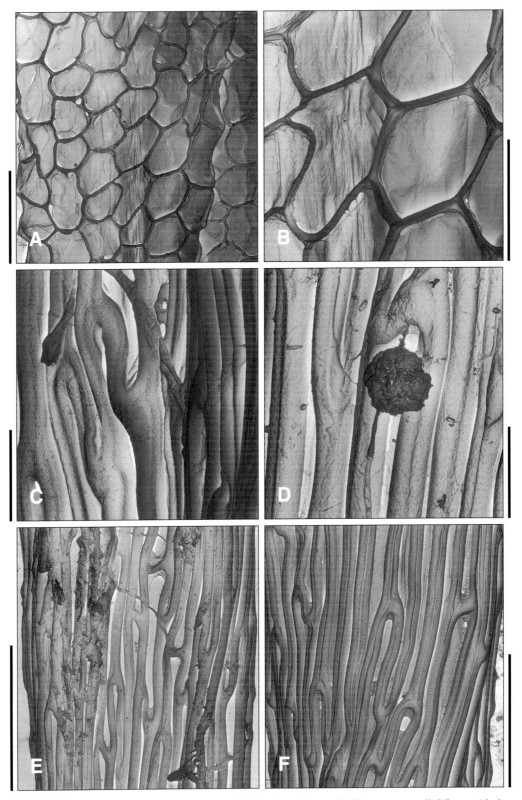

Fig. 10. Details of testa morphology (SEM). **A**, **B** *Dendrochilum pallidiflavens* var. *pallidiflavens* (*J. & M.S. Clemens* 32556, BO); **C** *D. cruciforme* var. *cruciforme* (*J. & M.S. Clemens* 32220, E); **D** *D. exasperatum* (*Carr* C. 3668, SFN 28029, SING); **E** *D. gibbsiae* (*J. & M.S. Clemens* 29235, E); **F** *D. haslamii* var. *haslamii* (*Collenette* 21535, L). **A**: scale bar equals 100 µm; **B**, **E**, **F**: scale bar equals 50 µm; **C**, **D**: scale bar equals 10 µm.

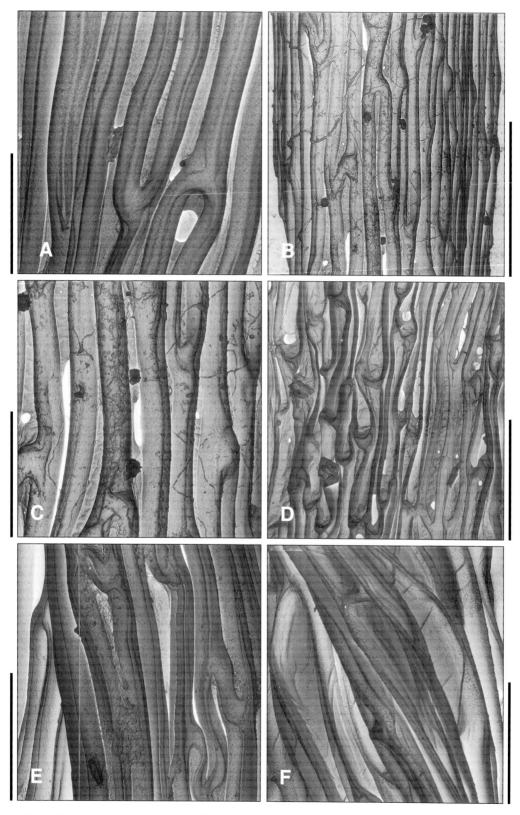

Fig. 11. Details of testa morphology (SEM). **A** *Dendrochilum haslamii* var. *haslamii* (*Collenette* 21535, L); **B**, **C** *D. corrugatum* (*Carr* 3128, SFN 27428, SING); **D** *D. dewindtianum* (*Carr* C. 3533, SFN 27562, SING); **E** *D. dulitense* (*Burtt & Martin* B. 4894, E); **F** *D. longifolium* (*Amdjah*, cult. Bogor, BO). **A**, **C**, **E**: scale bar equals 20 μm; **B**, **D**, **F**: scale bar equals 50 μm.

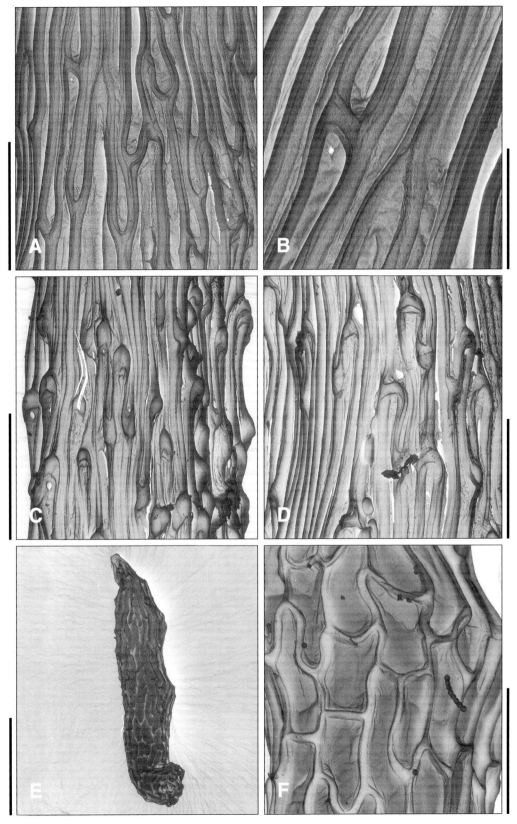

Fig. 12. Seed and details of testa morphology (SEM). **A**, **B**, *Dendrochilum muluense* (*Burtt & Martin* 5244, E); **C**, **D** *D. planiscapum* (*Chow & Leopold* SAN 74511, SAN); **E**, **F** *D. pubescens* (*de Vogel* 8888, L). **A**, **C**, **D**, **F**: scale bar equals 50 µm; **B**: scale bar equals 20 µm; **E**: scale bar equals 200 µm.

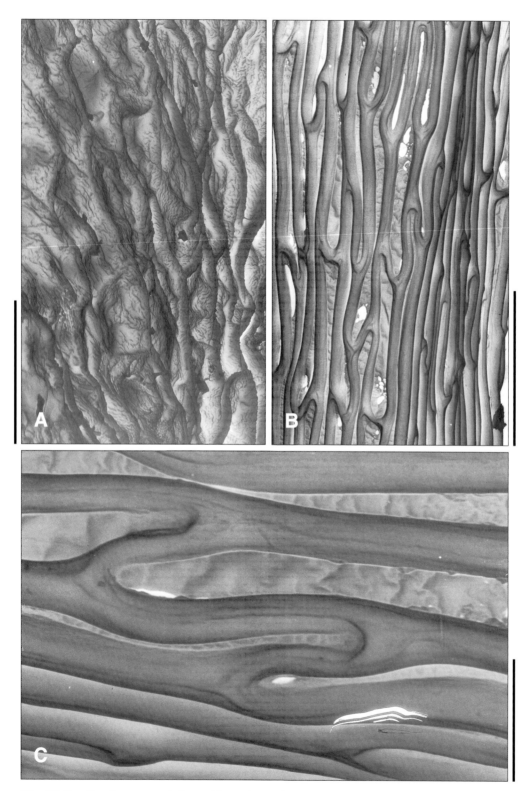

Fig. 13. Details of testa morphology (SEM). **A** *Dendrochilum simplex* (*Lee* S. 54798, K): **B**, **C** *D. tenompokense* var. *papillilabium* (*Dewol & Abas* SAN 89075, SAN). **A**, **B**: scale bar equals 50 µm; **C**: scale bar equals 10 µm.

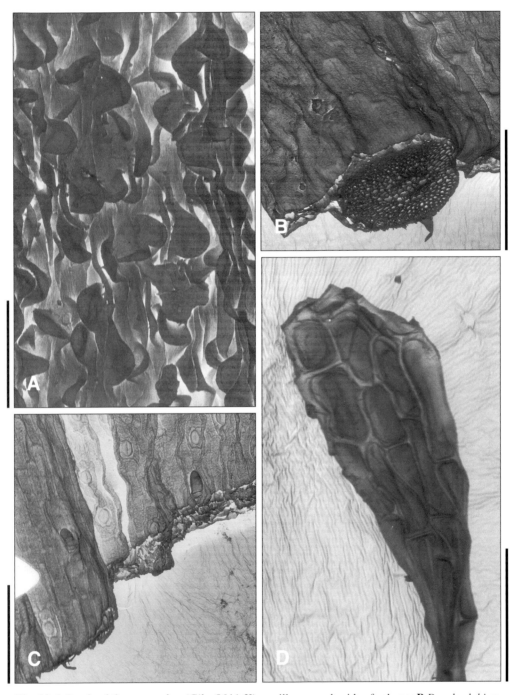

Fig. 14. A *Dendrochilum anomalum* (*Giles* 964A,K), papillae on underside of column; **B** *D. ochrolabium* (*Leiden* cult. (*Schuiteman et al.*) 933108, L), transverse section through abaxial surface of leaf showing prominent marginal nerve; **C** *D. ochrolabium* (*Leiden* cult. (*Schuiteman et al.*) 933108, L), portion of abaxial surface of leaf showing marginal nerve on left, trichomes and stomata; **D** *D. pallidiflavens* var. *pallidiflavens* (*Banyeng & Paie* S. 45050, L), elaiosome. **A**: scale bar equals 100 μm; **B & C**: scale bar equals 200 μm.; **D**: scale bar equals 50 μm.

thickening is moderate in *D. glumaceum*. A one-layered water-storing region below the upper epidermis has been observed in *D. glumaceum, D. kingii, D. longifolium* and *D. pallidiflavens*.

The epidermal mesophyll cells frequently contain various ergastic deposits. The most frequently occurring substance is calcium oxalate which occurs in the form of cubic crystals, styloids, raphides, druses, and bundles of needles. Calcium oxalate crystals are distributed in many taxa but are especially noticeable in section *Cruciformia*. Oil droplets may also be present in the epidermis or be restricted to the guard cells of the stomata.

The stomata, which are hypostomatic, i.e. only occuring on the lower (abaxial) surface of the leaf, are longitudinally orientated and are sometimes restricted to the bands of epidermis covering the intercostal areas. Rosinski (1992) gives the average density in these areas as 60 in *D. kingii* (sub syn. *D. sarawakense*).

Rosinski (1992) recognises two types of stomata in *Dendrochilum*. Type one occurs in leathery leaf surfaces, e.g., *D. pallidiflavens*. Here the stomata are tetra- or pentacytic, with large guard cells and large trapezoid lateral and polar subsidiary cells. Type two occurs in hard-textured leaf surfaces, e.g., *D. crassifolium* and *D. kingii*. Here the stomata are usually tetracytic, with rather small guard cells having a cuneate lumen, and narrow lateral, somewhat larger polar subsidiary cells.

The lamina contains a varying number of alternately larger and smaller longitudinal nerves arranged at about the same plane and containing collateral vascular bundles. The mid-nerve in several species of subgenus *Dendrochilum* appears less prominent than the laterals. The reverse is generally true of the other subgenera. In *D. ochrolabium* the two outermost longitudinal nerves nearest the margin of the lamina are as prominent as the mid-nerve (see Fig. 14B). Numerous small transverse nerves are present in a few species, e.g., *D. exasperatum* and *D. grandiflorum*. The nerves are often arranged at more than one level in the petiole. The mid-nerve may be orbicular or transversely elliptic in transverse section and is strengthened by longitudinal lower (abaxial) and upper (adaxial) sclerenchyma plates.

Trichomes on the leaf surface

All Bornean taxa have a variable number of sunken trichomes on the leaf surface. Each trichome is borne at a 60 degree angle to the lamina from a tiny epidermal depression and is composed of from one to three cells covered with a thin cuticle. A moderately thickened 'foot cell' extends from the base of the trichome through the epidermis and into the chlorenchyma. The epidermal depression is lined with 'subsidiary cells' that differ in size and shape from the surrounding epidermis. The lumen of the basal subsidiary cells is about twice the size of the others. The basal subsidiary cell walls are perforated and thin in subgenus *Dendrochilum*, while in subgenera *Acoridium* and *Platyclinis* they are perforated and thickened.

The distribution of sunken trichomes on the leaf surface, characteristic of the Coelogyninae, was also investigated by Rosinski (1992). Trichomes were found on both surfaces of the leaves of the taxa examined, but their density varied between species. The trichomes tended to be concentrated towards the base and apex, particularly on the lower surface.

Flowers

No anatomical investigation of the flowers of Bornean taxa has been undertaken and almost no palynological data are available. Pedersen (1997) reports noting a well-developed tectum in *D. glumaceum*.

CYTOLOGY

Chromosome numbers of 2n = 38, 40, 42 and 44 have been counted in subtribe Coelogyninae (Dressler 1993). Only a handful of dendrochilums have been investigated. Of those recorded from Borneo, *D. glumaceum* and *D. longifolium* have a diploid chromosome number of 2n = 40 (Hoffman 1929 and 1930, and Lim 1985 respectively).

CHEMISTRY

Little is known of the floral fragrances in *Dendrochilum*. Kaiser (1993) investigated *D. glumaceum* and found the scent to be strongly pervaded by anis aldehyde, geranyl acetate, and veratol (1, 2-dimothoxy benzene). Lüning (1964) reported an alkaloid content of about 0.001% in the floral fragrance of *D. glumaceum*, while Lawler & Slaytor (1970) reported an alkaloid content of 0.01% in New Guinean material of *D. longifolium*.

PHYLOGENETIC CONSIDERATIONS

De Vogel (1989) considered *Pholidota* Lindl. ex Hook. to be the most plesiomorphic genus of the Coelogyninae. He suggested that *Dendrochilum* may have evolved, "via its subgenus *Acoridium*", from *Pholidota* section *Acanthoglossum*, which is virtually confined to Borneo. Among other characters, the absence of stelidia and the simple rostellum certainly suggest that *Acoridium* is probably the most plesiomorphic subgenus of *Dendrochilum*. However, whether *Dendrochilum* is more closely related to *Acanthoglossum* than to other sections of *Pholidota* is debatable. *Acanthoglossum* is rather isolated within *Pholidota* on account of the presence in all six species of the section of a hamulus, i.e. a pollinium stalk derived from the top of the rostellum. In addition, the pedicel with ovary is more or less densely covered with minute hairs, which are often glued together by secretion, unlike other *Pholidota*. The presence of a hamulus is an apomorphic character state not present in *Dendrochilum*. Our knowledge of phylogenetic relationships, reproductive biology and, in many cases, species ranges in *Dendrochilum* is at most rudimentary. It is clear, however, that *Dendrochilum* and *Pholidota* are closely related and most likely represent phylogenetic sister groups sharing an immediate common ancestor. The two genera may be distinguished using the key below.

Pollinia entire. Pseudobulbs 1-leaved. Floral bracts persistent. Labellum often somewhat concave, but usually not saccate; if saccate (in some non-Bornean taxa), then the mid-lobe is usually smaller than the side lobes and the sac distinctly indented along the median line; in Bornean species of section *Mammosa* the hypochile and epichile are divided by two free, prominent calli .... *Dendrochilum*
Pollinia porate. Pseudobulbs 1- or 2-leaved. Floral bracts caducous or persistent. Labellum more or less saccate and with a prominent mid-lobe (sac rarely indented along the median line) .... *Pholidota*

Barkman (PhD dissertation, 1998) sampled 40 species of subgenus *Platyclinis*, mostly of Bornean origin, to evaluate the monophyly of sections *Cruciformia*, *Eurybrachium*, *Platyclinis*, and several informally recognised groups. Results based on combined parsimony analysis of nr DNA ITS 1 & 2 and *acc* D–*psa* I chloroplast spacer sequence variation indicate that none of the sections within subgenus *Platyclinis* recognised by Pedersen *et al.* (1997) are monophyletic. Interestingly, several previously recognised alliances are. Even more significant is the observation that *geographical location* is a better indicator of phylogenetic relationship than floral morphology.

24

The majority of the Bornean species of section *Eurybrachium* share a combination of covarying characters including a short, broad column with broad, rounded basal stelidia and an entire labellum which is usually as broad as long. These characters are also found in a few non-Bornean members of the section. Four of the fifteen species of section *Eurybrachium* represented in Borneo, however, have labellum side lobes and elongate stelidia indicating that it may not be monophyletic.

The endemic section *Cruciformia* would seem to be well circumscribed, having a more or less cruciform labellum with or without posterior side lobes, and distinct from other *Dendrochilum*. In section *Platyclinis*, on the other hand, which is the largest and most widespread within subgenus *Platyclinis*, there are few diagnostic features that unite the species. However, most do have a lobed labellum that is elastically attached to a column which usually has elongate, narrow stelidia. Several alliances consisting of two to several species, differentiated by similar floral morphologies, were informally recognised by Ames (1920) and Carr (1935). A formal subsectional classification of section *Platyclinis* has never been proposed, however.

Barkman found the phylogenetic estimates obtained from a consensus tree computed from fifteen equally parsimonious reconstructions were congruent with several lineages recognised by Ames (1920), Carr (1935) and Pedersen *et al.* (1997). Significantly, these appear to be *geographically* partitioned into separate monophyletic Bornean and Bornean/Philippine lineages.

Bornean *Eurybrachium* lineage

Molecular evidence supports a monophyletic lineage recognised by Carr (1935) including *D. alpinum, D. pterogyne* and *D. scriptum* together with the recently described *D. pseudoscriptum*. It is unclear if *D. alatum* and *D. corrugatum* are also allied to this lineage due to a lack of shared molecular characters. *D. cupulatum* and *D. joclemensii* do not appear to be closely related to other Bornean members of the section. *D. stachyodes* and *D. trusmadiense* were also found to form a monophyletic 'Sabah' lineage. Curiously incongruent with Pedersen *et al.* (1997) and the treatment presented here, *D. kamborangense* and *D. muluense* appear to be included within this lineage. They are, however, retained, for the purposes of this treatment, within section *Platyclinis*.

Section *Cruciformia*

Molecular analysis suggests that, within this section *D. exasperatum, D. gibbsiae* and *D. haslamii* are closely related. *D. cruciforme* and *D. grandiflorum*, on the other hand, do not appear to be closely related to each other or to the former three species.

Section *Platyclinis*

This, the largest section represented in Borneo, was also investigated by Barkman who found a lineage, already informally recognised by Carr (1935), including *D. angustilobum, D. lacteum* and *D. longirachis*. In addition, Barkman included *D. lancilabium*. To this group I would also add *D. imitator* and *D. longipes*. Surprisingly, *D. cupulatum* and *D. graminoides* also seem to have affinities with this lineage. The *D. lacteum* lineage and several members of section *Cruciformia* appear to form a monophyletic group.

High and mid elevation individuals of the widespread and variable *D. dewindtianum* were also analysed by Barkman. This species was found to be a sister to *D. tenompokense*, confirming Carr's (1935) view that they were related. The distinctive *D. planiscapum* is considered by Barkman to be sister to this clade. The clade containing the majority of section *Eurybrachium* appears to be nested within the clade consisting mostly of species from sections *Cruciformia* and *Platyclinis*. Barkman identified two species at the base of the Bornean clade, viz. *D. cruciforme* and *D. kingii* (section *Mammosa*).

Informal groupings (excluding *D. longifolium* and *D. pubescens* alliances) defined by Barkman (1998)

1. Core *Eurybrachium* (conforming, with the exception of *D. pseudoscriptum*, to Carr's section *Eurybrachion*):
   - *D. alatum*
   - *D. alpinum*
   - *D. corrugatum*
   - *D. pseudoscriptum*
   - *D. pterogyne*
   - *D. scriptum*
   - *D. joclemensii* (?)
2. 'Sabah' lineage:
   - *D. kamborangense*
   - *D. muluense*
   - *D. stachyodes*
   - *D. trusmadiense*
3. *D. gibbsiae* lineage (noted by Ames, 1920 and Carr, 1935):
   - D. exasperatum
   - D. gibbsiae
   - D. haslamii
   - Additional taxa: *D. dolichobrachium* and *D. hastilobum*.
4. *D. lacteum* lineage (noted by Carr, 1935):
   - *D. angustilobum*
   - *D. lacteum*
   - *D. lancilabium*
   - *D. longirachis*
   - Additional taxa: *D. imitator* and *D. longipes*
5. *D. dewindtianum* alliance (noted by Carr, 1935):
   - *D. dewindtianum*
   - *D. tenompokense*
   - Additional taxa: *D. crassifolium, D. geesinkii* and *D. murudense*
6. *D. longifolium* alliance:
   - *D. imbricatum*
   - *D. longifolium*
   - *D. oxylobum*
7. *D. pubescens* alliance:
   - *D. hamatum*
   - *D. pubescens*
   - *D. vestitum*

Core *Eurybrachium* lineage

The four species constituting the 'core' *Eurybrachium* lineage have an entire labellum, distinctive broad, rounded basal stelidia and globose, usually red or orange coloured pseudobulbs. These are *D. alpinum, D. pseudoscriptum, D. pterogyne* and *D. scriptum. D. alatum* and *D. corrugatum* also possess these characters, but their phylogenetic affinities to other species are not clear based on molecular data alone. *D. joclemensii* differs by its narrowly cylindrical green pseudobulbs and very obscurely lobed labellum. Similarly, *D. cupulatum*, although having a similar floral morphology, has green pseudobulbs that are narrowly cylindrical and widely spaced along the rhizome.

'Sabah' lineage

Barkman's concept of a 'Sabah' lineage, comprising *D. kamborangense, D. muluense, D. stachyodes* and *D. trusmadiense*, is incongruent with the sectional placement of these taxa by Pedersen *et al.* (1997). A suite of morphological characters are provided by Barkman, who states that they "might indicate relatedness to this lineage and hence serve to define a taxonomic group." The characters identified are a lobed labellum with serrate to erose margins, and non-basal, reduced, curved, finger-like stelidia on the column. It is hard to accept Barkman's concept here since, for example, *D. muluense*, described from Sarawak and occurring throughout Borneo, has more or less entire labellum lobes. Similarly, *D. trusmadiense* has a very shallowly lobed labellum which is neither serrate nor erose. It is also distinctive in having flowers that commence opening from the top of the inflorescence. The rare occurrence of rudimentary, non-basal stelidia in *D. stachyodes* may necessitate its future removal from section *Eurybrachium*.

A tentative lineage of taxa exhibiting morphological characters similar to those quoted by Barkman could include *D. flos-susannae, D. gracile, D. graminoides, D. ochrolabium, D. subulibrachium*, and possibly also *D. angustipetalum, D. lacinilobum* and *D. magaense*. Some of these taxa have much more prominent stelidia than those sampled by Barkman. It is clear that molecular analysis of this group of taxa is necessary in order to clarify whether real alliances exist, or if there is simply a convergence of homoplastic characters.

D. gibbsiae lineage

For the time being the taxa belonging to this alliance, as defined by Barkman, are retained in section *Cruciformia* (Pedersen *et al.*, 1997). Barkman suggests that any resemblance between the labellum morphology of *D. cruciforme* and *D. grandiflorum* to *D. exasperatum, D. gibbsiae* and *D. haslamii* may be superficial. Phylogenetic analysis suggests that *D. cruciforme* and *D. grandiflorum* may not be closely related to the '*D. gibbsiae* lineage', which is characterised by highly exaggerated side lobules of the terminal portion of the labellum, and tall, fleshy keels.

If these results are correct, a new section would need to be defined to include *D. exasperatum, D. gibbsiae* and *D. haslamii*. To these I would also provisionally include *D. dolichobrachium* and *D. hastilobum*. Section *Cruciformia* would then include *D. cruciforme*, but probably not *D. grandiflorum*. From their floral morphology I would consider *D. hosei*, and possibly *D. devogelii*, to be allied to *D. grandiflorum*. However, further sampling is necessary before relationships among these taxa can be clarified.

D. lacteum lineage

Members of this lineage all possess rather long, narrowly cylindrical, pencil-like pseudobulbs, and often slightly serrate petal margins. Labellum shape varies from entire (in *D. lancilabium*) to deeply three-lobed (e.g. in *D. longirachis*), while stelidia may be acute, truncate, or even bifid. *D. angustilobum* and *D. lacteum* are difficult to distinguish when sterile and Barkman found that they also have identical ITS 1 & 2 sequences.

D. dewindtianum alliance

D. dewindtianum is highly variable throughout its wide range which extends to Sumatra (as D. furfuraceum). Carr (1935) had already recognised a kinship between it and D. tenompokense and, indeed, both have similarly shaped pseudobulbs and close floral morphology. In D. tenompokense, however, the elevate lower margin of the stigmatic cavity is usually swollen and bilobed, or even prolonged, in var. papillilabium, into a bilobed flange. Both have narrow stelidia borne at the level of the stigma. Barkman's discovery that D. dewindtianum and D. tenompokense are related to D. planiscapum is surprising given the quite different floral morphology between them.

D. longifolium alliance

Ames (1920) considered his D. imbricatum to be allied to D. dewindtianum and D. oxylobum. Molecular analysis by Barkman has confirmed a close affinity between D. oxylobum and D. imbricatum, but not D. dewindtianum. It is clear that the widespread D. longifolium belongs here also. All three taxa have large pseudobulbs and leaves, similarly shaped labellum side lobes and keels, short, rather acute, non-basal stelidia, and an erose to irregularly lobed column apical hood.

D. pubescens alliance

D. hamatum, D. pubescens and D. vestitum appear to be closely related. The pubescent leaves and inflorescences of D. pubescens and, to a lesser extent D. vestitum, are distinctive among Bornean taxa. D. hamatum has a labellum and column structure closer to D. pubescens, but is virtually glabrous.

Some concluding observations

The revised subdivision of Dendrochilum proposed by Pedersen et al. (1997) was based on mostly floral morphological characters and, for practical reasons, is largely followed in this revision. However, as Chase & Palmer (1992, 1997) and Hapeman & Inoue (1997) point out, floral characters are often homoplastic. Homoplasty and the potential for convergence of several key characters used to define taxa in Dendrochilum, e.g., stelidia and labellum type, may possibly be higher than currently appreciated. These organs may evolve rapidly in response to pollinator availability and most probably play a significant role in specific pollination syndromes. As such, they may turn out to be of limited value at the subgeneric and sectional level.

As pointed out earlier, one of the most interesting observations resulting from Barkman's analysis is that generic subdivision within Dendrochilum may be correlated with geographical location. A basal subdivision of the major lineages was identified between Sumatra, Borneo and the Philippines. Several Sumatran and Philippine taxa assigned to different sections did appear to be closely related. Sampling of taxa from the islands of Indonesia and the Philippines remains too limited, however, to allow any strong conclusions to be drawn concerning the geographical subdivision hypothesis.

Dendrochilum is a species-rich and morphologically diverse genus, and consequently difficult to study given the current lack of accessibility to many taxa which would allow a broader survey to be undertaken. Further detailed molecular analysis is required which may have important implications for our understanding of the phylogenetic relationships within the genus. This could point the way forward to a possible future subdivision of section Platyclinis in particular. To quote Barkman: "additional taxon sampling in this large genus should allow finer resolution of species relationships."

BIOLOGICAL AND ECOLOGICAL ASPECTS

Habitat and ecophysiology

All five floristic altidudinal zones recognised in Malesia by van Steenis (1984) are represented in Borneo. These are: Tropical (0000–1000 m), Submontane (1000–1500 m), Montane (1500–2400 m), Subalpine (2400–4000 m) and Alpine (4000–4500 m). Altitudinal data are available for 78 Bornean species of *Dendrochilum*. These are represented by the figures given in the table below. No altitudinal data are available for *D. glumaceum, D. hamatum* or *D. pallidiflavens* var. *brevilabratum*. No species of *Dendrochilum* have been recorded from the Alpine zone in Borneo.

Table 1. Number and percentage of taxa represented in the floristic altitudinal zones in Borneo

	Total number of taxa	Percentage of total number of taxa	Number of taxa restricted to zone	Percentage of taxa restricted to zone	Number of non-endemic taxa	Number of sections restricted to zone	Total number of taxa in Borneo
Tropical Zone (0000–1000 m)	34	39.5%	15	17.4%	5	1	
Submontane Zone (1000–1500 m)	47	54.0%	12	13.7%	4	0	87
Montane Zone (1500–2400 m)	50	57.4%	13	15.1%	4	0	
Subalpine Zone (2400–4000 m)	15	17.4%	7	8.1%	1	0	

A few species have been recorded from transitional altitudes only, e.g., at 1500 metres which separates the submontane from the montane zone. Since overlap clearly occurs, these examples have been included in both zones. The figures in the table above show the submontane and montane zones to have the highest diversity of taxa.

Subgenus *Dendrochilum* is most diverse in the Tropical and Submontane zones, with only *D. pallidiflavens* var. *pallidiflavens* being represented in the Montane zone. The two species of section *Acoridium* recorded from Borneo are restricted to the Submontane and Montane zones. Section *Falsiloba* is probably restricted to the Montane zone, although there is a collection of very dubious provenance from the Tropical zone in Brunei. Section *Cruciformia* is evenly distributed from the Tropical to Montane zones, but is only represented by two species in the Subalpine zone. Section *Eurybrachium* is found in all zones, although it is only represented by *D. pterogyne* in the Tropical zone. It is most diverse in the Montane and Subalpine zones, each with nine and eight species respectively. All three species of section *Mammosa* are restricted to the Tropical zone. Section *Platyclinis* is well represented in all zones except the Subalpine where only five species have been recorded. One of these, *D. suratii*, is restricted to this zone. Four or five sections are found in all zones except the Subalpine where there are only three. The largest number of non-endemic taxa (71.4%) is found in the Tropical zone.

Some species occur in a variety of habitats and have a broad altitudinal range, while others appear very discerning in their requirements and are consequently very localised. *D. simplex* (section *Platyclinis*) is equally at home as an epiphyte in lower montane forest or as a rampant forest floor terrestrial in lowland kerangas forest. It has been collected at sites from around 400 metres to up to 2000 metres above sea level. Similarly, *D. gibbsiae* (section *Cruciformia*) is equally catholic in its tastes and occurs between *c.* 870 and 2400 metres above sea level.

Fig. 15. Lowland forest, Lambir Hills, Sarawak; 39.5% of the 87 taxa of *Dendrochilum* recorded from Borneo are found between sea level and 1000 metres altitude. (Photo: Hans P. Hazebroek)

Conversely, *D. acuiferum* (section *Eurybrachium*) is restricted to mostly ultramafic substrates between 2700 and 3000 metres on Mount Kinabalu, *D. havilandii* (section *Platyclinis*) prefers lowland and hill forest on limestone in Sarawak while *D. suratii* (section *Platyclinis*) has only been recorded as a twig epiphyte in mossy forest at between 2400 and 2500 metres on Mount Trus Madi.

It should be noted that some overlap in species zonation, especially between the Submontane and Montane zones, is to be expected. Future field observation will probably show that several taxa may have a broader altitudinal range than is currently known.

The altitudinal zonation of vegetation on Mount Kinabalu (4095 m) has been studied in some detail by Kitayama (1991) who recognised 21 vegetation map units using diagnostic canopy tree species to distinguish each zone. He recognises a fragmented upper-subalpine forest from about 3400 metres which extends to the tree line at about 3700 m. Above this level is a zone of alpine rock-desert with scattered communities of alpine scrub. As expected, Mount Kinabalu is the richest locality in Borneo for *Dendrochilum* diversity, with 32 (39.5%) out of the Bornean total of 81 species, (33 if Mount Tembuyuken is included) represented. Of these, 12 species and 2 varieties (9 species in section *Eurybrachium)* are endemic to the mountain.

Populations of *Dendrochilum* occur in habitats as diverse as coastal mangrove, e.g., *D. pallidiflavens*, and subalpine granite rock-faces, e.g., *D. stachyodes*. The majority of species, however, favour lower montane forest (approximately between 1200 and 2200 m) and especially mossy elfin forest, often along ridges, located in the humid cloud belt.

Most Bornean species of *Dendrochilum* are obligate epiphytes, though several may become semi-terrestrial in mossy forest. A few, such as *D. simplex*, are sometimes terrestrial at lower elevations. In the Subalpine zone on Mount Kinabalu, three species of section *Eurybrachium* (*D. alpinum, D. pterogyne* and *D. scriptum*) are facultatively epiphytic, also occurring as lithophytes on granite rocks. Here also are found *D. acuiferum* and *D. stachyodes*, which are unique among Bornean species in being obligately terrestrial or lithophytic.

Many species of *Dendrochilum* grow in exposed conditions in the canopies of ridge-top scrub, cliff-top kerangas forest or on windswept rocky slopes and consequently may suffer periodic drought and high rates of transpiration. Five xeromorphic adaptations facilitating survival during adverse conditions are found in Bornean *Dendrochilum*: specialised water-storage tissue in the pseudobulbs and, to a lesser extent, in the leaf; thick, rigid, fleshy leaves, shallowly v-shaped in transverse section, found in *D. pachyphyllum*; an often strongly developed cuticle limiting transpiration from the pseudobulbs and leaves; hypostomatic leaves which also limits transpiration; sunken trichomes which may be water absorbing. The trichome complexes are anatomically similar to those found on the leaves of the neotropical Pleurothallidinae which have been demonstrated to be absorptive by Pridgeon (1981).

Reproductive biology

Very little data are available on the pollination biology of *Dendrochilum*. From the few observations available it appears that various Coleoptera, Diptera and Hymenoptera are the most important pollen vectors. Pedersen (1995, 1997) speaks of a "syndrome for pollination by a variously specific selection of facultatively anthophilous insects". This syndrome would seem to apply to the majority of Bornean species, especially in section *Platyclinis* and subgenus *Dendrochilum*. Many of these have several characteristic features of the syndrome including clustered flowering shoots, long inflorescences with numerous small, scented flowers, a flat labellum with a shallow, usually linear nectary producing superficial nectar as a reward. Nectar is produced in a fine longitudinal furrow along the mid-line of the labellum in sections *Mammosa* and *Platyclinis*. The concave to saccate labellum and large obstructive

calli found in section *Mammosa* (especially *D. saccatum*) suggests a different adaptation to pollinators.

In common with most other Orchidaceae, the dust-like seed of *Dendrochilum* is almost exclusively adapted to wind-dispersal (anemochory). The seed of *D. pallidiflavens* (subgenus *Dendrochilum*), on the other hand, is relatively large and heavy compared with others in the genus, and the distal projection has an enlarged apex which functions as an oil-secreting elaiosome. Benzing & Clements (1991) also noted a presence of lipids in the testa. These seed characters are, rather, adaptations to myrmecochory in which ants are the vectors. A symbiotic association with dolichoderine ants referred to the genus *Iridomyrmex* (now subdivided into several genera) was first noted by Doctors van Leeuwen (1929a and b) in Javan specimens of *D. pallidiflavens* which were growing with a fern of the genus *Lecanopteris*. The swollen, hollow rhizomes of this 'ant fern' are used by ants as nest chambers (Ridley 1908b, Holttum 1954). The orchid seed is deposited inside the nests where germination is facilitated by the mineral-rich compost provided by the ants. The seeds of *D. pallidiflavens* are probably dispersed by wind in addition to ants and may be termed amphichorous, i.e. where dispersal by ants (myrmecochory) and wind (anemochory) are combined. Adaptation to ant dispersal has not been observed in other Bornean species of *Dendrochilum*, but future studies may show that a similar relationship exists in subgenus *Platyclinis* too. Associations in species, such as *D. gravenhorstii* (subgenus *Dendrochilum*), growing in kerangas forest developed over nutrient-poor podsolic soils may prove a fruitful area for study.

Examination of herbarium collections of *D. haslamii* var. *haslamii* (section *Cruciformia*) and *D. pterogyne* (section *Eurybrachium*) shows a consistenly high proportion of mature seed capsules which may suggest possible autogamy. This may be a response to the lack of suitable pollinators at high elevations.

Germination

Germination in *Dendrochilum* has only been studied in *D. glumaceum* in which the presence of a small, but distinct, cotyledon was identified by Pfitzer (1877, 1882). The development of a cotyledon is a very rare occurrence in the Orchidaceae and has only ever been observed in one other Asiatic orchid, *viz. Arundina graminifolia* (D. Don) Hochr. Despite its unusually advanced embryo, there is, in common with other orchids, no formation of endosperm in the seed of *D. glumaceum*.

BIOGEOGRAPHY

Borneo, with 81 species of *Dendrochilum*, is the second richest area for *Dendrochilum* after the Philippines where 89 species have been recorded (Pedersen, Wood & Comber 1997). The percentage of endemism is almost as high (90.1%), with 73 species occurring nowhere else. In the Philippines this figure is 94%. Available collections suggest that a large proportion of these endemics are of local occurrence, many restricted to a single mountain. Some local endemics are restricted to the Subalpine zone which, in Borneo, is only found on Mount Kinabalu. Others, such as *D. grandiflorum*, occur in the Submontane and Montane zones on the mountain but have never been collected from equivalent altitudes elsewhere. *D. trusmadiense*, restricted to the Montane zone on Mount Trus Madi, has showy flowers and would be easily spotted in the neighbouring Crocker Range. Other more diminutive species,

Fig. 16 (opposite). Lambir Hills, Sarawak. The gentler slopes behind the cliffs supports kerangas vegetation dominated by smallish trees with an open canopy often rich in orchids, including several species of *Dendrochilum*. Tall, lush dipterocarp forest, where fewer species of orchid are found, can be seen around the base of the cliffs. (Photo: Hans P. Hazebroek)

such as *D. microscopicum* from Mount Kemul and *D. suratii* from Mount Trus Madi, may simply have been overlooked elsewhere. A statement by Ames (1920) may, however, still hold true for many of these narrow endemics today: "If it were not for the repeated discovery of some of these species in their original stations, and for their absence from collections made elsewhere, the suggestion that deductions as to restricted distribution are premature at this time might be seriously entertained".

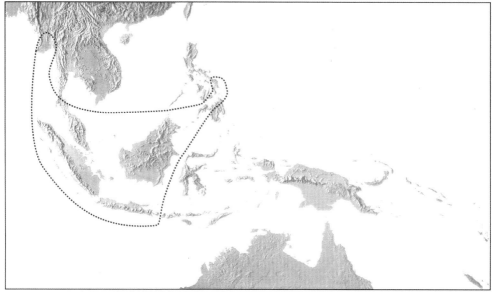

Fig. 17. (above) The geographical range of *Dendrochilum* subgenera *Acoridium* (broken line) and *Platyclinis* (solid line). **Fig. 18.** (below) The geographical range of *Dendrochilum* subgenus *Dendrochilum.*

Bornean taxa can be more or less gathered into one or more of four distribution groups:

1. *Local endemics*: those restricted to a single mountain, part of a mountain range or district.
2. *Regional endemics*: those restricted to two or more adjacent mountain ranges without notable disjunctions.
3. *Island endemics*: those restricted to Borneo but exhibiting more disjunct and/or wide distributions than in regional endemics.
4. *Non-endemics*: those also occurring outside Borneo.

Distribution patterns of endemics

The ranges of Bornean endemics can be compared using small geographical areas termed 'range units' (Vermeulen 1993). The range of each species or variety is equivalent to the sum of range units. The range units identified after study of a large number of extant collections show a heterogenous pattern of distribution at the regional and national level.

Local endemics

Following analysis of the ranges of all of the taxa within this group, 21 different range units of local endemism were recognised accommodating 38 taxa (see Table 2):

Brunei
1. Belait District, Bukit Batu Patam.

Kalimantan
2. Kalimantan Barat.
3. Mount Kemul, Kalimantan Timur.
4. Kalimantan Timur.

Sabah
5. Mount Kinabalu massif, including Mount Tembuyuken.
6. Mount Trus Madi.
7. Mount Lumaku.
8. Crocker Range, excluding the above mountains.
9. Sipitang District.

Sarawak
10. Mount Murud.
11. Mount Mulu.
12. Mount Dulit.
13. Mount Matang.
14. Tama Abu Range (Mount Temabok), upper Baram valley.
15. Northern Hose Mountains.
16. Lanjak Entimau Protected Forest.
17. Limbang District.
18. Undup in Sri Aman District.
19. Baram District
20. Mount Penrissen

North Natuna Archipelago
21. Natuna Besar (Bunguran Island).

Table 2. The range of local endemic taxa expressed in range units and generalised ranges.

Subgenus / Section / Taxon	1	2	3	4	5	6	7	8	9	10	11	12	13	14	15	16	17	18	19	20	21	Generalised range
Acoridium																						
Acoridium																						
auriculilobum									9													Sabah
Dendrochilum																						
pallidiflavens var. brevilabratum																			19			Sarawak
pallidiflavens var. oblongum	1																					Brunei
Platyclinis																						
Cruciformia																						
cruciforme var. longicuspe					5																	Sabah
devogelii									9													Sabah
grandiflorum					5																	Sabah
haslamii var. haslamii					5																	Sabah
haslamii var. quadrilobum											11											Sarawak
hastilobum																	17					Sarawak
hosei															15							Sarawak
Eurybrachium																						
acuiferum					5																	Sabah
alatum					5																	Sabah
alpinum					5																	Sabah
corrugatum					5																	Sabah
lewisii											11											Sarawak
microscopicum			3																			Kalimantan
pseudoscriptum					5																	Sabah

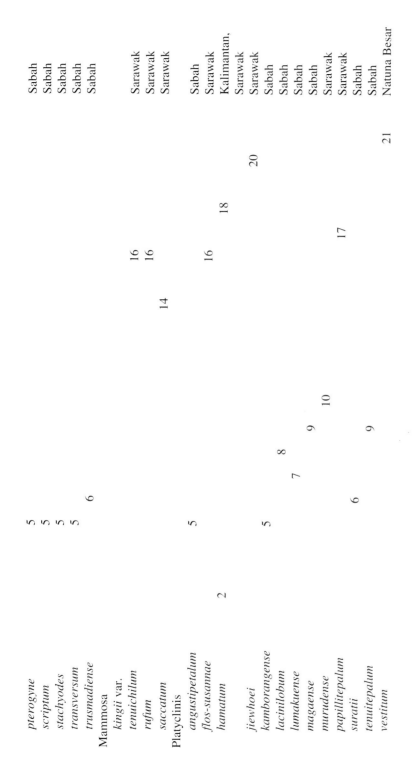

37

Regional endemics

Comparison of the ranges of all taxa within this group produced a total of 13 range units accommodating 14 taxa (see Table 3):

1. *Sarawak*: lowlands and hills, especially around Bau, Bidi and Kuap.

2. *Sarawak*: Mount Matang, Mount Santubong and Mount Serapi.

3. *Sarawak*: Mount Dulit and Mount Murud; Sabah: Sipitang District.

4. *Sarawak*: Tama Abu Range (Mount Temabok), upper Baram Valley, and Hose Mountains; Brunei: Belait District.

5. *Sarawak*: Mount Dulit and Hose Mountains; Brunei: Mount Retak.

6. *Sarawak*: Mount Dulit and Mount Batu Buli.

7. *Sarawak*: Limbang District; Brunei: Belait District; Sabah: Sipitang District.

8. *Brunei*: Mount Retak; Kalimantan Barat: Tilung Hill (Bukit Tilung).

9. *Brunei*: Temburong District; Sabah: Mount Trus Madi.

10. *Sabah*: Mount Kinabalu and Crocker Range.

11. *Sabah*: Crocker Range and Mount Monkobo.

12. *Sabah*: Mount Kinabalu, Mount Alab and Maliau Basin.

13. *Kalimantan Timur*: Mount Beratus, Kutai.

Table 3. The range of regional endemic taxa expressed in range units and generalised ranges.

Subgenus / Section / *Taxon*	1	2	3	4	5	6	7	8	9	10	11	12	13	Generalised range
Subgenus														
Acoridium														
Acoridium														
hologyne			3											Sarawak, Sabah
Falsiloba														
pandurichilum							7							Sarawak, Brunei, Sabah
Platyclinis														
Cruciformia														
cruciforme var. *cruciforme*												12		Sabah
dolichobrachium													13	Kalimantan
Eurybrachium														
minimiflorum						6								Sarawak
Platyclinis														
angustilobum										10				Sabah
dulitense					5									Sarawak, Brunei
globigerum		2												Sarawak
gracile var. *bicornutum*									9					Brunei, Sabah
havilandii	1													Sarawak
johannis-winkleri								8						Brunei, Kalimantan
pubescens				4										Sarawak, Brunei
sublobatum			3											Sarawak
tenompokense var. *papillilabium*											11			Sabah

39

Island endemics

Comparison of the ranges of all taxa within this group produced a total of 9 especially rich primary range units accommodating 25 taxa (see Table 4):

Sabah
1. Mount Kinabalu Massif (including Mount Tembuyuken)
2. Crocker Range (including Mount Alab, Mount Anginon, Mount Lumaku, Mount Trus Madi)
3. Southwest Sabah, Sipitang District (Maga River/Long Pa Sia area and Rurun River headwaters).

Sarawak
4. Mount Murud and Tama Abu Range.
5. Gunung Mulu National Park.
6. Mount Batu Lawi and adjacent areas of Limbang District.
7. Mount Dulit.
8. Western Sarawak Mountains (including Mount Penrissen, Mount Pueh, Mount Rumput, Mount Santubong).

Kalimantan Timur
9. Apo Kayan area.

Table 4. The range of island endemic taxa occurring within primary range units; generalised range on the right.

Taxon	1	2	3	4	5	6	7	8	9	Generalised range
Subgenus										
Section										
Taxon										
Platyclinis										
Cruciformia										
exasperatum	1		3				7			Sabah, Sarawak
gibbsiae	1	2		4	5				9	Kalimantan, Sabah, Sarawak
Platyclinis										
anomalum	1	2	3				7		9	Kalimantan, Sabah, Sarawak
crassifolium	1	2	3	4						Sabah, Sarawak
crassilabium	1				5	6			9	Kalimantan, Sabah, Sarawak
geesinkii	1					6			9	Kalimantan, Sarawak
gracilipes		2	3	4	5		7			Brunei, Sabah, Sarawak
graminoides		2		4						Brunei, Sabah, Sarawak
imbricatum	1		3	4		6			9	Kalimantan, Sabah, Sarawak
imitator	1		3		5					Sabah, Sarawak
integrilabium		2	3				7			Sabah, Sarawak
lacteum		2						8		Sabah, Sarawak
lancilabium	1	2		4	5					Kalimantan, Sabah, Sarawak
longipes				4	5				9	Kalimantan, Sarawak
longirachis	1	2	3	4			7		9	Kalimantan, Sabah, Sarawak
mucronatum								8	9	Kalimantan, Sarawak
muluense		2		4	5	6			9	Brunei, Kalimantan, Sabah, Sarawak
ochrolabium		2	3					8		Sabah, Sarawak
oxylobum		2								Kalimantan, Sabah, Sarawak
pachyphyllum			3			6				Sabah, Sarawak
planiscapum		2	3							Sabah
simplex	1	2		4		6		8		Kalimantan, Sabah, Sarawak
subintegrum	1		3							Sabah, Sarawak
subulibrachium					5	6				Kalimantan, Sarawak
tenompokense var.										
tenompokense	1	2			5					Sabah, Sarawak

41

An additional group of 19 secondary range units, from where a fewer number of island endemic taxa have been recorded, were also recognised. These accommodate 13 taxa (see Table 5):

Brunei
1. Mount Pagon and Mount Retak.

Kalimantan
2. Raya Hill (Bukit Raya)
3. Kotawaringin
4. Liangangang
5. Long Petak
6. Mount Kemul
7. Talaj River (Sungai Talaj)
8. Serawai, Uut Labang River

Sabah
9. Mount Monkobo
10. Mount Tawai
11. Nabawan area
12. Sandakan District
13. Tenom District

Sarawak
14. Bario area
15. Belaga District (including Batu Laga)
16. Hose Mountains (including Bukit Batu)
17. Kapit District (Entulah/Balleh Rivers)
18. Kuching area
19. Tama Abu Range (Mount Temabok), upper Baram Valley

Fig. 19. (Following pages) Gunung Mulu National Park, Mount Mulu, Sarawak, showing upper montane forest on sandstone; type locality of *D. haslamii* var. *quadrilobum*, *D. lewisii* and *D. muluense*; other species found on the mountain include *D. gracilipes*, *D. lancilabium*, *D. longipes*, *D. subulibrachium* and *D. tenompokense* var. *tenompokense*. (Photo: Hans P. Hazebroek)

Table 5. The range of island endemic taxa occurring within secondary range units; generalised range on the right.

Subgenus / Section / *Taxon*	Secondary range units																			Generalised range
	1	2	3	4	5	6	7	8	9	10	11	12	13	14	15	16	17	18	19	
Dendrochilum																				
gravenhorstii							7				11	12		14						Kalimantan, Sabah, Sarawak
Platyclinis																				
Cruciformia																				
gibbsiae										10					15					Kalimantan, Sabah, Sarawak
Platyclinis																				
gracilipes	1														15	16			19	Brunei, Sabah, Sarawak
graminoides	1									10										Brunei, Sabah, Sarawak
integrilabium								8							15				19	Kalimantan, Sabah, Sarawak
lancilabium						6														Kalimantan, Sabah, Sarawak
longipes		2				6														Kalimantan, Sarawak
muluense	1																			Brunei, Kalimantan, Sabah, Sarawak
ochrolabium			3														17			Sabah, Sarawak
oxylobum											11							18		Kalimantan, Sabah, Sarawak
simplex				4					9		11					16	17			Kalimantan, Sabah, Sarawak
subintegrum															15					Sarawak
subulibrachium					5														19	Kalimantan, Sarawak

43

Non-endemics

A total of eight taxa occur outside Borneo, which can be placed into one of three subcategories:

1. Species mainly distributed within Borneo

D. dewindtianum. A common and variable plant on Mount Kinabalu, this species is also recorded from Mount Alab and Mount Trus Madi in Sabah and from Mount Dulit and Mount Murud in Sarawak. Outside of Borneo it has been recorded, as *D. furfuraceum* J.J. Sm., from West Sumatra.

D. galbanum. Restricted, so far as is known, to Mount Murud in Borneo, it has also been found on Mount Ketambe in Mount Leuser National Park, Aceh, Sumatra.

2. Species mainly distributed outside of Borneo

D. glumaceum. This species has only been recorded once in Borneo, at Kotawaringin in Central Kalimantan, but is otherwise widespread in the Philippines.

Fig. 20. Gunung Mulu National Park, Mount Benarat/Melinau Gorge, Sarawak. The largest undisturbed areas of limestone vegetation on alluvial soils in Sarawak are in the Melinau river drainage at the bases of Mount Api and Mount Benarat. (Photo: Hans P. Hazebroek)

D. gracile var. *gracile*. The typical variety of *D. gracile* is recorded in Borneo from western Sarawak, eastern Sabah and Mount Beratus in East Kalimantan. It is also widespread in Peninsular Malaysia, Sumatra, Java and the Lesser Sunda Islands.

D. kingii var. *kingii*. The typical variety is recorded from numerous sites in Sarawak and Sabah, but only from Sangkulirang in East Kalimantan and Keribung River in Kalimantan (province unknown). Outside of Borneo it is found in Peninsular Malaysia and Palawan Island in the Philippines.

D. longifolium. This species is recorded in Borneo from Sarawak, Sabah and Kalimantan. Outside of Borneo it is widespread, occurring from Myanmar (Burma) in the west to New Britain (part of the Bismarck Archipelago, Papua New Guinea) in the east.

D. pallidiflavens var. *pallidiflavens*. The typical variety is widespread throughout Borneo. Two varieties are endemic to Borneo. The typical variety is widespread outside Borneo where its range extends from Myanmar (Burma) in the west to the Philippines in the east.

3. Species distributed in Malaysia only

D. crassum. Originally described from Perak in Peninsular Malaysia, this species is also found in the mountains of Sabah.

Distribution patterns of *Dendrochilum* in Borneo

Study of the endemism of *Dendrochilum* within Borneo shows that the regional endemics, which make up 14.9% of the total number of taxa, are a heterogenous group comprising species with a narrow distribution, and others that are more widespread. The island endemics, which represent 30.2% of the total, is also a heterogenous group comprising locally common species with a somewhat continuous distribution, e.g., *D. muluense* and *D. simplex*, and rarer species with a more disjunct distribution, e.g., *D. geesinkii* and *D. ochrolabium*.

All except one species of subgenus *Platyclinis* section *Eurybrachium* and most of section *Cruciformia* are narrowly distributed local endemics, the majority occurring only on Mount Kinabalu. Most of subgenus *Platyclinis* section *Platyclinis* are either local or island endemics. It is interesting that all except one of the island endemics belong to section *Platyclinis*.

It should be stressed that distributional trends, especially at the regional and island level, often reflect where the greatest concentration of collecting activity has taken place and not necessarily areas of greatest diversity. Montane areas of southern and east-central Sarawak and large portions of upland Kalimantan, which are probably rich in *Dendrochilum*, still remain where little or no collecting has taken place. On the other hand, areas such as Mount Kinabalu in Sabah or Mount Batu Lawi, Mount Mulu and Mount Murud in Sarawak, which have received more attention from scientists, are also areas of high *Dendrochilum* diversity.

Centres of *Dendrochilum* diversity in Borneo

Two centres of high diversity, at specific and sectional level, each containing a large number of endemics, can be recognised in Borneo:

1. *Mount Kinabalu Diversity Centre* (KDC). The Kinabalu Massif is the richest area with a total of 34 species represented. Of these, 14 are local endemics, 3 are regional endemics, 15 are national endemics, and 2 are non-endemic.
2. *Eastern Sarawak Diversity Centre* (ESDC). This region includes Sarawak's highest mountain, Mount Murud (2438 m), Mount Mulu, the Tama Abu Range, Mount Batu Buli, Mount Dulit and other mountainous areas. A total of 30 species are represented, of which 3 are local endemics, 4 are regional endemics, 20 are national endemics, and 3 are non-endemic.

47

Many national endemic species of *Dendrochilum* are scattered throughout other regions of Borneo, but diversity is much lower and restricted to isolated pockets. Potentially rich areas such as the many mountains along the Kalimantan/Sarawak border, e.g., Batu Tiban Hill (Bukit Batu Tiban) (1749 m), Mount Cemaru (1681) and Mount Liangpran (2240 m), and much of the Crocker Range between Mount Kinabalu and Keningau in Sabah remain to be investigated. Undescribed taxa almost certainly occur in these and other areas and, in addition, the ranges of many existing species will no doubt be expanded.

In Borneo areas of low diversity correspond to the coastal fringes (including offshore islands), peat-swamp forests, lowland dipterocarp and, to a lesser extent, hill forests, particularly in Kalimantan.

Taxonomic correlations of distribution patterns in Borneo

No subgenus or section is endemic to either the Kinabalu Diversity Centre (KDC) or Eastern Sarawak Diversity Centre (ESDC). Subgenus *Acoridium* is found in the ESDC and neighbouring Sipitang District in Sabah, but is absent from the KDC. Its endemic section *Falsiloba* is absent from both diversity centres. The predominantly West Malesian subgenus *Dendrochilum* (no sections recognised) occurs in the KDC, ESDC and other primarily low elevation localities. Of subgenus *Platyclinis*, section *Cruciformia* occurs in the KDC, ESDC and is scattered elsewhere. Section *Eurybrachium* is mostly restricted to the KDC but with outlying sites on Mount Trus Madi in Sabah, Mount Batu Buli and Mount Dulit in Sarawak and Mount Kemul in Kalimantan Timur. Only *D. lewisii* is represented in the ESDC. Section *Mammosa* is only represented in the ESDC and in scattered low elevation localities elsewhere. The largest section, *Platyclinis*, is well represented and widespread in the KDC and ESDC and elsewhere, including areas of low diversity, throughout Borneo.

Fig. 21. The geographical range of *Dendrochilum* subgenus *Acoridium* sections *Acoridium* (broken line) and *Falsiloba* (solid line).

Fig. 22. (above) The geographical range of *Dendrochilum* subgenus *Platyclinis* sections *Cruciformia* (solid line) and *Eurybrachium* (broken line). **Fig. 23.** (below) The geographical range of *Dendrochilum* subgenus *Platyclinis* sections *Mammosa* (broken line) and *Platyclinis* (solid line).

Diversity at sectional and specific level occurs in the KDC, ESDC and many areas of lower diversity, while subgeneric diversity is highest in the ESDC and adjacent areas of Southwest Sabah.

Table 6. The number of taxa representing each of the Bornean subgenera and sections of *Dendrochilum* in the Kinabalu Diversity Centre (KDC) and Eastern Sarawak Diversity Centre (ESDC). E = endemic to Borneo; P = taxa with their centre of diversity in the Philippines; PM, B & P = taxa distributed in Peninsular Malaysia, Borneo and the Philippines (Palawan only); S & B = taxa with their centre of distribution in Sumatra and Borneo; W = widespread taxa.

Subgenus/section	KDC	ESDC	Total range
Acoridium	0	2	P
Acoridium	0	1	P
Falsiloba	0	1	E
Dendrochilum	2	2	W
Platyclinis	33	36	W
Cruciformia	6	4	E
Eurybrachium	10	3	S & B
Mammosa	0	2	PM, B & P
Platyclinis	17	27	W

Orographical and climatic correlations

The two centres of greatest diversity recognised above are predictably located in the areas of highest relief on the island. The pockets of local endemism show a strong tendency to occupy prominent mountains such as Mount Kinabalu, Mount Murud and Mount Trus Madi, or parts of mountain ranges such as the Crocker Range in Sabah and Hose Mountains in Sarawak. The reverse was found to be the case for local endemicity in the Philippines (Pedersen 1997).

The distribution patterns of *Dendrochilum* in Borneo are clearly a reflection of local temperature (especially minimum values), height of cloud-base, aspect and degree of exposure, and forest type, especially the type of canopy and the degree of light penetration within it. Many species seem to favour ridge-top shrubbery and mossy elfin-forest which experience variable and often harsh conditions ranging from bright sunlight to dense cloud cover or drying winds. These localities consequently experience a large fluctuation in temperature.

An interesting correlation exists between the degrees of endemism and the tendency for restriction of taxa to higher elevations. Local and island endemics are particularly well represented above 1500 metres, while regional endemics are evenly represented below 1000 metres and above 1500 metres (see Table 7).

Table 7. The percentage of *Dendrochilum* taxa within each of the distributional categories represented below 1000 m a.s.l. and above 1500 m. a.s.l. in Borneo. 'N' is the number of taxa within each category for which altitudinal data are available. 'T' is the total number of taxa within each category.

Distributional category	below 1000 m	above 1500 m	N	T
Local endemics	27 %	64.8 %	37	38
Regional endemics	50 %	50 %	14	14
Island endemics	32.4 %	40.5 %	37	37
Non-endemics	71.4 %	57.1 %	7	8

The Origin of high-elevation taxa in the Kinabalu Diversity Centre

Mount Kinabalu (4095 m), the highest mountain between the Himalayas and Irian Jaya (Indonesian New Guinea), is an isolated granitic batholith which, at *c.* 1.5 million years old, is comparatively young in geological terms. Geographical isolation and recent geological origin are reflected by a diverse, insular flora with a high percentage of endemicity. Wood *et al.* (1993) reported a 10 : 1 ratio of endemic : non-endemic orchid species above an elevation of 2700 m.

It is now generally accepted that extensive migration of alpine Malesian floras took place during the Pleistocene when the limits of the altitudinal zones of taxa were depressed and hence the opportunity for range expansion increased. Some of the present day high- and mid-elevation endemic taxa of *Dendrochilum* may have evolved in Borneo after the Pleistocene. These neoendemics have continued to inhabit small and isolated mountain islands amid a sea of lowland forest. Consequently, the opportunity to widen their range through migration has always been limited by available time and their montane environment compared with their lowland counterparts. Many taxa are further restricted from migration by their preference for ultramafic habitats and associated mycorrhizal flora. Most ultramafic localities are isolated and, rather like islands, act as barriers to dispersal. An obvious example is the large number of endemic species of section *Eurybrachium* restricted to such localities on Mount Kinabalu. In addition, there are no neighbouring mountains of comparable elevation upon which new populations could establish via wind-blown seed. The relatively few exclusively low elevation Tropical zone taxa may or may not have evolved earlier. Within *Dendrochilum*, however, about 93% of species are endemic to a single island and it is unlikely that the majority of this large genus could have evolved since the end of the Pleistocene. This high level of endemicity, especially among species inhabiting lower and mid altitude elevations, may seem surprising since orchid seeds are dust-like and presumably able to be carried long distances by the wind. Their lack of food reserves and susceptibility to desiccation, though, may explain why long distance dispersal is very often unsuccessful.

Dendrochilum, represented by 32 species on Mount Kinabalu, (excluding Mount Tembuyuken) is one of the most species-rich genera on the mountain. In addition, it is biogeographically very interesting because of the high percentage of endemism present above 2000 m. Section *Eurybrachium*, for example, is represented by nine endemic species on Mount Kinabalu, seven of which are most abundant at sites above 2600 m. Of these, *D. alpinum, D. pterogyne* and *D. stachyodes* are commonest above about 3200 m.

Section *Eurybrachium* is widespread today throughout Sundaland, i.e. Sumatra, Java, Borneo and the southern Philippines. This suggests that the ancestor, or ancestors, of modern upper montane endemics were either capable of long distance dispersal during periods of island isolation, or were formerly more widespread when land bridges were present. It is interesting that there are striking morphological similarities between, for example, *D. alpinum* and *D. pterogyne* and species from Sumatra, the Philippines and areas of lower elevation in Borneo.

Biogeographical hypotheses in a phylogenetic context and the evidence for vertical radiation

Van Steenis (1965) and Stapf (1894) postulated that certain taxa were either centric, having evolved *in situ*, or eccentric, having " migrated from remote centres of development" (van Steenis 1965). Despite its relative isolation, van Steenis (1965) suggested that intimate montane affinities exist between Mount Kinabalu and Sulawesi, Sumatra, and several islands of the Philippines. These islands form part of the Sunda shelf and were probably joined to one another when sea levels dropped during the Pleistocene (Roe 1965, Wood & Lamb 1994).

Evolution of high altitude taxa through vertical altitudinal radiation of lowland congeners was hypothesised by van Steenis (1965). The Kinabalu endemic, *Leptospermum recurvum* Hook.f. (Myrtaceae) is thought by Lee & Lowry (1980), for example, to have arisen from a low elevation population of *L. flavescens* Sm. This pattern of speciation was refuted, however, by Patton & Smith (1992) who, while studying Andean mice, found a vicariant or horizontal radiation pattern of speciation. Which of these patterns of speciation offers the best explanation for present day endemicity among upper montane *Dendrochilum* on Mount Kinabalu?

Barkman tested centric and eccentric hypotheses of ancestry by comparing expected branching patterns of ancestry with observed phylogenetic branching patterns. The probable monophyletic origin of core *Eurybrachium* taxa corresponding to Carr's section *Eurybrachion*, including *D. alpinum, D. pterogyne, D. scriptum* and *D. transversum*, were investigated. Montane taxa from neighbouring islands, with a mostly similar floral morphology, were compared with possible lower elevation ancestors from Mount Kinabalu, the Crocker Range and Mount Trus Madi. Areas outside of Borneo chosen were Mount Kemiri (Leuser) and Danau (Lake) Toba in Sumatra, and Mount Apo in the Philippines. In total 24 taxa belonging to various sections of subgenus *Platyclinis* were sampled from Borneo, Sumatra and the Philippines. A species of *Pholidota* was chosen as an outgroup in order to root the tree. The taxa sampled are listed below:

Section *Eurybrachium* from Borneo:
  D. alpinum Carr
  D. cupulatum J.J. Wood
  D. joclemensii Ames
  D. pseudoscriptum T.J. Barkman & J.J. Wood
  D. pterogyne Carr
  D. scriptum Carr
  D. stachyodes (Ridl.) J.J. Sm.
  D. transversum Carr
  D. trusmadiense J.J. Wood

Section *Eurybrachium* from Sumatra:
    D. dewildei J.J. Wood & J.B. Comber
    D. karoense J.J. Wood
    D. ovatum J.J. Sm.
Section *Eurybrachium* from the Philippines:
    D. graciliscapum Ames
Section *Cruciformia* (endemic to Borneo):
    D. exasperatum Ames
    D. gibbsiae Rolfe
    D. grandiflorum (Ridl.) J.J. Sm.
    D. haslamii Ames
Section *Platyclinis* from Borneo:
    D. dewindtianum W.W. Sm. (highland and mid-elevation individuals)
    D. kamborangense Ames
    D. muluense J.J. Wood
    D. tenompokense Carr
Section *Platyclinis* from the Philippines:
    D. arachnites Rchb.f.
    D. cobbianum Rchb.f.
    D. glumaceum Lindl.

Twenty seven consensus trees were estimated with the same likelihood score in the unconstrained analysis. The Kishino–Hasegawa test (Kishino & Hasegawa, 1989), which uses the mean and variance of support score differences between competing trees, was used to assess the significance of these differences in the Bornean, Sumatran and Philippine origin hypotheses. The results obtained provided little evidence to support a close relationship between the Kinabalu endemics and species from Sumatra or the Philippines. In addition, *D. cupulatum* and *D. joclemensii* did not appear to be closely related to high level endemics, in spite of a similar floral morphology. The molecular evidence provided by Barkman demonstrates that the Kinabalu high endemics and those from neighbouring islands are unrelated, and any gross morphological or ecological similarities are purely coincidental. It can therefore be inferred that the high elevation Kinabalu endemics are centric in origin, having evolved *in situ*, and belong to a single lineage. This corroborates the suggestion proposed by van Steenis (1965) that montane endemics have arisen from widespread lower elevation taxa. A similar pattern of centric evolution has been identified within *Dacrycarpus* (Podocarpaceae) by de Laubenfels (1988) and the orchid genus *Calanthe* (Chan & Barkman, 1997). It is notable that Holloway (1996) also refers to many 'duplexes', or high elevation species with lower elevation sister species, in several genera of Kinabalu butterflies and moths.

Barkman identified a pattern of high elevation taxa nested within or sister to a group of lower elevation species. Clearly, lower elevation ancestors would need to be able to adapt in order to successfully colonise an alien habitat formed by the uplift of the mountain repeatedly within the last 1.5 million years. Various factors, such as lower temperature, higher ultraviolet light levels and water stress, lower nutrient availability, and different or absence of pollinators would have to be overcome. It would be interesting to discover which morphological characters, that covary with elevation in a phylogenetic context, facilitate adaptation to a high-elevation environment.

Horizontal high-elevation vicariant radiation on Mount Kinabalu

Some high elevation endemics on Mount Kinabalu have poorly resolved relationships and could not have arisen via vertical radiation diversification. Sympatric monophyletic taxa, including *D. alpinum, D. pseudoscriptum, D. pterogyne, D. scriptum* and *D. transversum*, with ancestry from a lower elevation congener, may have diversified at high elevations via vicariant radiation. Kitayama (1997) has reported several angiosperm genera to be represented by multiple congeners on Mount Kinabalu. Here diversification of sympatric congeners is explained by competitive niche exclusion whereby a newly evolved species is only able to enter a community if it colonises a new, unused niche or is able to exclude competing species from its current niche.

Kitayama's model would predict that high elevation species of *Dendrochilum* currently occupy different niche spaces. Barkman considers niche differentiation in the lineage including *D. pterogyne* and *D. scriptum* to be related to pollination rather than space. These primarily epiphytic species, which often occupy the same trunk or branch of a tree, have different reproductive characters. The flowers of *D. pterogyne* are unscented and held upright, while those of *D. scriptum* have a rather unpleasant, musty odour and are held upside down on mostly erect inflorescences. Here the two species utilise the same ecological niche, but partition the reproductive niche by using the same pollinator (or pollinators) at different times. This avoids competition for the same reproductive niche space. Other taxa may partition space in which to grow but utilize similar reproductive space. Barkman also mentions the possibility that, if these taxa have recently evolved and are still occupying the same vegetative space, this system may not be at equilibrium: "Sufficient time may not have elapsed for them to have undergone sufficient competition to exclude one another or undergo some process of niche differentiation via character displacement."

TAXONOMY

Dendrochilum Blume

Bijdr.: 398–399, Tab. 2 (1825). - Pedersen, Wood & Comber 1997: 15. - Lectotype: *Dendrochilum aurantiacum* Blume (designated by Bechtel *et al.* 1981: 127).

Very small to medium-sized, sympodial herbs with an epiphytic, lithophytic, or terrestrial habit of growth. *Roots* borne from the rhizome and sometimes from the bases of the pseudobulbs, thin to fleshy, usually more or less branched. *Rhizome* short to elongate, usually more or less branched, initially covered with imbricate, scarious cataphylls which soon disintegrate into fibres and finally disappear. *Pseudobulbs* clustered to widely separated on the rhizome, long-lived, each consisting of a single internode, subspherical to terete, initially covered with imbricate cataphylls which soon disintegrate into fibres which finally disappear. *Leaves* 1 per pseudobulb, long-lived, convolute or conduplicate, articulate, distinctly petiolate (rarely subsessile); petiole channelled to terete; blade dorsiventrally complanate to terete, elliptic to filiform, rounded to acuminate, thin-textured to leathery or rigid, 1-nerved or containing a prominent mid nerve and 2–10 distinct (and several to many indistinct) parallel nerves (small transverse nerves are also present in a few species). *Inflorescences* racemose, synanthous (to hysteranthous) or heteranthous, differentiated into a (sometimes very short) peduncle and a rachis; flowering commencing from the basal, central or top part of the rachis, or all flowers opening more or less simultaneously; peduncle stout to very slender, frequently curved in the latter case, sometimes tightly enclosed by the unexpanded leaf as a whole or squeezed into a median slit of the otherwise expanded leaf; rachis erect to pendent, few- to many-flowered, lax to dense, nearly always quadrangular in transverse section and furrowed along two opposite sides, usually separated from the peduncle by one or more imbricate, adpressed, sterile non-floriferous bracts; flowers distichously alternating, but the rachis axis sometimes spirally twisted so as to produce an inflorescence with flowers facing all sides (quaquaversal); floral bracts persistent, minute to large and conspicuous, glumaceous, more or less scarious (rarely membranous), semi-orbicular to lanceolate, rounded to acuminate, quite entire to distinctly erose, in most species containing many conspicuous nerves. *Indumentum* of the vegetative parts (and sometimes the sepals, petals, and ovary) present to a various extent and consisting of minute, scattered brown or blackish trichomes. *Flowers* less than 2 cm across (rarely a little larger), long-lasting, developing subsimultaneously within an inflorescence, non-resupinate or resupinate. Sepals and petals free (lateral sepals connate in the Philippine *D. exiguum*), straight to gradually curved (never bent in a right angle), flat to somewhat concave, linear to suborbicular, rounded to acuminate, usually 1- or 3-nerved; sepals always quite entire; petals entire or sometimes erose to fimbriate or somewhat serrate. *Labellum* narrowly and either firmly or elastically attached to the column-foot when present, or to the base of the column, straight to somewhat (but never sigmoidly) curved at the base, sessile or shortly

55

clawed, entire or variously 3-lobed (very rarely bilobed), smooth to finely papillose, flat to somewhat concave (rarely saccate with the sac usually being distinctly indented along the mid-line), disc usually ornamented with relatively small (rarely prominent) keels or calli and often provided with a nectar secreting, longitudinal, median furrow. *Column* sub- or semiterete to somewhat clavate, sometimes provided with two stelidia and sometimes at the top prolonged into an apical wing or hood; foot absent or short (rarely long), flat (very rarely deeply concave); anther-cap terminal to somewhat ventral, strongly incumbent, flexibly attached by a short connective, biloculate to imperfectly quadriloculate, dehiscing ventrally by slits, usually with a variously shaped wart on top; pollinia 4, superposed, relatively hard, entire, slenderly clavate to subspherical, provided with very short to somewhat elongate caudiculae; stigma entire, concave, margin often elevate, very rarely extended into a ventral flange; rostellum entire and flat, rarely deeply bilobed and channelled (in subgenus *Pseudacoridium* only), producing a minute, orbicular to quadrate, and extremely fragile viscidium but no stipes. *Capsule* uniloculate, ellipsoid to spherical with 3 jugae and 3 valvae with median keel and often with the extant, shrivelled column forming a small beak, dehiscing by longitudinal slits; seed fusiform, with a smooth appearance and prosenchymatic central testa cells, or filiform with a somewhat spiny appearance and nearly isodiametric central testa cells; in the latter case a tiny, clavate elaiosome may be present at the distal end of the seed.

KEY TO THE SUBGENERA OF DENDROCHILUM

1. Column at the top prolonged into an apical wing or hood which distinctly (in the Philippine *D. auriculare* indistinctly) exceeds the anther; stelidia present or absent. Labellum variously shaped but never transversely E-shaped 3
   Column sometimes with an apical hood but never distinctly prolonged beyond the anther; stelidia always absent. Labellum cupulate or pandurate (in Bornean species), or often transversely E-shaped (in Philippine species) 2

2. Rostellum deeply bilobed, distinctly channelled .... *Pseudacoridium* (Philippines only)
   Rostellum entire, flat (lateral margins sometimes recurved or slightly incurved) ....
   .... *Acoridium* (p. 56)

3. Inflorescence synanthous (to nearly hysteranthous). Rhizome short to elongate (if elongate then usually pendent). Labellum entire to distinctly 3-lobed. Seed fusiform, 0.9–2.4 mm long; central testa cells prosenchymatic, with the junctions of the anticlinal walls not protruding .... *Platyclinis* (p. 79)
   Inflorescence heteranthous. Rhizome elongate, creeping. Labellum entire (more or less pandurate). Seed filiform, 4.4–6.7 mm long; central testa cells isodiametric, with the junctions of the anticlinal walls conspicuously protruding .... *Dendrochilum* (p. 64)

SUBGENUS ACORIDIUM

Dendrochilum subgenus **Acoridium** (*Nees & Meyen) Pfitzer & Kraenzl.* in Engler, *Pflanzenreich* 4, 50, 2 B 7: 87, 111 (1907) - Butzin 1974: 254, 255; 1984–1986: 944; de Vogel 1989: 103; Rosinski 1992: 92, 96, 101, 201.

Acoridium Nees & Meyen, *Nova Acta Acad. Caesar. Leop. Carol.* 19, Suppl. 1: 131 (1843) - Endlicher 1843: 59, Böckeler 1879: 158–159; Bentham 1883b: 1043; Rolfe 1904: 219; Ames 1922: 80; Smith 1934: 201, 211; Ames 1937: 66–68; Schweinfurth 1959: 518; Farr *et al.* 1979; Dressler 1981: 215; Burns-Balogh 1989: 35, 39; Bechtel *et al.* 1992: 47; Gunn *et al.* 1992: 222; Webster 1992: 1.4, 4.14; Greuter *et al.* 1993: 10. - Holotype: *Acoridium tenellum* Nees & Meyen.

Acoridium sect. *Euacoridium* Ames, *Proc. biol. Soc. Wash.* 19: 143 (1906) *nom. illeg.* - Type not designated.

Dendrochilum sect. *Acoridium* (Nees & Meyen) Ames, *Orchidaceae* 2: 78–79 (1908) - Ames 1908: 5; Kränzlin 1908: 38; L.O. Williams 1951: 281–295.

Rhizome short, pseudobulbs more or less clustered. *Leaves* convolute or conduplicate. *Inflorescences* synanthous or heteranthous. *Labellum* firmly attached to the column, entire to distinctly 3-lobed (often transversely E-shaped in Philippine species). *Column* sometimes with an apical hood but never distinctly prolonged beyond the anther; foot and stelidia absent. *Rostellum* entire, flat (lateral margins sometimes recurved or slightly incurved).

KEY TO THE SECTIONS OF SUBGENUS *ACORIDIUM* IN BORNEO

Column distinctly shorter than dorsal sepal; foot absent. Labellum sessile, entire or distinctly 3-lobed, without 2 prominent, wing-like keels reminiscent of side lobes .... *Acoridium* (p. 57)
Column subequal to dorsal sepal; short foot present. Labellum distinctly clawed, entire, with 2 prominent, suberect, wing-like keels reminiscent of side lobes .... *Falsiloba* (p. 62)

Section **Acoridium**

Leaves conduplicate, dorsiventrally complanate to terete, often very narrow and xeromorphic. *Inflorescences* synanthous, peduncle free or for some distance adherent to or tightly enclosed in the subtending leaf at the time of flowering. *Flowering* commencing from the bottom of the rachis, very rarely from the top of the rachis (only in the Philippine *D. oliganthum* and occasionally in the Philippine *D. williamsii*). *Petals* quite entire to finely erose. *Labellum* sessile to shortly clawed, entire to 3-lobed, never provided with two prominent, wing-like keels reminiscent of side lobes. *Column* distinctly shorter than the dorsal sepal; foot absent.

DISTRIBUTION. Borneo and the Philippines.

KEY TO THE SPECIES OF SUBGENUS *ACORIDIUM* SECTION *ACORIDIUM* IN BORNEO

Pseudobulbs 2.8–3 cm long. Inflorescences with flowers borne 2–2.5 mm apart. Labellum with auriculate side lobes, disc with a fleshy, transverse basal ridge. Sepals 4 mm long .... 1. *D. auriculilobum* J.J. Wood
Pseudobulbs 10–14.5 cm long. Inflorescences with flowers borne 1 mm apart. Labellum entire, triangular-ovate to obliquely subquadrate, disc lacking basal ridge. Sepals 2–2.5 mm long .... 2. *D. hologyne* Carr

1. DENDROCHILUM AURICULILOBUM

Dendrochilum auriculilobum J.J. Wood in Wood and Cribb, *Checklist orch. Borneo*: 165, fig. 22, H & J (1994). Type: Malaysia, Sabah, Sipitang District, Rurun River headwaters, *c.* 1700 m, December 1986, *Vermeulen & Duistermaat* 1057 (holotype L!, herbarium material only, isotype K!, spirit material only). **Fig. 24.**

DESCRIPTION. *Terrestrial. Roots* 1 mm in diameter, sparingly branched. *Cataphylls* 3.5–8 cm long, pale fawn speckled pale brown, enclosing young pseudobulbs and basal part of peduncle, becoming fibrous. *Pseudobulbs* cylindrical, 2.8–3 × 0.6–0.8 cm, 2–3 cm apart on rhizome. *Leaf*: petiole sulcate, 5–6 cm long; blade linear-lanceolate, acute, thin-textured, 20–25 × 1.2–1.3 cm. *Inflorescences* densely many-flowered, flowers arranged in two ranks, borne 2–2.5 mm apart, opening from proximal part of rachis; peduncle filiform, wiry, 20–25 cm long; non-floriferous bracts about 6, imbricate, lowermost up to 1 mm long; rachis curving, 18 cm long; floral bracts broadly ovate, obtuse, margin hyaline, prominently nerved, ramentaceous, involute, entirely enclosing pedicel with ovary, 3–3.2 × 4–4.5 mm. *Flowers* somewhat fragrant, very pale greenish, labellum green. *Pedicel with ovary* straight, 2.3–2.4 mm long. *Sepals* and *petals* spreading, 3-nerved. *Dorsal sepal* ovate-elliptic, acute, 4 × 1.7–1.8 mm. *Lateral sepals* ovate, acute, 4 × 2 –2.1 mm. *Petals* elliptic, acute, 3.9–4 × 1.7–1.8 mm. *Labellum* 2.5 mm long, 3 mm wide across side lobes, sessile, concave, cup-like; side lobes auriculate, acute, 1.1 × 0.2–0.3 mm; mid-lobe oblong-ovate, obtuse, with a small obtuse apical mucro, 1.5–1.6 × 1.5 mm, 3-nerved, with a fleshy, transverse basal ridge. *Column* oblong, entire, 0.8 × 1 mm; stelidia and apical hood absent; anther-cap cucullate, apiculate, *c.* 0.6–0.7 × 0.6–0.7 mm.

DISTRIBUTION. Borneo.

SABAH. Sipitang District, Long Pa Sia area, 22 November 1986, *Phillipps et al.* SNP 2948, 2969 & 2970 (Sabah Parks Herbarium, Kinabalu Park!). Sipitang District, Rurun River headwaters, December 1986, *Vermeulen & Duistermaat* 1057 (holotype L! isotype K!).

HABITAT. Low and open mossy ridge forest with a dense undergrowth of bamboo and rattan palms. *c.* 1700 m.

D. auriculilobum is an interesting species which is known only from a few collections from Southwest Sabah. Closely related to *D. hologyne* Carr, which was described from Sarawak but is also recorded from Sipitang District in Southwest Sabah, it is distinguished by the much shorter, thicker pseudobulbs, laxer inflorescences, larger flowers and labellum with auriculate side lobes and a fleshy transverse basal ridge.

The specific epithet is derived from the Latin *auriculatus*, with ear-like appendages, and *lobus*, lobe, referring to the side lobes of the labellum.

2. DENDROCHILUM HOLOGYNE

Dendrochilum hologyne Carr in *Gdns. Bull. Straits Settl.* 8: 89–90 (1935). Type: Malaysia, Sarawak, Marudi District, Dulit Ridge, *Synge* S. 513 (holotype SING!, isotypes K!, L!). **Fig. 25, plate 7D.**

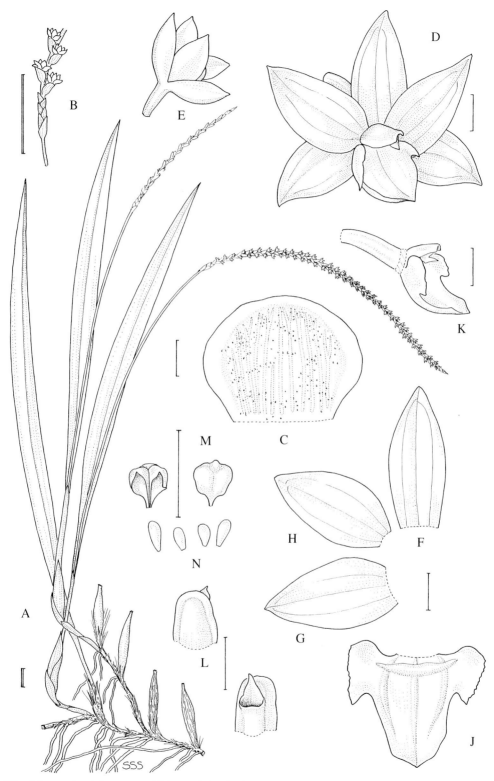

Fig. 24. *Dendrochilum auriculilobum*. **A** habit; **B** lower portion of inflorescence; **C** floral bract; **D** flower, front view; **E** flower, side view; **F** dorsal sepal; **G** lateral sepal; **H** petal; **J** labellum, front view; **K** pedicel with ovary, labellum and column, side view; **L** column apex, front and back views; **M** anther-cap, front and back views; **N** pollinia. **A–N** from *Vermeulen & Duistermaat* 1057 (holotype). Scale: single bar = 1 mm; double bar = 1 cm. Drawn by Susanna Stuart-Smith.

DESCRIPTION. Terrestrial or partially epiphytic. *Rhizome* long-creeping, clump-forming, internodes 0.7–2.5 cm long, 2–3 mm in diameter. *Roots* 2–4 borne from each node, branching distally, 0.5–2 mm in diameter. *Cataphylls* 3–6, tubular, obtuse to acute, imbricate, smooth to finely rugulose, 0.5–11 cm long, pale brown, speckled darker brown. *Pseudobulbs* narrowly cylindrical, pencil-like, rugulose, erect, (8–)11–15 × 0.3–0.4 cm. *Leaf:* petiole sulcate to canaliculate, 1.3–2.3 cm long; blade narrowly linear-lanceolate, ligulate, acute, abruptly narrowed at base, thin-textured, (16–)18–25 × 0.85–1.4 cm, distinct nerves 5–6, mid-nerve raised and prominent on abaxial surface, numerous calcium oxalate bodies usually present. *Inflorescences* borne from apex of almost mature pseudobulbs, as long as or longer than leaves, densely many-flowered, flowers arranged in two ranks, each flower borne 2 mm apart, opening from proximal part of rachis; peduncle filiform, 12.5–20 cm long; rachis quadrangular, concave alternately on each side above the flower, 6–12 cm long; non-floriferous bracts 3–7; floral bracts broadly ovate or triangular-ovate, obtuse, 2–2.7 × 3.2 mm, glabrous. *Flowers* unscented, sepals and petals very pale green or very pale ochre, brownish at base, labellum pale green, often ochre or apricot in centre and at base, column brownish. *Pedicel with ovary* straight, 1.8–1.9 mm long. *Sepals* 3-nerved. *Dorsal sepal* oblong-elliptic, narrowly obtuse, 1.5–1.6(–2) × 0.8–0.9 mm. *Lateral sepals* oblong-ovate, obtuse to subacute, sometimes somewhat falcate, often apically carinate on reverse, 2.2–2.5 × 0.9–1.1 mm. *Petals* narrowly oblong, acuminate or acute, 0.9–2.2 × 0.5–0.6 mm, 1-nerved. *Labellum* entire, concave with erect margins, triangular-ovate or obliquely subquadrate, obtuse to acute, 1.8–2 × 2 mm when flattened, 3-nerved, provided inside in the lower half above base with 3 obscure rounded keels, the median keel short, tubercular, the outer keels divaricate from base and almost incurved at a right angle, much dilate in the upper half, with the apex nearly contiguous and reaching a little beyond middle of labellum. *Column* entire, 0.6–1 mm long; stelidia and apical hood absent; rostellum suberect, triangular; stigmatic cavity semiorbicular; anther-cap minute, cucullate.

DISTRIBUTION. Borneo.

SARAWAK. Lawas District, Mount Murud, Oct. 1992, *Mjöberg* 65 (AMES!). Marudi District, Mount Dulit, Dulit Ridge, 6 Sept. 1932, *Synge* S. 415 (K); & 17 Sept. 1932, *Synge* S. 513 (holotype SING!, isotypes K!, L!).

SABAH. Sipitang District, Long Pa Sia area, 22 November 1986, *Phillipps et al.* SNP 2949 (National Parks Herbarium, Kinabalu Park!). Sipitang District, Maligan to Long Pa Sia trail, Dec. 1986, *Vermeulen & Duistermaat* 905 (K! L!). Sipitang District, ridge between Maga River headwaters and Malabid River headwaters, Dec. 1986, *Vermeulen & Duistermaat* 1011 (K! L!). Sipitang District, ridge east of Maga River, *c.* 1.5 km south of confluence with Pa Sia River, 17 Oct. 1986, *de Vogel* 8339 (L!). Sipitang District, Ulu Long Pa Sia, 8 km north west of Long Pa Sia, 24 Oct. 1985, *Wood* 646 (K!).

HABITAT. Mossy forest; podsol forest; ridge-top forest with *Agathis borneensis* Warb., small rattan palms, etc.; open low stunted mossy forest 5–10 metres high, with dense undergrowth of terrestrial orchids and other herbs, on narrow sandstone ridges. 1400–1600 m.

Carr compared *D. hologyne* with *D. fuscescens* Schltr. and *D. lamellatum* J.J. Sm., both Sumatran species belonging in section *Eurybrachium*. It differs from *D. auriculilobum* by the longer pseudobulbs, denser inflorescences and entire labellum.

The specific epithet is derived from the Greek *holo*, entire, whole, and *gyn* or *gyno*, female or pertaining to female organs, and refers to the column which lacks stelidia and an apical hood.

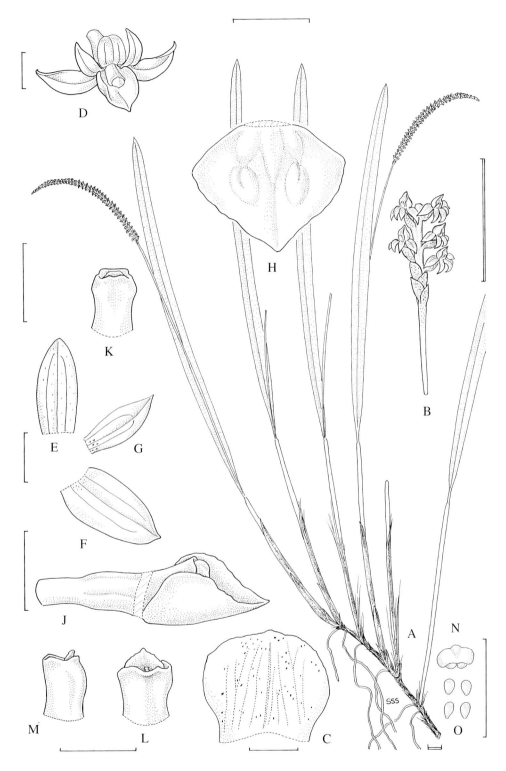

Fig. 25. *Dendrochilum hologyne*. **A** habit; **B** lower portion of inflorescence; **C** floral bract; **D** flower, oblique view; **E** dorsal sepal; **F** lateral sepal; **G** petal; **H** labellum, front view; **J** pedicel with ovary, labellum and column, side view; **K** column, front view; **L** column, back view; **M** column, side view; **N** anther-cap, front view; **O** pollinia. **A & B** from *de Vogel* 8339, **C–O** from *Wood* 646. Scale: single bar = 1 mm; double bar = 1 cm. Drawn by Susanna Stuart-Smith.

Section **Falsiloba**

Section **Falsiloba** *J.J. Wood & H. Ae. Peders.* in *Opera Bot.* 130: 25 (1997). Type species: *Dendrochilum pandurichilum* J.J.Wood.

Leaves conduplicate, dorsiventrally complanate, narrow. *Inflorescences* synanthous; peduncle free from the subtending leaf at the time of flowering. *Flowering* commencing from the top of the rachis. *Petals* quite entire. *Labellum* distinctly clawed, without side lobes but proximally with two prominent, suberect, wing-like keels reminiscent of side lobes. *Column* subequal to the dorsal sepal, with a short foot.

DISTRIBUTION. Borneo.

In general the keels and calli normally present on the labellum in *Dendrochilum* are of limited taxonomic importance above species level. In the monospecific section *Falsiloba* (and section *Mammosa* discussed later), however, they are of diagnostic value. The labellum lacks side lobes but is provided with two prominent, proximal, wing-like keels reminiscent of side lobes.

3. DENDROCHILUM PANDURICHILUM

Dendrochilum pandurichilum J.J. Wood in Wood & Cribb, *Checklist orch. Borneo*: 190–191, fig. 23 A–C (1994). Type: Malaysia, Sabah, Sipitang District, Mount Lumaku, 1500 m, December 1963, *J.B. Comber* 102 (holotype K!). **Fig. 26, plate 12F & G.**

DESCRIPTION. Clump-forming epiphyte. *Rhizome* branching, 1–2 cm long. *Roots* filiform, wiry, flexuose, sparsely branched. *Cataphylls* 3, tubular, acute, pale brown, 3–8 mm long. *Pseudobulbs* crowded on rhizome, ovoid-elliptic, 5.7 × 2.3 mm, yellowish. *Leaf*: petiole sulcate, 1–3 mm long; blade narrowly ligulate, obtuse and mucronate, 0.8–3.5 × 0.2–0.4 cm. *Inflorescences* laxly 3- to 8-flowered, flowers borne 2 mm apart, opening from distal part of rachis; peduncle filiform, naked, 1–3.5 cm long; rachis 0.6–1.8 cm long; non-floriferous bracts absent; floral bracts subulate, acute, 2–3 mm long. *Flowers* pale orange. *Pedicel with ovary* clavate, 1.8 mm long. *Sepals* and *petals* 1-nerved. *Dorsal sepal* ovate-elliptic, subacute, 3 × 1–1.1 mm, concave, cucullate, median nerve prominent and dorsally carinate. *Lateral sepals* triangular-ovate, subacute, 2.9 × 1.5–1.6 mm, somewhat concave. *Petals* linear, acute, slightly falcate, 2.9–3 × *c.* 0.8–0.9 mm. *Labellum* pandurate, 2.1–2.8 mm long, 2.4–2.5 mm wide across keels; side lobes absent; mid-lobe trilobulate, 2–2.1 mm wide when flattened, outer lobules obtuse, median lobule tooth-like, subacute; disc with prominent oblong, rounded wing-like keels each 0.8 × 1 mm, looking, at first sight, like side lobes. *Column* narrow, arcuate, 3 mm long (when straightened); apical hood entire; foot present; stelidia absent; anther-cap cucullate.

DISTRIBUTION. Borneo.

SARAWAK. Limbang District, Pa Mario River, Ulu Limbang, route to Mount Batu Lawi, October 1987, *Awa & Lee* S. 50731 (K!, L!, SAR!). Limbang District, trail from Bario to Batu Lawi, Batu Buli, 14 March 1998, *Leiden cult. (Vogel et al.)* 980090 (K!, L!).

BRUNEI. Belait District, Badas Forest Reserve, 17 January 1992, *Leiden* cult. *(de Vogel)* 911260A (K!, L!).

SABAH. Sipitang District, Mount Lumaku, December 1963, *J.B. Comber* 102 (holotype K). Sipitang District, Northwest of Long Pa Sia to Long Semado trail, on Sarawak border, 24 October 1986, *de Vogel* 8569 (L!).

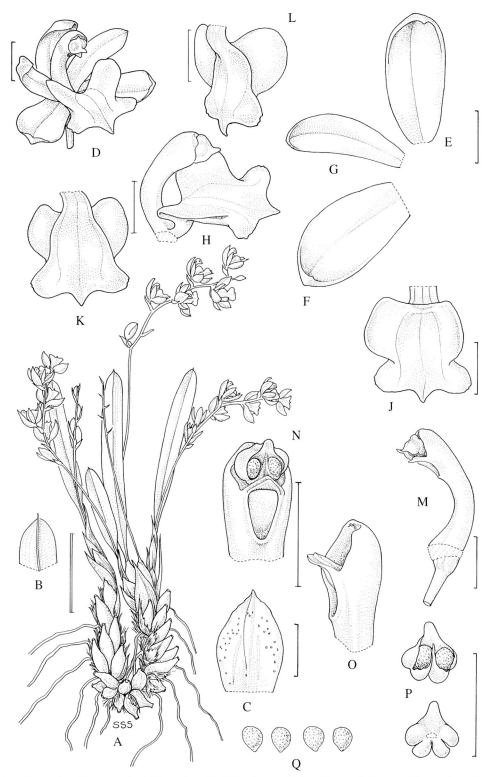

Fig. 26. *Dendrochilum pandurichilum.* **A** habit; **B** leaf apex; **C** floral bract; **D** flower, oblique view; **E** dorsal sepal; **F** lateral sepal; **G** petal; **H** labellum and column, oblique view; **J** labellum, front view; **K** & **L** labellum, back views; **M** pedicel with ovary and column, side view; **N** column, front view; **O** column, anther-cap removed, side view; **P** anther-cap, interior and back views; **Q** pollinia. **A** from *J.B. Comber* 102 (holotype), **B–Q** from *Leiden* cult. (*de Vogel*) 911260A. Scale: single bar = 1 mm; double bar = 1 cm. Drawn by Susanna Stuart-Smith.

HABITAT. Lower montane forest; riparian forest; lowland kerangas forest. (10–)1500–1900 m.

It is curious that *D. pandurichilum* has been found in lowland kerangas forest at only 10 metres above sea level in Brunei, whereas all other collections have been made in lower montane forest. The Brunei collection (*Leiden* cult. 911260A) was said to be made in forest about 30 metres in height growing on pure white sand and containing numerous large *Agathis* trees and with an undergrowth of slender pole trees. The provenance is questionable. *De Vogel* 8569 from Sabah, although sterile, is almost certainly this species. The data label records the flower buds, since lost, as pink. It was collected in primary ridge forest up to 20 metres high on sandstone with an understorey of climbing bamboos, rattan palms and an abundance of small trees.

The specific epithet is derived from the Latin *panduratus*, fiddle-shaped, and the Greek *chilus*, lipped, referring to the distinctive and elegant shape of the labellum.

SUBGENUS DENDROCHILUM

Dendrochilum subgenus **Dendrochilum**
Butzin 1974: 254; 1984–1986: 944; de Vogel 1989: 103; Wood & Comber 1995: 57.

Dendrochilum sensu Benth., *J. Linn. Soc., Botany* 18: 195 (1881), non Blume. - Pfitzer 1882: 152, 153, 156; Bentham 1883a: 506–507; J.D. Hooker 1886–1890: 782; Pfitzer 1888–1889: 180–181; Stein 1892: 21, 213–214; Ridley 1896: 287; 1907: 84; Constantin 1913: 31; Ridley 1924: 81; Hsieh 1955: 240; Wirth & Withner 1959: 172; Chadefaud & Emberger 1960: 485, Fig. 784(3); Hiroe 1971: 95.
Dendrochilum sect. *Eudendrochilum* J.J.Sm., *Recl. Trav. bot. néerl.* 1: 58 (1904), *p.p., nom. illeg.* - Smith 1905a: 161; Ames 1908: 5; Smith 1934: 204, 212; Mansfeld 1937: 673; 1955: 71; Holttum 1957: 233; 1964: 233; Schlechter 1982: 238. Type not designated.
Dendrochilum sect. *Dendrochilum*. - Seidenfaden & Wood 1992: 188.
Dendrochilum subgen. *Eudendrochilum* (J.J. Sm.) Pfitzer & Kraenzl. in Engler, *Pflanzenreich* 4, 50, 2 B 7: 87 (1907), *nom. illeg.* - Rosinski 1992: 92, 97–98, 104, 201.
Dendrochilum sensu Ames, *Orchidaceae* 7: 80 (1922), *p.p.*, non Blume. - Smith 1934: 200, 211; Greuter *et al.* 1993: 325.

DESCRIPTION. *Rhizome* elongate, creeping, pseudobulbs widely separated. *Leaves* conduplicate. *Inflorescences* heteranthous. *Labellum* elastically attached to the column, entire (often more or less pandurate). *Column* distally prolonged into an apical wing which distinctly exceeds the anther; stelidia present. *Rostellum* entire, flat. *Seeds* filiform; central testa cells nearly isodiametric; junctions of the anticlinal testa cell walls protruding from the testa.

DISTRIBUTION. Myanmar (Burma), Thailand, Peninsular Malaysia, Singapore, Sumatra, Bangka, Java, Lesser Sunda Islands, Borneo and the Philippines.

Species delimitation within subgenus *Dendrochilum* in Borneo has proven to be the most problematic in this study. Seidenfaden (1986), in his treatment of the Thai species, conceded that "the identification of these few plants has been cumbersome and the results unsatisfactory". He goes on to say that "I am not certain about the correctness of the identification."

There is certainly little doubt that a thorough investigation of the species occurring outside of Borneo, and in particular Sumatra which is the centre of diversity and poorly collected, is required before a broader understanding of the subgenus can be gained. In Borneo two species, *viz. D. crassum* Ridl. and *D. gravenhorstii* J.J. Sm., appear well-defined and vary very little. The reverse is true, however, of the group of Bornean taxa grouped around Blume's *D. pallidiflavens.* From the material at hand it seems that there exists a cline of variation in several key characters, notably habit, leaf index, rachis indumentum, labellum shape and papillosity, number of keels, stelidia length and apical hood shape. I cannot honestly find any consistent differences separating these Bornean taxa at specific level. I have, therefore, reluctantly decided to recognise only one variable species, *viz. D. pallidiflavens* Blume, which is the earliest available epithet. Certainly, further study of living material in the field throughout Borneo and neighbouring islands, especially Sumatra, is required before a clearer picture of species delimitation across the range of the subgenus can be gained. Of the original group of Bornean taxa, only *D. brevilabratum* (Rendle) Pfitzer is maintained, albeit at varietal level. A parallel example of similar variability within Bornean taxa of the *Coelogyninae* is found in *Chelonistele sulphurea* (Blume) Pfitzer (de Vogel 1986).

KEY TO THE SPECIES OF SUBGENUS *DENDROCHILIUM* IN BORNEO

1. Labellum with a few narrow teeth along the margins of the basal auriculate portion, disc with a raised, somewhat compressed, fleshy flange each side at the base and 3 smooth, fleshy keels, the outer terminating on the top portion of the blade. Leaves linear, ligulate .... 4. *D. gravenhorstii* J.J. Sm.

    Labellum without basal teeth, although sometimes minutely erose, often somewhat pandurate, fleshy basal flanges absent, keels 2, less often 3, minutely papillose, usually terminating midway along the blade. Leaves commonly narrowly elliptic, oblong-elliptic or elliptic 2

2. Top portion of labellum relatively short, keels 2, very prominent, broad and flange-like, protruding beyond the base of the blade. Sepals distinctly incurved, fleshy. Apical hood of column entire .... 5. *D. crassum* Ridl.

    Top portion of labellum more elongate and prominent, keels 2 or 3, rather low, less prominent and usually narrower, not protruding beyond the base of the blade. Sepals spreading, less incurved, thinner-textured. Apical hood of column variably shaped, entire to bifid. .... 6. *D. pallidiflavens* Blume

4. DENDROCHILUM GRAVENHORSTII

Dendrochilum gravenhorstii J.J. Sm. in *Bull. Jard. Bot. Buitenz.* 3, 2: 28–29 (1920). Type: Indonesia, Kalimantan Barat, Upper Kapuas, Talaj River, 1916, *Gravenhorst* s.n., cult. hort. Bogor (holotype BO, isotype L!). **Fig. 27.**

DESCRIPTION. Terrestrial or epiphytic. *Rhizome* elongate, at least up to 40 cm long, *c.* 2.5 mm in diameter, internodes 1–1.3 cm long, branching, glossy, clothed in numerous reddish-brown sheaths when young. *Roots* 0.2–0.8 mm in diameter, branching distally, wiry, smooth. *Cataphylls* 4–6, 0.3–2.5 cm long, ovate-elliptic, acute to acuminate, glossy chestnut-brown,

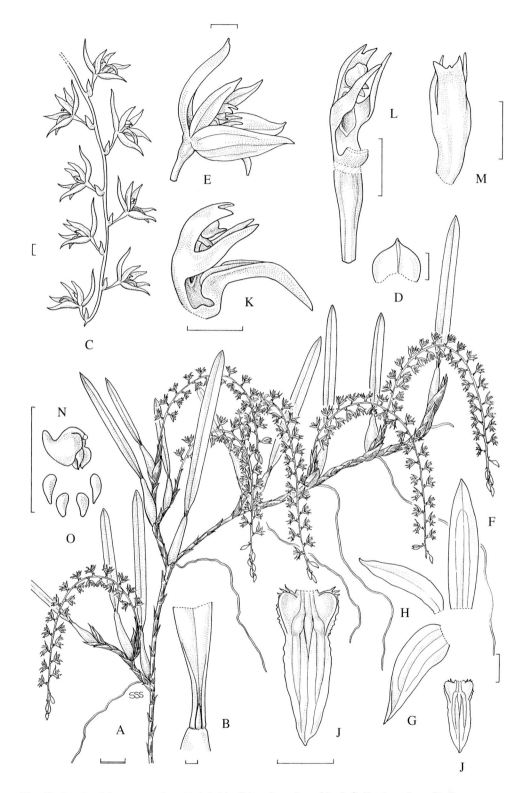

Fig. 27. *Dendrochilum gravenhorstii.* **A** habit; **B** basal portion of leaf; **C** distal portion of inflorescence; **D** floral bract; **E** flower, side view; **F** dorsal sepal; **G** lateral sepal; **H** petal; **J** labellum, front view; **K** labellum and column, side view; **L** pedicel with ovary and column, oblique view; **M** column, back view; **N** anther-cap, side view; **O** pollinia. **A & B** from *Awa et al.* S. 50416, **C–O** from *Gravenhorst* s.n. (holotype). Scale: single bar = 1 mm; double bar = 1 cm. Drawn by Susanna Stuart-Smith.

sometimes speckled. *Pseudobulbs* borne 1.3–4.5(–7) cm apart, often at an acute angle to the rhizome, 1.3–2.5 × 0.55–0.75 cm, obliquely oblong-ovoid, becoming multisulcate, glossy, pale green to ochre. *Leaf:* petiole 2–3 mm long, cuneate, canaliculate; blade 4.5–11.5 × 0.5–0.8(–0.9) cm, linear, ligulate, obtuse to acute, rigid, coriaceous, main nerves 5, median nerve somewhat canaliculate, glossy green. *Inflorescences* gently decurved, subdensely many-flowered, flowers borne 2–3 mm apart; cataphylls 8–9, 0.3–3 cm long, ovate-elliptic, obtuse to acute, imbricate, pale brown; peduncle erect to porrect, 1–2 cm long; rachis quadrangular, arcuate, (4–)5–8 cm long; floral bracts ovate-orbicular, obtuse to shortly acute, 0.8–1 × 1.75–1.8 mm. *Flowers* sweetly scented, very pale green to pale yellow, labellum yellowish-green to ochre with a whitish margin, stelidia white. *Pedicel with ovary* 2–2.25 mm long, clavate to slightly sigmoid, pale green. *Sepals* 3-nerved. *Dorsal sepal* 4.4–4.5 × 1.3 mm, narrowly oblong, apiculate. *Lateral sepals* 4 × 1.6 mm, obliquely oblong-ovate, subfalcate, acute. *Petals* 3.7 × 0.8 mm, obliquely lanceolate to linear-lanceolate, acute, 1-nerved. *Labellum* 2.7 mm long, 1.25 mm wide at base, 0.8 mm wide at centre, subentire, shortly clawed, decurved, apex often recurved, lanceolate, dilated and auriculate below, auriculate basal portion with a few narrow basal teeth, erose above, 3-nerved; disc with a raised, somewhat compressed, fleshy flange each side at the base, keels 3, arising on the dilated basal portion, thick, fleshy, outer keels terminating on distal portion of blade, median keel shorter and narrower, terminating on proximal portion of blade. *Column* gently curving, 2 mm long; foot distinct, truncate; apical hood quadrangular, 2- to 4-toothed; stelidia arising opposite stigmatic cavity, obliquely linear-ligulate to slightly sigmoid, acute, slightly longer than apical hood; stigmatic cavity ovate to oblong, lower margin elevate, sometimes reflexed, rotund-truncate; rostellum prominent, transversely suborbicular, acute; anther-cap 0.5 mm wide, cucullate, triangular, acuminate or acute. *Capsule* 5–7 × 4–5 mm, ovoid, floral remains persistent.

DISTRIBUTION. Borneo.

SARAWAK. Marudi District, route to Pa Ukat, Bario, 27 July 1985, *Awa et al.* S.50416 (AAU, K!, L, MEL, SAR, SING).

SABAH. Sandakan District, Mile 87.5, Telupid Hap Seng logging area, 22 June 1976, *Leopold & Taha* SAN 83505 (K!, L!, SAN, SAR). Keningau to Sepulot road, 6 km past Nabawan, near old airstrip, June 1986, *Vermeulen & Lamb* 325 (L!).

KALIMANTAN BARAT. Upper Kapuas, Talaj River, 1916, *Gravenhorst* s.n., cult. hort. Bogor (holotype BO, isotype L!).

HABITAT. Kerangas forest; podsolic dipterocarp/*Dacrydium* forest on very wet sandy soil. 100–1000 m.

D. gravenhorstii is easily distinguished from other Bornean members of subgenus *Dendrochilum* by the unusually ornate labellum.

The specific epithet honours C.A. Gravenhorst (born 1884), the head of a Danish oil factory laboratory who travelled to the Dutch East Indies to investigate the properties of oil-bearing seeds. He made several collections of living orchids in Kalimantan which he sent to J.J. Smith at Buitenzorg (Bogor) Botanic Gardens.

5. DENDROCHILUM CRASSUM

Dendrochilum crassum Ridl. in *J. Linn. Soc., Botany* 32: 288 (1896). Type: Peninsular Malaysia, Perak, Hermitage Hill, *Ridley* s.n. (type not located). **Fig. 28, plate 2B.**

DESCRIPTION. Scrambling terrestrial, epiphyte or lithophyte. *Rhizome* elongate, up to 60 cm or more long, 0.3–0.5 cm in diameter, internodes 0.5–3 cm long, branching, terete, orange-coloured. *Roots* branching, smooth to minutely papillose, *c.* 1–2 mm in diameter. *Cataphylls* 4 to several, 0.5–2.5(–4) cm long, ovate-elliptic, acute, dark brown, sometimes speckled. *Pseudobulbs* (1–)2–5(–8) cm apart, cylindrical or fusiform, smooth to ridged, (1.5–)3–4 × 0.5–0.8 cm, olive-green to orange-coloured. *Leaf*: petiole sulcate, 0.4–0.8(–1) cm; blade oblong-elliptic to elliptic, obtuse to acute, coriaceous, 4–9.5 × 1.1–4 cm, main nerves 7–9. *Inflorescence* decurved, subdensely many-flowered, flowers borne 2–3 mm apart; cataphylls 4–5, triangular-ovate, acute, 0.5–2 cm long; peduncle dilated at base, 1.5–2 cm long, enclosed by cataphylls; rachis quadrangular, 8–16 cm long; floral bracts ovate, acute or apiculate, 2.5–3 mm long. *Flowers* sweetly scented, sepals and petals apple-green, edged whitish, or entirely creamy white, labellum pale green, brighter green towards base, column pale green, anther-cap cream. *Pedicel with ovary* narrowly clavate, gently curved, 3.5–4 mm long. *Sepals* spreading, incurved, fleshy, minutely papillose, 3-nerved, sparsely covered with brown trichomes, especially on reverse. *Dorsal sepal* oblong, obtuse, apiculate, 4.9–5 × 2 mm. *Lateral sepals* oblong, obtuse, apiculate, slightly subfalcate, 4–4.4 × 2 mm. *Petals* oblong–spathulate, obovate, obtuse, apiculate, fleshy, 1-nerved, directed forward and hiding the column, 4 × 2 mm. *Labellum* entire, subpandurate, broadly ovate distally, obtuse and apiculate, straight, fleshy, minutely papillose, 2 mm long, 1–1.1 mm wide at base, 0.8–0.9 mm wide distally, disc with 2 prominent, erect, fleshy, flange-like basal keels terminating near middle of blade. *Column* 1.8–1.9 mm long, gently curving; foot short; apical hood entire, oblong, obtuse, cucullate; stelidia borne opposite stigmatic cavity, lanceolate, falcate, acute, shorter than apical hood, 0.8–0.9 mm long; stigmatic cavity quadrate, lower margin thickened; rostellum broadly ovate, obtuse; anther-cap ovate-cordate, minutely papillose, *c.* 0.6 × 0.6–0.7 mm.

DISTRIBUTION. Peninsular Malaysia & Borneo.

SABAH. Locality unknown, *Argent* 111, cult. hort. RBG Edinburgh, no. C14753 (E!). Near Kundasang, June 1994, *Barkman* 7 (K!, Sabah Parks Herbarium, Kinabalu Park). Mount Kinabalu, Minitinduk Gorge, 21 March 1933, *Carr* 3172, SFN 26668 (K!, SING!). Ulu Kimanis, 16 Oct. 1986, *Chan* 12/86 (SING!). 17.6 km west of Keningau, 1984, *Clements* 3226, cult. hort. RBG Kew, EN 385–84.03893 (K!). Mount Trus Madi, 20 May 1986, *Joseph et al.* SAN 113509 (K!, L, SAN, SAR). Tambunan District, Sinsuron road, km 64, April 1984, *Lamb* s.n. (K!). Crocker Range, Mount Alab, Sinsuron road, km 65–66 from Kota Kinabalu to Tambunan, 12 May 1985, *Lamb & J.B. Comber* in *Lamb* AL 345/85 (K!). Near Nabawan, 1986, *Leiden* cult. (*Vermeulen*) 26602 (L!). Crocker Range, Keningau to Kimanis road, Dec. 1986, *Vermeulen & Duistermaat* 693 (K!, L!). Crocker Range, Kimanis road, 10 Oct. 1985, *Wood* 584 (K!). Penampang District, Crocker Range, Sinsuron road, 30 Oct.1985, *Wood* 733 (K!).

HABITAT. Sandstone and shale outcrops beside road, associated with *Arundina graminifolia* (D. Don) Hochr., *Gahnia* spp., *Lycopodium* spp., *Melastoma* spp., *Nepenthes fusca* Danser, ferns, grasses, etc., fully exposed to sun; cliffs; hill forest; dipterocarp/*Dacrydium* forest 10–20 metres high on podsolic fine white sandy soil. 400–1600 m.

A distinctive species recognised by the strongly incurved, fleshy, sparsely furfuraceous sepals and petals, and very prominent fleshy keels on the proximally broad labellum. The rhizomes and pseudobulbs become stained with orange pigment in open, exposed habitats.

Fig. 28. *Dendrochilum crassum*. **A** habit; **B** portion of inflorescence; **C** close-up of trichomes on rachis; **D** floral bract; **E** flower, side view; **F** dorsal sepal; **G** lateral sepal; **H** petal; **J** labellum, front and oblique views; **K** pedicel with ovary, labellum and column, side view; **L** pedicel with ovary and column, back view; **M** column, front and side views; **N** anther-cap, back view; **O** pollinia. **A** from *Vermeulen & Duistermaat* 693, **B–O** from *Wood* 733. Scale: single bar = 1 mm; double bar = 1 cm. Drawn by Susanna Stuart-Smith.

The specific epithet is from the Latin *crassus*, thickened, and refers to the fleshy nature of the flowers.

6. DENDROCHILUM PALLIDIFLAVENS

Dendrochilum pallidiflavens Blume, *Bijdr.*: 399, plate 10, 52, left (1825). Type: Indonesia, Java, Pantjar, *Blume* 1939 (holotype L).

DESCRIPTION. Epiphyte, sometimes lithophytic or a scrambling terrestrial. *Rhizome* elongate, often up to 1 m long, 3–4 mm in diameter, often 5–6 mm in diameter below a pseudobulb and branch, internodes very variable in length, 0.2–2.6 cm long, shortest on the smaller distal branches, repeatedly branching, especially distally, often somewhat fractiflex, reddish-brown or yellowish, striate, rooting profusely from nodes, clothed, when young, with ovate, acute to acuminate, greyish-brown sheaths *c.* 0.2–2.5 cm long, which soon disintegrate. *Roots c.* 0.2–2 mm in diameter, branching, branches few or profuse and forming a dense mass, elongate, wiry, smooth. *Cataphylls* 2–5, *c.* 0.6–3 cm long, ovate-elliptic, acute to acuminate, greyish-brown, often speckled. *Pseudobulbs* borne 1–12 cm apart, often at an acute angle to the rhizome, often densely crowded on the short, distal rhizome branches, 1.3–3.7(–4.5) × 0.3–0.5(–1) cm, cylindrical to fusiform, sometimes ovoid to elliptic, smooth to striate, wrinkled when dry, green, yellowish-green, bright yellow or reddish-brown. *Leaf*: petiole 0.2–1.8 cm long, sulcate; blade 3–7.5 (–13.5) × 0.5–2.5(–4.3) cm, linear-lanceolate, narrowly elliptic, oblong-elliptic, sometimes broadly elliptic, obtuse to subacute, less often acute, thinly coriaceous, sometimes thicker-textured and stiff, main nerves 5–9, with numerous small transverse nerves, glossy green or yellowish-green. *Inflorescences* often paired, gently decurved, subdensely many-flowered, flowers borne (1.5–)2–3 mm apart; cataphylls *c.* 5, *c.* 0.6–1.8 cm long, ovate, acute, imbricate, pale brown, speckled; peduncle erect to porrect, 0.3–2 cm long; rachis quadrangular, straight to arcuate, 5–14 cm long; floral bracts ovate, obtuse, acute or shortly acuminate, 1.5–2 mm long, shorter than or nearly as long as pedicel with ovary. *Flowers* sweetly scented or unscented, sepals and petals creamy yellow, greenish yellow, pale yellowish white, pale lemon-yellow or pale ochre-yellow, often paler at base, labellum pale yellow with a pale green centre, or green at base, pale yellow or greenish white distally, or bright green with paler margins, keels sometimes bright green, column pale green with whitish stelidia and apical hood, fruit capsule yellowish. *Pedicel with ovary* 1.3–3.5 mm long, slender, straight or geniculate, pale green or reddish-green. *Sepals* 3-nerved, minutely papillose. *Dorsal sepal* (2–)3.2–4(–6) × 1–1.2(–1.5) mm, narrowly oblong-ligulate, or oblong-subspathulate, obtuse, subacute or shortly apiculate. *Lateral sepals* (2–)3–4(–6) × 0.9–1.4 (–1.5) mm, oblong-elliptic, often subfalcate and slightly oblique, obtuse, subacute or shortly apiculate. *Petals* 3–4.5(–5) × 0.6–1.1 mm, linear, obliquely ligulate, or subspathulate, obtuse to subacute, minutely papillose, 1-nerved. *Labellum* 1.5–2 × 0.3–1.25 mm, entire, with or without a short claw, decurved, linear-oblong to oblong-subspathulate, or rhomboid-spathulate, with parallel sides or narrowed centrally and pandurate, obtuse to subacute, 3-nerved, minutely papillose, disc with 2 minutely papillose keels usually extending from near the base to and terminating near or at the centre of the labellum, sometimes with a shorter, lower median keel. *Column* curved, minutely papillose, 1.2–2.5 mm long; foot distinct, upcurved, truncate; apical hood variable in shape, entire, truncate, obtuse or acute, to shallowly bifid or deeply bifid; stelidia usually arising opposite stigmatic cavity, narrowly linear, often falcate, acute to acuminate, sometimes erose, shorter than to longer than apical hood, rarely with a tooth in the sinus between base of hood and stelidia; stigmatic cavity oblong to ovate-cordate, lower margin sometimes thickened; rostellum prominent, oblong-ovate to cordate, obtuse to subacute; anther-cap ovate, cucullate, obtuse. *Capsule* 0.7–0.8 × 0.8–0.9 mm, globose.

KEY TO THE VARIETIES OF *D. PALLIDIFLAVENS*

1. Sepals 5.5–6 mm long. Rachis virtually glabrous. Labellum 3-keeled, median keel shorter and less prominent. Stelidia long-acuminate, equal to or longer than apical hood. Apical hood bifid .... var. *brevilabratum* (Rendle) J.J. Wood
    Sepals 2–3(–4) mm long. Rachis usually sparsely to densely covered with brownish or blackish trichomes, less often virtually glabrous. Labellum usually 2-keeled, less often with a shorter, less prominent median keel. Stelidia variable in length, but often shorter than apical hood. Apical hood variable in shape, entire to bifid 2
2. Labellum subspathulate, pandurate or narrowly linear-oblong, papillose, especially on keels, 2-keeled, rarely 3-keeled. Column somewhat minutely papillose .... var. *pallidiflavens*
    Labellum oblong or narrowly ovate-oblong, distinctly papillose, 3-keeled. Column distinctly minutely papillose .... var. *oblongum* J.J. Wood

a. var. pallidiflavens. Figs. 29–33, plate 12D & E; 18–24.

Dendrochilum conopseum Ridl. in *Trans. Linn. Soc. London, Bot.*, ser. 2, 4: 236 (1894). Type: Malaysia, Sabah, Mount Kinabalu, Marai Parai Spur, 1700, *Haviland* s.n. (holotype SAR, photograph K!), **syn. nov.**

Bulbophyllum pteriphilum Rolfe in *Kew Bull.* 1894: 391 (1894). Type: Peninsular Malaysia, Penang, *Curtis* s.n. (holotype K!).

Dendrochilum album Ridl. in *J. Linn. Soc., Botany* 32: 287 (1896). Types: Peninsular Malaysia, Perak, Maxwell Hill, June 1895, *Ridley* s.n. (syntype K!); Larut Hills, *Ridley* s.n. (syntype not located).

Dendrochilum ellipticum Ridl. in *J. Straits Branch Roy. Asiat. Soc.* 39: 77 (1903). Type: Singapore, Sumbawang, *Ridley* 6536 (holotype ?SING), **syn. nov.**

Dendrochilum aurantiacum Blume var. *pallidiflavens* (Blume) J.J. Sm. in *Recl. Trav. bot. néerl.* 1: 60 (1904).

Dendrochilum micranthum Schltr. in *Bull. Herb. Boissier* 2, 6: 303 (1906). Type: Indonesia, Kalimantan Timur, Long Gombeng, *Schlechter* 13561 (holotype B†), **syn. nov.**

Dendrochilum album Ridl. var. *acutifolium* Pfitzer & Kraenzl. in Engler, *Pflanzenreich* 4, 50, 2 B 7: 88 (1907). Type: Thailand, Phang-nga, *Curtis* 2901 (holotype SING).

Dendrochilum pteriphilum (Rolfe) Pfitzer in Engler, *Pflanzenreich* 4, 50, 2 B 7: 89 (1907).

Dendrochilum intermedium Ridl. in *J. Straits Branch Roy. Asiat. Soc.* 50: 135 (1908). Type: Malaysia, Sarawak, Kuching District, Mount Matang, June 1907, *Hewitt* s.n. (holotype SING!, isotype K!), **syn. nov.**

Dendrochilum spathulatum Ridl. in *J. Straits Branch Roy. Asiat. Soc.* 50: 134 (1908). Types: Peninsular Malaysia, Pahang, Tahan River, *Ridley* s.n. (syntype ? SING); Indonesia, Sumatra, *Ridley* s.n. (syntype ? SING); Peninsular Malaysia, Sungai Kelantan, 1898, *Ridley* s.n. (syntype SING!); Peninsular Malaysia, Siak, *Ridley* s.n. (syntype ? SING), **syn. nov.**

Dendrochilum weberi Ames in *Philipp. J. Sci., Botany* 8: 410 (1913). Type: Philippines, Mindanao, Cabadbaran, *Weber* 59 (holotype AMES, isotype K!).

Dendrochilum bancanum J.J. Sm. in *Bull. Jard. Bot. Buitenzorg* 3, 1: 95 (1919), *nom. nud.*

Dendrochilum bulbophylloides Schltr. in *Notizbl. bot. Gart. Mus. Berl.* 8: 16 (1921). Type: Malaysia, Sarawak, Kuching District, Gunung Mattan (=Matang), Dec. 1866, *Beccari* 3036 (holotype B†, isotypes FI!, *Carr* drawing and photograph K!, L!), **syn. nov.**

Dendrobium ridleyi Ames in Merr., *Bibl. Enum. Born. Pl.*: 164 (1921), *nom. superfl.*

? *Dendrochilum beyrodtianum* Schltr., *nom. nud.*; Butzin: 57 (1979).

Dendrochilum brevilabratum (Rendle) Pfitzer var. *petiolatum* J.J.Wood in *Kew Bull.* 39(1): 78, fig. 3 (1984). Type: Malaysia, Sarawak, Marudi District, Gunung Mulu National Park, between Sungai Berar and Sungai Mentawai, 15 March 1978, *Nielsen* 675 (holotype AAU!, isotype K!), **syn. nov.**

DISTRIBUTION. Myanmar (Burma), Thailand, Peninsular Malaysia, Singapore, Sumatra, Bangka, Java, Lesser Sunda Islands, Borneo and the Philippines.

SARAWAK. Base camp at Sadok Hill, 14 Oct. 1982, *Banyeng & Paie* S.45050 (L!, SAR). Marudi District, between Bario & Pa Umor, 14 Feb. 1995, *Beaman* 11261 (K!). Kuching District, Mount Matang, Dec. 1866, *Beccari* 3036 (holotype of *D. bulbophylloides* FI; *Carr* drawing and photograph K!; isotype L!). Lawas District, Ba Kelalan, 23 Aug. 1955, *Brooke* 10525 (BM!, G!, L!, SING!). Balingian, Tunggal Hill, Stuan Forest Reserve, 10 March 1960, *Brunig* S.12015 (SAR!). Marudi District, Gunung Mulu National Park, near Berar Hill, 16 March 1981, *Collenette* 2349 (K!). Limbang District, Tg. Long Amok, Ensungei River, 10 Sept. 1980, *George et al.* S. 42814 (K!, L!, SAR). Kuching District, Mount Matang, June 1907, *Hewitt* s.n. (holotype of *D. intermedium* SING!; isotype K!). Marudi District, Gunung Mulu National Park, around camp at Long Tapin on Tutoh River, 29 March 1978, *Jermy* 13894 (K!). Locality unknown, *Lobb* s.n. (K!). Belaga District, Batu Laga, Kapit, 29 Aug. 1984, *Mohtar* S. 48084 (AAU, K, L!, MEL, SAR, SING). Marudi District, Gunung Mulu National Park, along rentice between Berar River and Mentawai River, 15 March 1978, *Nielsen* 675 (holotype of *D. brevilabratum* var. *petiolatum* AAU!; isotype K!). Marudi District, Pa Lungan, Kelabit Highlands, 12 April 1970, *Nooteboom & Chai* 02097 (L!, SAR!). Kapit District, Salong Hill, Ulu Sampurau waterfall, Melinau, 23 Aug. 1967, *Paie* S. 26585 (L!, SAR). Song District, summit of Bakak Hill, between Ulu Sg. Janan, Katibas Song & Ulu Sg. Yong, 12 March 1975, *Paie* S. 36322 (KEP!, L!, SAR, SING!). Locality unknown, collected through Sarawak Museum for Bureau of Science, Manila, Philippines, *Sarawak Museum native collector* 903 (AMES!). Marudi District, Mount Dulit, 5 Aug. 1932, *Synge* S. 100 (K!). Marudi District, Mount Dulit, near Long Kapa, Ulu Tinjar, 11 Aug. 1932, *native collector* in *Synge* S. 175 (K!, L!, SING!). Marudi District, Gunung Mulu National Park, Ubung River, *Warwick* 144A, cult. Edinburgh Botanic Garden C 14815 (E!). Kapit District, Nanga Berkakap, Melatai River, Batang Balleh, 17 April 1985, *Yii* S. 48403 (K!, KEP!, L!, MEL, SAR, SING!).

BRUNEI. Belait District, Ulu Ingei, Batu Patam Hill, 9 June 1989, *Boyce* 295A (K!). Temburong District, Selapon, banks of Selapon River east of village, 20 Nov. 1990, *Dransfield* 6979 (BRUN!, K!). Tutong District, Ramba subdistrict, Ulu Tutong, down valley to south-west of helicopter pad, LP 239, 8 May 1992, *Johns* 7555 (K!). Temburong District, Temburong River near Kuala Belalong and slightly up Belalong River, July 1989, *Jongejan* cult. (*de Vogel*) 3976 (K!, L!). Temburong District, ridge between Kuala Belalong and upper Temburong River, 9 July 1989, *de Vogel* 8953 (BRUN!, L). Temburong District, Temburong River near Kuala Belalong and slightly up Belalong River, July 1989, *Leiden* cult. (*de Vogel*)

Fig. 29. *Dendrochilum pallidiflavens* var. *pallidiflavens*. **A & B** habits; **C** floral bract; **D** flower, front view; **E** dorsal sepal; **F** lateral sepal; **G** petal; **H** labellum, front view; **J** labellum, side view; **K** labellum and column, side view; **L** ovary and column, side view; **M & N** column, oblique views; **O** column, side view; **P** anther-cap, back view; **Q** pollinia. **A, C–J, L, P & Q** from *Wood* 749, **B** from *Wood* 817, **K, M–O** from *Lamb* AL 1128/89. Scale: single bar = 1 mm; double bar = 1 cm. Drawn by Susanna Stuart-Smith.

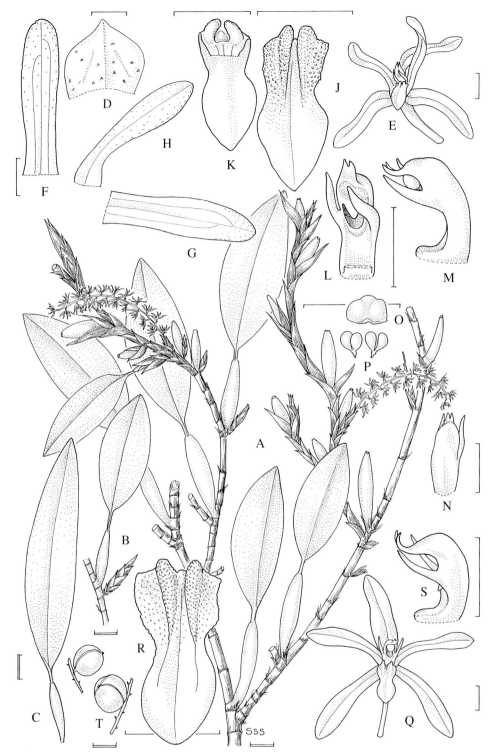

Fig. 30. *Dendrochilum pallidiflavens* var. *pallidiflavens*. **A & B** habits; **C** pseudobulb and leaf; **D** floral bract; **E** flower, front view; **F** dorsal sepal; **G** lateral sepal; **H** petal; **J** labellum, front view; **K** labellum, back view; **L** column, anther-cap removed, front view; **M** column, side view; **N** column, back view; **O** anther-cap, back view; **P** pollinia; **Q** flower, front view; **R** labellum, front view; **S** column, anther-cap removed, side view; **T** fruit capsules. **A** from *J. & M.S. Clemens* 40633, **B** from *George et al.* S. 42814, **C** from *J. & M.S. Clemens* 26899, **D–P** from *Carr* C. 3186, SFN 26759, **Q–S** from *Haviland* 1381, **T** from *J. & M.S. Clemens* 32556. Scale: single bar = 1 mm; double bar = 1 cm. Drawn by Susanna Stuart-Smith.

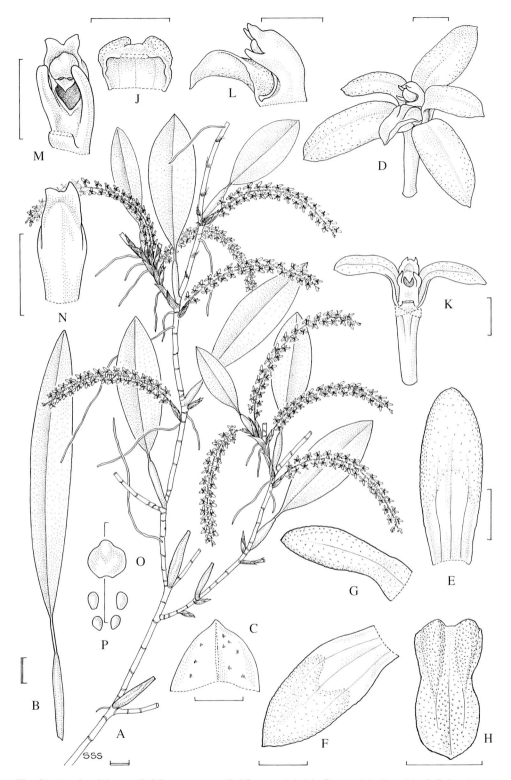

Fig. 31. *Dendrochilum pallidiflavens* var. *pallidiflavens*. **A** habit; **B** pseudobulb and leaf; **C** floral bract; **D** flower, oblique view; **E** dorsal sepal; **F** lateral sepal; **G** petal; **H** labellum, front view; **J** base of labellum, back view; **K** pedicel with ovary, petals and column, anther-cap removed, front view; **L** labellum and column, side view; **M** column, oblique view; **N** column, back view; **O** anther-cap, back view; **P** pollinia. **A** from *Jermy* 13894, **B–P** from *Nielsen* 675 (holotype of *D. brevilabratum* var. *petiolatum*). Scale: single bar = 1 mm; double bar = 1 cm. Drawn by Susanna Stuart-Smith.

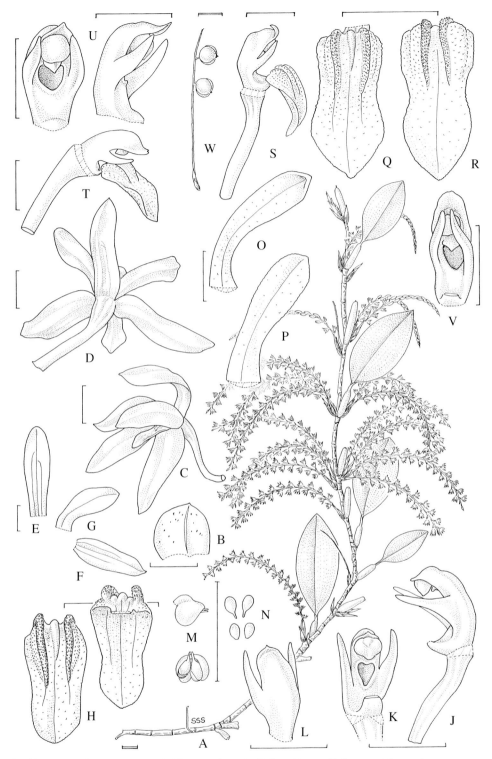

Fig. 32. *D. pallidiflavens* var. *pallidiflavens*. **A** habit; **B** floral bract; **C** flower, oblique view; **D** flower, back view; **E** dorsal sepal; **F** lateral sepal; **G** petal; **H** labellum, front and back views; **J** pedicel with ovary and column, side view; **K** column, front view; **L** column, back view; **M** anther-cap, side and interior view showing pollinia; **N** pollinia; **O** petal; **P** petal, **Q** labellum, front view; **R** labellum, front view; **S & T** pedicel with ovary, labellum and column, side views; **U** column, front and side views; **V** column, front view; **W** infructescence. **A–C, E–N** from *Hewitt* s.n. (holotype of *D. intermedium*), **O, R, S & V** from *Collenette* 2349, **P, Q, T, U & W** from *Boyce* 295A. Scale: single bar = 1 mm; double bar = 1 cm. Drawn by Susanna Stuart-Smith.

30322 (K, L). Temburong District, Kuala Belalong, 22 July 1988, *Wong* WKM 297 (BRUN!, K!, L!, SING!).

SABAH. Mount Kinabalu: Mahandei River, March 1933, *Carr* 3186, SFN 26759 (K!, SING!); Kaung, May 1933, *Carr* 3412, SFN 27315 (K!, SING!); Dallas, 31 Oct. 1931, *J. & M.S. Clemens* 26899 (BM!, BO!, E!, HBG!, K!, L!); Penibukan, 24 Jan. 1933, *J. & M.S. Clemens* 31337 (BM!); Marai Parai, 3 April 1933, *J. & M.S. Clemens* 32556 (BM!, BO!, E!, L!); Penibukan, 9 Oct. 1933, *J. & M.S. Clemens* 40633 (AMES!, BM!, E!, K!, L!); Kaung, Aug. 1892, *Haviland* 1381 (K!); Marai Parai, *Haviland* s.n. (holotype of *D. conopseum* SAR, photograph K!). Nabawan, Syt. Benawood, Maadun River, 26 May 1987, *Krispinus* SAN 119385 (K!, SAN!). Crocker Range, Ulu Kimanis, 1989, *Lamb* AL 1128/89 (K!). Tambunan District, Ingaran, between Tambunan and Mount Trus Madi, 17 March 1969, *Nooteboom* 1334 (L!, SAN!). Keningau to Sepulot road, 6 km past Nabawan, near old airstrip, June 1986, *Vermeulen & Lamb* 316 (L!), *Vermeulen & Lamb* 340 (L!) & 9 Oct. 1986, *de Vogel* 8141 (L!). Near Nabawan, 13 Oct. 1985, *Wood* 596 (K!) & 21 May 1988, *Wood* 749 (K!). Crocker Range, Keningau to Kimanis road, 30 May 1988, *Wood* 817 (K!).

KALIMANTAN BARAT. Liangangang, *Hallier* 2647 (BO, K!). Ketapang, Mount Palung National Park, Cabang Panti Research Site, 17 March 1997, *Laman et al.* TL 870 (AMES, BO, K!).

KALIMANTAN TENGAH. Sampit, 7 Sept. 1940, *Buwalda* 7728 (BO!). Sintang, HPH km. 70, 13 April 1994, *Church et al.* 898 (L!).

KALIMANTAN TIMUR. Near Balikpapan, 13 Aug. 1974, *Darnaedi* D.467 (BO!). Between Long Bawan and Panado, foothills of Mount Tapa Sia, 23 July 1981, *Geesink* 9198 (L!). Mount Njapa, Krubung, Berau, 10 Jan. 1981, *Kato & Wiriadinata* B.5834 (L!). Sinar Baru, north of Long Bawan, Apo Kayan, 4 Aug. 1981, *Kato et al.* B.10627 (L!). Near Long Keluh, *c.* 50 km south of Tanjung Redeb, Berau, 1 Sept. 1981, *Kato et al.* B.11884 (L!). Loa Djanan, west of Samarinda, 13 April 1952, *Kostermans* 6690 (BO!, L!, SING!).

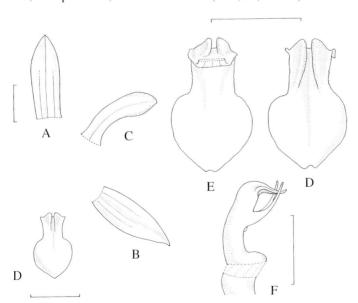

Fig. 33. *Dendrochilum pallidiflavens* var. *pallidiflavens*. **A** dorsal sepal; **B** lateral sepal; **C** petal; **D** labellum, front view; **E** labellum, back view; **F** column, side view. **A–F** from *Beccari* 3036 (isotype of *D. bulbophylloides*). Scale: single bar = 1 mm. Drawn by Susanna Stuart-Smith (after Vermeulen ined.).

HABITAT. Lowland mixed dipterocarp forest on alluvium in river valleys and on ridges on red clay soils over sandstone and shales, including Elaeocarpaceae, *Hopea* spp., *Myristica* spp. and *Shorea* spp.; hill forest on sandstone and limestone; *Agathis* forest; wet kerangas forest *c.* 10 to 20 metres high including *Dacrydium* spp., Dipterocarpaceae, *Garcinia* spp. and *Syzygium* spp., with an understorey of *Rhododendron longiflorum* Lindl. and *R. malayanum* Jack, rattans, etc., and a field layer of ferns, mosses, orchids, *Nepenthes ampullaria* Jack, etc., developed over white sandy soil, with pools of stagnant brown water; riverine forest to *c.* 30 metres high developed on alluvial soil over shale outcrops; lower montane mossy forest, sometimes on ultramafic substrate; steep roadside cuttings on shale, associated with *Lycopodium* spp., *Melastoma* spp., *Nepenthes fusca* Danser, *Rhododendron* spp., ferns, etc. in full sun; on trees among paddyfields; epiphytic on *Rhodomyrtus* spp. in lowland dipterocarp forest; recorded as a terrestrial in kerangas forest where it sometimes covers the forest floor in open areas. Near sea level– 1700 m.

The habit of var. *pallidiflavens* is variable and may be open and lax or dense and very congested depending on habitat (see Plates 18–24).

b. var. **brevilabratum** (Rendle) J.J. Wood **comb. et stat. nov. Fig. 34.**

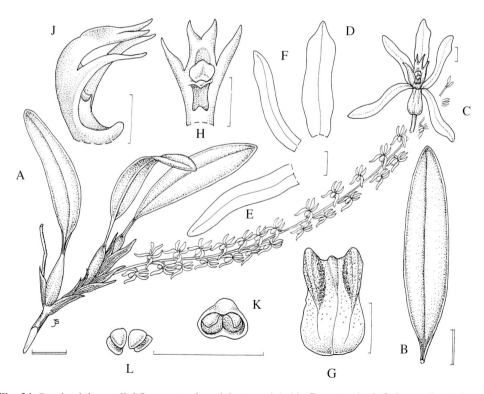

Fig. 34. *Dendrochilum pallidiflavens* var. *brevilabratum*. **A** habit; **B** mature leaf; **C** flower, front view; **D** dorsal sepal; **E** lateral sepal; **F** petal; **G** labellum, front view; **H** column, front view; **J** column, side view; **K** anther-cap, interior view showing pollinia; **L** pollinia. **A–L** from *Hose* 52 (holotype). Scale: single bar = 1 mm; double bar = 1 cm. Drawn by Judi Stone.

Platyclinis brevilabrata Rendle in *J. Bot., Lond.* 39: 173 (1901). Type: Malaysia, Sarawak, Baram District, *Hose* 52 (holotype BM!, isotype E!).

Acoridium brevilabratum (Rendle) Rolfe in *Orchid Rev.* 12: 220 (1904).

Dendrochilum brevilabratum (Rendle) Pfitzer in Engler, *Pflanzenreich* 4, 50, 2 B 7: 89 (1907).

DISTRIBUTION. Borneo.

SARAWAK. Baram District, 25 Oct. 1894, *Hose* 52 (holotype BM!, isotype E!).

HABITAT. Unknown. Altitude unknown.

The varietal epithet is from the Latin *brevis*, short and *labellum*, lip.

c. var. **oblongum** J.J.Wood **var. nov.**, a varietate typica labello 3-carinato magis distincte papilloso oblongo vel anguste oblongo distinguitur. Typus: Brunei, Belait District, Ulu Ingei, Bukit Batu Patam, 12 June 1989, *Boyce* 319 (holotypus K!, herbarium and spirit material). **Fig. 35.**

Flowers off-white, crystalline, column and base of labellum pale green, or yellow, sepals stained green proximally, petals stained white promixally, labellum lime-green, column yellowish green. *Pedicel with ovary* 3 mm long. *Dorsal sepal* 4.2–4.4 × 1.5 mm. *Lateral sepals* 4 × 1.5 mm. *Petals* 3.5 × 1.1 mm. *Labellum* 1.6–1.7 × 1–1.1 mm. *Column* 1.6 mm long.

DISTRIBUTION. Borneo.

BRUNEI. Belait District, Ulu Ingei, Bukit Batu Patam, 9 June 1989, *Boyce* 295 (BRUN!) & 12 June 1989, *Boyce* 319 (holotype K!).

HABITAT. Mixed dipterocarp forest on valley bottoms, ridge tops and steep slopes, on sandstone, recorded growing on branches and trunks. 55 m.

The two collections from Brunei cited above differ from others examined primarily by the three-keeled, oblong or narrowly ovate-oblong, distinctly papillose labellum, hence the varietal epithet.

SUBGENUS PLATYCLINIS

Dendrochilum subgenus **Platyclinis** (*Benth.*) *Pfitzer* in Engler, *Pflanzenreich* 4, 50, 2 B 7: 87, 91–92 (1907). - Butzin 1974: 254, 255; 1984–1986: 944; de Vogel 1989: 103; Rosinski 1992: 92, 101, 104, 201.

Platyclinis Benth., *J. Linn. Soc., Botany* 18: 295 (1881). - Hemsley 1881: 656; Bentham 1883a: 496; B.S. Williams 1885: 544; J.D. Hooker 1886–1890: 708; Möbius 1887: 561–563; Pfitzer 1888–1889: 128; Veitch 1889: 79–80; Stein 1892: 19, 519; Meinecke 1894: 142; B.S. Williams 1894: 679; Royal Gardens, Kew 1896: 189; Burbidge 1900: 127, 128; de Dalla Torre & Harms 1900–1907: 98; Siebert 1901: 132–133; Watson 1903: 431–432; Rendle 1904: 346; Royal Botanic Gardens, Kew 1904: 189; Zörnig 1904: 737; Ridley 1907: 26; Velenovsky 1907: 349; Bernard 1909: 50; Wehmer 1911: 119, Constantin 1913: 31; Frolik 1913: 46; Miethe 1913: 437–438; Bailey 1916: 2710; Ridley 1924: 24; Solereder & Meyer 1930: 153, 158; Thurgood 1931: 174–175; Curtis 1950:

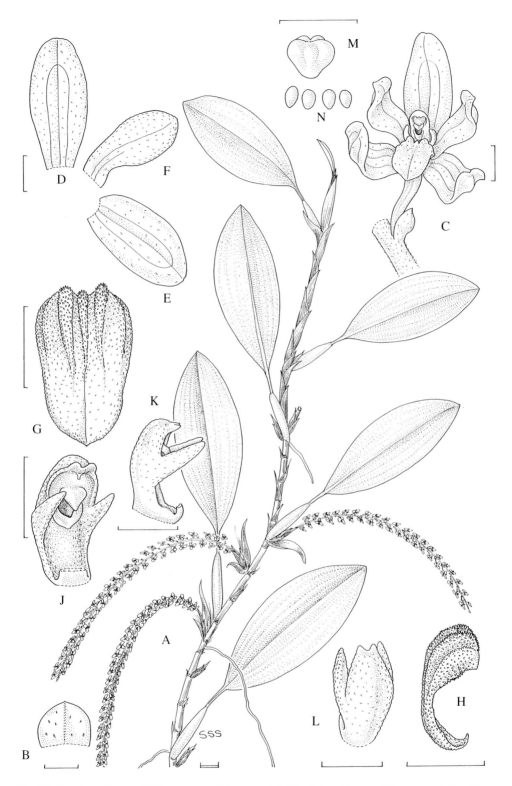

Fig. 35. *Dendrochilum pallidiflavens* var. *oblongum*. **A** habit; **B** floral bract; **C** flower with floral bract and portion of rachis, front view; **D** dorsal sepal; **E** lateral sepal, **F** petal; **G** labellum, front view; **H** labellum, side view; **J** column, anther-cap removed, oblique view; **K** column, anther-cap removed, side view; **L** column, back view; **M** anther-cap, back view; **N** pollinia. **A–N** from *Boyce* 319 (holotype). Scale: single bar = 1 mm; double bar = 1 cm. Drawn by Susanna Stuart-Smith.

159–160; Arnold 1953: 166; Hsieh 1955: 265; Blowers 1957: 182; Wirth & Withner 1959: 172; Chadefaud & Emberger 1960: 1174; Northen 1962: 223, 230; Sander 1969: 158; Hiroe 1971: 75; Hu 1975: 139; Farr *et al.* 1979: 1357. Lectotype: *Platyclinis abbreviata* (Blume) Benth. ex Hemsl.

Dendrochilum sect. *Platyclinis* (Benth.) J.J. Sm., *Recl. Travl. bot. néerl.* 1: 61 (1904). - Smith 1905: 162; Ames 1908: 5–6; Holttum 1957: 228–229; 1964: 228–229; Seidenfaden & Wood 1992: 184; Wood *et al.* 1993: 25.

Acoridium sect. *Platyclinis* (Benth.) Ames, *Proc. biol. Soc. Wash.* 19: 143 (1906). - Type not designated.

Dendrochilum sensu Ames, *Orchidaceae* 7: 80 (1922) *p.p.*, non Blume. - Smith 1934: 200, 211; Greuter *et al.* 1993: 325.

Rhizome short or elongate (if elongate then usually pendent). *Pseudobulbs* clustered or widely separated. *Leaves* convolute or conduplicate. *Inflorescences* synanthous. *Labellum* firmly or elastically attached to the column, entire to distinctly 3-lobed (never transversely E-shaped). *Column* distally prolonged into an apical wing or hood which distinctly (in the Philippine *D. auriculare* and the Sumatran *D. pholidotoides* indistinctly) exceeds the anther; foot and stelidia present or absent; rostellum entire, flat. *Seeds* fusiform; central testa cells prosenchymatic; junctions of the anticlinal testa cell walls not protruding from the testa.

KEY TO THE SECTIONS OF SUBGENUS *PLATYCLINIS* IN BORNEO

1. Labellum divided into a distinctly saccate hypochile and a flat epichile by two free, prominent calli; firmly attached to the column. Column strongly incurved; foot absent .... *Mammosa* (p. 142)
   Labellum not divided into a hypochile and epichile by two prominent calli; firmly or elastically attached to the column. Column straight to somewhat (rarely strongly) incurved; foot present or absent 2

2.ˑ Labellum more or less distinctly cruciform; firmly attached to the column. Column slender (relatively stout in *D. haslamii*); foot absent; stelidia always basal and subequal to the column proper, usually narrowly subspathulate, never absent .... *Cruciformia* (p. 81)
   Labellum entire to variously 3-lobed but never cruciform; firmly or elastically attached to the column. Column stout or slender; foot present or absent; stelidia very variable, sometimes absent 3

3. Labellum firmly attached to the column, entire to obscurely 3-lobed (rarely distinctly 3-lobed). Column comparatively stout and straight; foot absent ..... *Eurybrachium* (p. 109)
   Labellum elastically attached to the column, entire to distinctly 3-lobed. Column comparatively slender (stout in *D. glumaceum*), more or less incurved; foot present, usually short .... *Platyclinis* (p. 154)

Section **Cruciformia**

Section **Cruciformia** *J.J. Wood & H. Ae. Peders.* in *Opera Bot.* 130: 30 (1997). Type species: *Dendrochilum cruciforme* J.J. Wood.

Labellum firmly attached to the column, making a right angle to the latter; more or less distinctly cruciform, neither replicate nor coiled-up, not divided into a hypochile and epichile by two prominent calli. *Column* shorter than the dorsal sepal, slender (relatively stout in *D. haslamii*), generally slightly incurved, not tapering from the base upwards; foot absent; stelidia present, always basal and subequal to the column proper, usually narrowly spathulate.

DISTRIBUTION. Borneo.

KEY TO THE SPECIES OF SUBGENUS *PLATYCLINIS* SECTION *CRUCIFORMIA*

1. Mature leaf-blades 5.3–5.7 cm wide, elliptic. Inflorescence rather rigid, the rachis fleshy, terete. Disc of labellum bearing a fleshy transverse basal callus and two separate crest-like keels above at either side which, although lying very close to, are not united with the basal callus .... 7. *D. hosei* J.J. Wood
    Mature leaf-blades much narrower, rarely up to 3.5 cm wide, narrowly linear, ligulate or elliptic. Inflorescence rarely rigid, (if so then rachis thin and quadrangular), often arcuate and rather graceful, the rachis narrow, often quadrangular. Disc of labellum bearing two keels united at base to form a U-shaped structure 2

2. Mature leaf-blades (0.7–)0.9–2.6 (rarely to 3.5) cm wide 3
    Mature leaf-blades 0.1–0.6 cm wide (usually between 0.1 and 0.4 cm wide) 5

3. Leaf-blades oblong-elliptic, obtuse, 1.8–2.6 cm wide, with finely reticulate venation ....
    .... 8. *D. exasperatum* Ames
    Leaf-blades linear-lanceolate to elliptic, acute to acuminate, 0.7–2.6(–3.5) cm wide, with or without finely reticulate venation 4

4. Inflorescence densely many-flowered, each flower borne 2–3 mm apart on rachis. Flowers small, 4–5 mm across, transparent yellow, the disc with a brown, yellow-brown or citron blotch around the distal part of the keels, often extending to around the base of the labellum lobules, or reduced to two separate blotches, one at the base of each mid-lobe side lobule. Pseudobulbs closely caespitose, without small transverse wrinkles. Leaves without finely reticulate venation, containing sparsely or densely distributed crystalline calcium oxalate bodies (clearly visible in herbarium material), generally linear-lanceolate or oblong-lanceolate, 9–25 × 0.7–1.8 cm. Labellum appearing 5-lobed, the side lobes reduced, the mid-lobe subulate, the side lobules variable in length and shape, usually oblong-falcate, obtuse, retrorse, ascending. Stelidia equalling or longer than the column apex .... 9. *D. gibbsiae* Rolfe
    Inflorescence laxly few- to several-flowered, each flower borne up to 5 mm apart on rachis. Flowers relatively large, 1 cm across, flesh-pink to pinkish-brown, the labellum with darker keels, without a darker central blotch. Pseudobulbs less closely caespitose, often borne 1–2 cm apart on rhizome, with small transverse wrinkles (especially apparent on herbarium material). Leaves with fine reticulate venation, generally shorter and broader, often around 2 cm wide (occasionally up to 3.5 cm wide), calcium oxalate bodies absent. Labellum pandurate, tridentate, the side lobes absent, the mid-lobe with 3 acute lobules. Stelidia slightly shorter than or equalling the column apex ....
    .... 10. *D. grandiflorum* (Ridl.) J.J. Sm.

5. Side lobes of labellum distinct, obtuse or acute to acuminate, sometimes falcate 6
   Side lobes of labellum absent, or obscure and rounded, never falcate 7

6. Labellum 1.5–1.6 mm long, the mid-lobe transversely oblong to narrowly hastate, the side
   lobes small, keel-like, obtuse to subacute, sometimes somewhat falcate. Sepals and
   petals 2–3 mm long .... 11. *D. dolichobrachium* (Schltr.) Schltr.
   Labellum 3.8–4 mm long, the mid-lobe broadly hastate, the side lobes well developed,
   triangular, acute. Sepals and petals 4–4.1 mm long .... 12. *D. hastilobum* J.J. Wood

7. Labellum cruciform. Stelidia shorter than column apex or, if longer, then the labellum
   apex longly subulate-acuminate. Pedicel with ovary usually 0.5–1(–2) mm long. Disc
   usually with a dark purple-brown central blotch, apex of keels dark purple-brown ....
   .... 13. *D. cruciforme* J.J. Wood
   Labellum pandurate, or expanded above the middle into large lanceolate or oblong,
   falcate, acute or obtuse divaricate lobules. Pedicel with ovary 3 mm long. Disc and
   keels without purple-brown pigmentation 8

8. Labellum pandurate, not expanded above the middle into divaricate lobules. Stelidia 2–2.1
   mm long, ligulate, decurved at the apex, hamate. Flowers pale green ....
   .... 14. *D. devogelii* J.J. Wood
   Labellum expanded above the middle into large lanceolate or oblong, falcate, acute or
   obtuse divaricate lobules, or rarely lobules almost straight to subfalcate, brown or deep
   yellow. Sepals and petals obtuse or subacute, strongly reflexed. Stelidia equalling or
   slightly longer than column apex, not hamate .... 15. *D. haslamii* Ames

7. DENDROCHILUM HOSEI

Dendrochilum hosei J.J .Wood in Wood & Cribb, *Checklist orch. Borneo*: 179, fig. 23 D&E
(1994). Type: Malaysia, Sarawak, Kapit District, northern Hose Mountains, base of ridge
leading to Batu Hill, 1 Dec. 1991, *Leiden* cult. (*de Vogel*) 913960, originally cited by Wood &
Cribb 1994 as *de Vogel* 1244 (holotype L!, isotype K!, SAR, both spirit material only). **Fig.
36.**

DESCRIPTION. Epiphyte. *Rhizome* not collected. *Roots c.* 1.8 mm in diameter. *Cataphylls*
4–5, ovate-elliptic, acute, up to 8 cm long, persistent, disintegrating and becoming fibrous.
Pseudobulbs elliptic, up to 5 × 2 cm, slightly wrinkled. *Leaf*: petiole deeply sulcate, 5–6 cm
long; blade elliptic, acute, cuneate at base, tough, coriaceous, 22–25 × 5.3–5.7 cm.
Inflorescences rather rigid, laxly many-flowered, flowers borne 5–6 mm apart, those along
central portion of rachis opening first; peduncle tough, 22 cm long, 2–2.8 mm wide; non-
floriferous bracts absent; rachis ± quadrangular, shallowly sulcate, fleshy, 14 cm long, 2–3 mm
wide; floral bracts triangular-ovate, acute, 3–5 × 2 mm. *Flower* colour not recorded. *Pedicel
with ovary* clavate, gently curved, *c.* 1.8 mm long. *Sepals* and *petals* 3-nerved. *Dorsal sepal*
oblong-elliptic to ovate-elliptic, apiculate, 4 × 1.9 mm. *Lateral sepals* ovate-elliptic, apiculate,
4 × 1.9 mm. *Petals* elliptic, acute, 3.5–3.6 × 1.6–1.7 mm. *Labellum* pandurate, margins
sometimes slightly erose towards base, 2.9–3 mm long, 4 mm wide across side lobes; side
lobes obscure, oblong, margins rather uneven to erose; mid-lobe with erect, oblong-ovate,
obtuse, slightly falcate lateral lobules and a tooth-like, apiculate, mucronate central lobule;

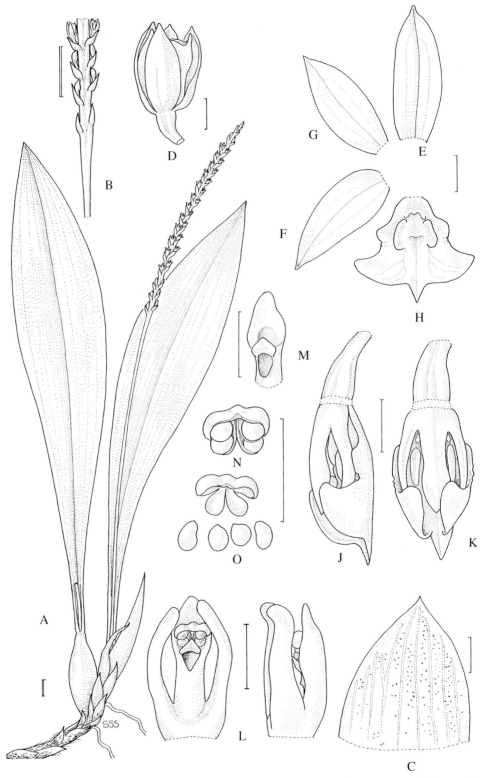

Fig. 36. *Dendrochilum hosei.* **A** habit; **B** lower portion of inflorescence; **C** floral bract; **D** flower, side view; **E** dorsal sepal; **F** lateral sepal; **G** petal; **H** labellum, front view; **J** pedicel with ovary, labellum and column, side view; **K** pedicel with ovary, labellum and column, viewed from above; **L** column, front and side views; **M** column apex, oblique view; **N** anther-cap, interior views, with and without pollinia; **O** pollinia. A–O from *Leiden* cult. (*de Vogel*) 913960 (holotype). Scale: single bar = 1 mm; double bar = 1 cm. Drawn by Susanna Stuart-Smith.

disc with a fleshy basal callus and two separate fleshy, crest-like keels either side which almost touch the basal callus at their base. *Column* slightly dorsally carinate, 2.1 mm long; foot absent; apical hood ovate, entire; stelidia basal, ligulate, obtuse, decurved at apex, slightly longer than apical hood, 2 mm long; stigmatic cavity oblong or triangular-oblong; rostellum broadly ovate, obtuse; anther-cap cucullate.

DISTRIBUTION. Borneo.

SARAWAK. Kapit District, northern Hose Mountains, base of ridge leading to Batu Hill, 1 Dec. 1991, *Leiden* cult. (*de Vogel*) 913960 (holotype L!, isotype K!).

HABITAT. Not recorded. 1200 m.

The labellum shape and column structure of this rather unattractive species bear a striking resemblance to the widespread *D. gibbsiae* Rolfe. The broad leaves of *D. hosei* are quite different however, and similar to those of *D. longifolium* Rchb.f., while the rather rigid inflorescence is not unlike *D. longipes* J.J. Sm. The two keels on the disc are aligned very close to, but are not united with the transverse basal ridge. In many species, including *D. gibbsiae*, these are united to form a u- or m-shaped structure.

The specific epithet refers to the Hose Mountains in Sarawak which were named after the Rev. George F. Hose (1838–1922), Bishop of Singapore, Labuan and Sarawak, from 1881 until 1908.

8. DENDROCHILUM EXASPERATUM

Dendrochilum exasperatum Ames, *Orchidaceae* 6: 50 (1920). Type: Malaysia, Sabah, Mount Kinabalu, Marai Parai Spur, 1915, *J. Clemens* 396 (holotype AMES!). **Figs. 37 & 38, plate 3C & D.**

DESCRIPTION. Epiphyte, rarely terrestrial. *Rhizome* 1–8 cm long, *c.* 4 mm in diameter. *Roots* produced in a large, dense mass, smooth, *c.* 1 mm in diameter. *Cataphylls* usually 4, ovate-elliptic, acute to acuminate, 1–5 cm long, finely nerved, reddish brown, unspotted, soon becoming fibrous. *Pseudobulbs* closely caespitose, narrowly cylindrical or fusiform, somewhat flattened, 1–6 × 0.4–0.6 cm, finely wrinkled when dry. *Leaf:* petiole sulcate, 0.5–4 cm long; blade oblong-elliptic to elliptic, obtuse and minutely mucronate, sometimes somewhat constricted *c.* 2–2.5 cm below apex, 5–16 × (1.1–)1.5–3.5 cm, main nerves 7(–9), with numerous fine transverse nerves, thin and parchment-like to somewhat leathery-textured, frequently with numerous small irregular calcium oxalate bodies similar to *D. gibbsiae*, particularly noticeable in dried material. *Inflorescences* gently curving, many-flowered, subdense, buds appearing curved and acuminate; peduncle terete, 5–15 cm long; rachis quadrangular, 8–25 cm long; non-floriferous bract solitary, 4–9 mm long; floral bracts oblong-ovate to ovate, obtuse, apiculate, longer than pedicel with ovary, 2–6 × 2 mm, nerves prominent and raised on abaxial surface. *Flowers* borne 2–4 mm apart, variously described as: greenish with some ochre, labellum brown in centre; pale reddish ochre, labellum dark red in centre, side lobes reddish brown, mid-lobe pinkish, tip white; pale greenish yellow, with a triangular yellowish brown spot at base of each mid-lobe side lobule; creamy yellow, dark speck on labellum; sepals pale ochre-red, petals yellowish, labellum ochre, with 2 brown central spots; pale brown, tips of upper sepals and labellum appendages reddish brown; with little or no scent. *Pedicel with ovary* clavate, 1.8–3 mm long. *Dorsal sepal* narrowly oblong-elliptic, acute, 6 × 1.5 mm, nerves 3. *Lateral sepals* slightly obliquely oblong-elliptic, somewhat falcate, acute, 4.6–5 × 1.7–1.8 mm, sometimes overlapping distally, nerves 3. *Petals*

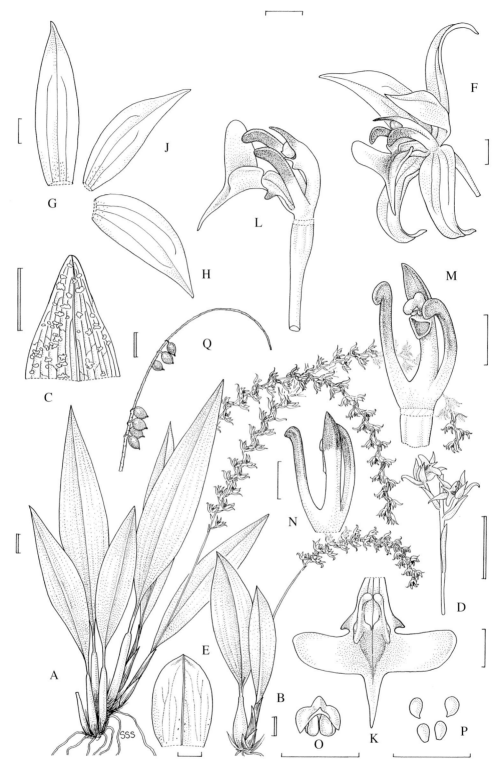

Fig. 37. *Dendrochilum exasperatum*. **A & B** habits; **C** leaf apex, adaxial surface, showing calcium oxalate crystal bodies; **D** lower portion of inflorescence showing solitary non-floriferous bract; **E** floral bract; **F** flower oblique view; **G** dorsal sepal; **H** lateral sepal, **J** petal; **K** labellum, front view; **L** pedicel with ovary, column and labellum, side view; **M** column, oblique view; **N** column, back view; **O** anther-cap, front view; **P** pollinia; **Q** infructescence. **A** from *Vermeulen & Duistermaat* 668 & 1038, **B** from *J. & M.S. Clemens* 50246, **C–Q** from *Vermeulen & Duistermaat* 668. Scale: single bar = 1 mm; double bar = 1 cm. Drawn by Susanna Stuart-Smith.

narrowly elliptic, acute, somewhat falcate, 4–5 × 1.1–1.5 mm, nerves 2–3. *Labellum* 5-lobed, 2.5–3 mm long, 1.8–2 mm wide across mid-lobe; nerves 3; side lobes reduced, triangular, rounded; mid-lobe cruciform, expanded into spreading to erect oblong, rounded to truncate, sometimes slightly falcate side lobules and a triangular, acuminate, cuspidate, horizontal to decurved terminal lobule; side lobules often porrect, 1–1.1 × 0.9–1 mm, nerves 4; terminal lobule 1–1.1 mm long; disc with 2 lamellate keels joined at base to form a horseshoe-shaped U, arising from base and terminating at base of mid-lobe, with a low septum level with side lobes between. *Column* slender, 2.8–3 mm long, very obscurely dorsally carinate; foot absent; apical hood elongate, oblong-triangular, obtuse to acute, entire, 1 mm long; stelidia basal, ligulate, apex obtuse, rather spathulate or hamate, slightly shorter than apical hood, 2 mm long; rostellum small, triangular, acute; anther-cap ovate, cucullate, glabrous, 0.4–0.5 × 0.6 mm. *Capsule* ovoid-elliptic, 5–7 × 3.5–4 mm, lines of dehiscence 3.

DISTRIBUTION. Borneo.

SARAWAK. Marudi District, Mount Dulit, Dulit Ridge, 6 Sept. 1932, *Synge* S. 436 (K).

SABAH. Mount Kinabalu, Kemburongoh, 20 Dec. 1995, *Barkman* 212 (Sabah Parks Herbarium, Kinabalu Park!). Mount Kinabalu, Pig Hill, 29 Dec. 1995, *Barkman* 214 (Sabah Parks Herbarium, Kinabalu Park!). Mount Kinabalu, Marai Parai Spur, 1915, *J. Clemens* 396 (holotype AMES). Mount Kinabalu, Dallas, 18 Oct. 1931, *J. & M.S. Clemens* 26784A (BM!) & 20 Nov. 1931, *J. & M.S. Clemens* 27256 (BM!, BO!, E!, K!). Mt Kinabalu, Kemburongoh, 1993, *Carr* 3750 (SING!) & 13 Nov. 1931, *J. & M.S. Clemens* 27186 (BO!). Mount Kinabalu, above Lumu-Lumu, Jan. 1932, *J. & M.S. Clemens* 30173 (AMES!). Mount Kinabalu, Penibukan, 4 Jan. 1933, *J. & M.S. Clemens* 30640 (BM!), 7 Jan. 1933, *J. & M.S. Clemens*

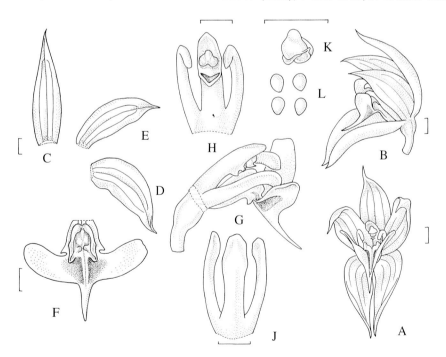

Fig. 38. *Dendrochilum exasperatum*. **A** flower, front view; **B** flower, side view; **C** dorsal sepal; **D** lateral sepal; **E** petal; **F** labellum, front view; **G** pedicel with ovary, labellum and column, side view; **H** column, front view; **J** column, back view; **K** anther cap, oblique view; **L** pollinia. **A–L** from *Barkman* 212. Scale: single bar = 1 mm. Drawn by Susanna Stuart-Smith.

30799 (BM!) &10 & 15 Jan. 1933, *J. & M.S. Clemens* s.n. (BM!). Mount Kinabalu, Tenompok, Aug. 1933, *Carr* 3668, SFN 28029 (K!, SING!), 2 Nov. 1933, *Carr* SFN 28049 (SING!). Mount Kinabalu, Tenompok Orchid Garden, 9 Nov. 1933, *J. & M.S. Clemens* 50235 (BM!, mixed collection with *D. crassifolium* and *D. kamborangense)* & 50246 (AMES!, BM!, K!, L!) & 9 Nov. 1933, *J. & M.S. Clemens* 50256 (SING!). Tenom District, Sapong, Mount Anginon, April 1965, *H.F. Comber* 4037 (K!). Sipitang District, Mount Lumaku, 1963–4, *J.B. Comber* 122 (K!). Mount Alab, 1987, *Leiden* cult. (*Vermeulen*) 26484 (K!, L!). West Coast Zone/Interior Zone, Crocker Range, Keningau to Kimanis road, Dec. 1986, *Vermeulen & Duistermaat* 668 (L!) & *Vermeulen & Duistermaat* 690 (L!). Sipitang District, Rurun River headwaters, Dec. 1986, *Vermeulen & Duistermaat* 1038 (L!).

HABITAT. Lower montane ridge forest on sandstone soils; mossy forest, recorded growing in 60% sunlight high up in canopy; low and rather open, wet, somewhat podsolic forest, with a very dense undergrowth of *Pandanus* spp. and rattan palms; cleared, steeply sloping roadside on sandstone and shale outcrops partially covered with grass and small bushes, fully exposed to the sun. Recorded by Barkman as a low growing trunk epiphyte on *Leptospermum javanicum* Blume. 900–2400 m.

D. exasperatum is a close ally of the more common *D. gibbsiae*, with which it shares similar labellum and column characters. The leaves however, are generally shorter, broader, more obtuse and have a distinctive fine transverse venation, while the floral bracts and flowers are larger and the labellum side lobes are less well developed.

The specific epithet is from the Latin *exasperatus*, covered with short hard points or roughened and presumably refers to the surface of the leaves which frequently contain numerous small calcium oxalate bodies producing a rough texture, particularly in dried material.

9. DENDROCHILUM GIBBSIAE

Dendrochilum gibbsiae Rolfe in *J. Linn. Soc., Botany* 42: 147–148 (1914). Type: Malaysia, Sabah, Mount Kinabalu, Marai Parai Spur, 2100 m, Feb. 1910, *Gibbs* 4087 (holotype BM!, isotype K!). **Figs. 39 & 40, plate 4A–C.**

D. kinabaluense Rolfe in *J. Linn. Soc., Botany* 42: 148 (1914). Type: Malaysia, Sabah, Mount Kinabalu, Marai Parai Spur, 2100 m, Feb. 1910, *Gibbs* 4085 (holotype BM!, isotype K!).
D. quinquelobum Ames, *Orchidaceae* 6: 63–64, pl. 82, II, 2 (1920). Type: Malaysia, Sabah, Mount Kinabalu, Kiau, 30 Nov. 1915, *J. Clemens* 361 (holotype AMES!, isotypes BM!, BO!, K!, SING!).

DESCRIPTION. Epiphyte or terrestrial. *Rhizome* 1–10 (rarely up to 16) cm long, 0.2–0.3 cm in diameter. *Roots* produced in a large, dense mass, smooth, 0.5–1.8 mm in diameter. *Cataphylls* 3–4, 0.5–2.5 cm long, finely nerved, greyish brown or reddish brown, sometimes finely speckled, persistent and taking a long time to become fibrous. *Pseudobulbs* closely caespitose, rarely distant and up to 2.5 cm apart, cylindrical, narrowly oblong, ovoid-oblong or narrowly fusiform, 1.2–4.8 × 0.3–1 cm, smooth or slightly wrinkled. *Leaf*: petiole sulcate, (1–)1.5–3 cm long; blade linear-lanceolate or oblong-lanceolate, subacute or acute, (3.5–)10–28 × (0.3–)0.6–1.8(–2.1) cm, main nerves 5–7, with numerous parallel secondary nerves, transverse nerves absent, thin-textured, containing scattered to densely distributed crystalline calcium oxalate bodies, often covering the entire surface, variable in shape, from

Fig. 39. *Dendrochilum gibbsiae*. **A–C** habits; **D** floral bract; **E** flower, oblique view; **F** pedicel with ovary, labellum and column, front view; **G** dorsal sepal; **H** lateral sepal; **J** petal; **K** labellum, front view; **L** column with anther-cap removed, oblique view; **M** column, side view; **N** column, back view; **O** anther-cap, oblique view; **P** pollinia; **Q–U** labelli, front views; **V** infructescence. **A** from *Carr* C. 3620, SFN 27884, **B** from *Carr* C. 3134, SFN 26745; **C–P** from *Wood* 578; **Q** from *Gibbs* 4085 (holotype of *D. kinabaluense*), **R** from *J. Clemens* 361 (holotype of *D. quinquelobum*), **S** from *Bailes & Cribb* 722, **T** from *Gibbs* 4087 (holotype), **U** from *Wood* 621, **V** from *Carr* SFN 27970. Scale: single bar = 1 mm; double bar = 1 cm. Drawn by Susanna Stuart-Smith.

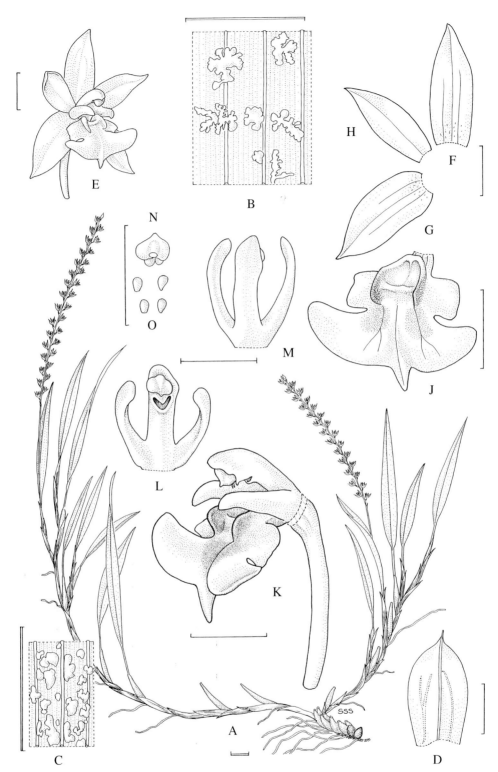

Fig. 40. *Dendrochilum gibbsiae*. Variant from Mount Mulu. **A** habit; **B & C** close-up of portion of abaxial surface of leaf showing irregular calcium oxalate crystal bodies; **D** floral bract; **E** flower, oblique view; **F** dorsal sepal; **G** lateral sepal; **H** petal; **J** labellum, front view; **K** pedicel with ovary, labellum and column, side view; **L** column, front view; **M** column, back view; **N** anther-cap, back view; **O** pollinia. **A** from *Argent* s.n., RBGE no. 781696 & *Argent & Jermy* 988; **B** from *Burtt & Martin* B. 5299; **C** from *J.A.R. Anderson* S. 30838; **D–O** from *Argent* s.n., RBGE no. 781696. Scale: single bar = 1 mm; double bar = 1 cm. Drawn by Susanna Stuart-Smith.

simple to intricate and dentritic in shape, particularly noticeable in dried material. *Inflorescence* an ascending to gently curving, many-flowered, narrow subdense raceme, flowers borne 2–3 mm apart; peduncle terete, (3.5–)8–13 cm long; rachis quadrangular, (5.5–)10–16(–25) cm long; non-floriferous bracts absent; floral bracts ovate-triangular or oblong-elliptic, acute, 1.2–3 × 0.1–1.2 mm, spreading. *Flowers* unscented, or scented of oranges, variously described as: cream, flesh coloured, salmon, creamy green, pale yellowish green, pale greenish, greenish yellow, creamy yellow, yellow, straw-coloured or pale khaki, pedicel pink, labellum pale green, orange-brown near base, pale ochre at apex, or orange, with 2 pink, purple, maroon, brown or orange-brown spots at base of mid-lobe, calli yellowish green, brown at tip, column pale green or brownish green, reddish at base, stelidia greenish white, anther-cap cream. *Pedicel with ovary* narrowly clavate, 1–3 mm long. *Dorsal sepal* oblong-elliptic, acute, 2.2–4 × 0.8–1.1 mm, concave, minutely papillose-hairy at base inside, nerves 3. *Lateral sepals* oblong-ovate, acute, 2–4 × 1.2–1.5 mm, minutely papillose-hairy inside, nerves 3. *Petals* narrowly elliptic or oblong-elliptic, acute, 1.6–3.6 × 0.8–1 mm, minutely papillose-hairy inside, nerves 1–2. *Labellum* 5-lobed, 1.6–3 mm long, 2.5–4 mm wide across mid-lobe; nerves 3, median prominent; slightly decurved at side lobes; side lobes oblong, obtuse, truncate or slightly erose, sometimes reduced to triangular, falcate teeth, 0.2–1 × 0.2–0.6 mm, spreading or erect; mid-lobe cruciform, expanded into spreading or retrorse, oblong and rounded to oblong-falcate, obtuse side lobules and a cuspidate, acuminate, deflexed or ascending terminal lobule; side lobules 1–2 × 0.5–1.1 mm, nerves 3; terminal lobule 0.8–1 mm long; disc with 2 fleshy lamellate keels joined at base to form a horseshoe-shaped U, arising from just above base and terminating at base of mid-lobe, and usually with an obscure, slightly raised transverse septum level with side lobes between. *Column* slender, 1.4–2 mm long; foot absent; apical hood short, narrowly ovate, obtuse or subacute; stelidia basal, oblong-linear, apex obtuse, hamate, (1–)1.5–2.1 mm long, equalling or longer than apical hood; rostellum small, ovate-triangular, obtuse or subacute; anther-cap ovate, cucullate, minute, glabrous. *Capsule* ovoid, 5 × 4.5–5 mm.

DISTRIBUTION. Borneo.

SARAWAK. Marudi District, Gunung Mulu National Park, Mount Api, Ulu Melinau, Tutoh, northeast flank of mountain, 30 Sept. 1971, *J.A.R. Anderson* S. 30838 (A, E!, K!, L!, SAR, SING). Marudi District, Gunung Mulu National Park. cult. Edinburgh Botanic Garden, 5 Sept. 1988, *Argent* s.n., RBGE no. 781696 (E!). Marudi District, Gunung Mulu National Park, Mount Api, Ridge Camp, 14 April 1978, *Argent & Jermy* 988 (E!, K!). Belaga District, Linau-Balui divide, Nawai/Balui Rivers, 5 Sept. 1978, *Burtt* B. 11460 (E!). Lawas District, route from Ba Kelalan to Mount Murud, below Camp III, 29 Sept. 1967, *Burtt & Martin* B. 5299 (E!, SAR). Marudi District, Gunung Mulu National Park, Mount Api, Ulu Melinau, 7 Sept. 1970, *Chai* S. 30098 (A, BO, K!, KEP!, L!, SAN, SAR, SING!). Simunjan District, Mount Ampungan, southeast of Serian, 1993, *Leiden* cult. (*Schuiteman et al.*) 933273 (L!). Marudi District, Gunung Mulu National Park, Mount Api, trail to limestone pinnacles, 23 March 1998, *Leiden* cult. (*Vogel et al.*) 980299 (K!, L!). Kapit District, northern Hose Mountains, base of ridge leading to Batu Hill, 1 Dec. 1991, *de Vogel* 1245 (L!). Marudi District, proposed Mount Murud National Park, 11 Sept. 1982, *Yii* S. 44402 (K!, L, SAR).

BRUNEI. Temburong District, Mount Pagon, 1992, *de Vogel* 2043 (K!, L!).

SABAH. Sinsuron (Sensuron) road, cult. Royal Botanic Garden, Edinburgh, 4 May 1989, *Argent* s.n., RBGE no. 801344 (E!). Ranau District, Mount Kinabalu, Pinosuk Plateau, golf course area, 1983, cult. Kew, *Bailes & Cribb* 722 (K!). Ranau District, Mount Kinabalu, Pig Hill, *Barkman* 51 (Sabah Parks Herbarium, Kinabalu Park). Ranau District, Mount Kinabalu, Pinosuk Plateau, golf course site near East Mesilau River, 13 Oct. 1983, *Beaman* 7239 (K!).

Ranau District, Mount Kinabalu, Pig Hill, 25 May 1984, *Beaman* 9889 (K!). Mount Kinabalu, Minitinduk River, 25 March 1933, *Carr* 3134, SFN 26745 (SING!). Mount Kinabalu, near Bundu Tuhan, July 1933, *Carr* 3620, SFN 27884 (K!, SING!). Mount Kinabalu, Tenompok, 11 Sept. 1933, *Carr* 3695 (SING!) & 29 Sept. 1933, *Carr* 3702 (SING!). Mount Kinabalu, Dallas, 20 Oct. 1933, *Carr* 3729 (SING!) & *Carr* 3745 (SING!). Mount Kinabalu, Marai Parai, 21 Aug. 1933, *Carr* SFN 27544 (SING!), SFN 27550 (SING!) & SFN 27970 (SING!). Mount Kinabalu, Kadamaian River, 14 Oct. 1933, *Carr* SFN 28034 (SING!). Ulu Kimanis, 16 Oct. 1986, *C.L. Chan* 17/86 (K!). Mount Kinabalu, Kiau, 30 Nov. 1915, *J. Clemens* 361 (holotype of *D. quinquelobum* AMES!, isotypes BM!, K!, SING!). Mount Kinabalu, Dallas, 18 Oct. 1931, *J. & M.S. Clemens* 26784 (BM!), 28 Oct. 1931, *J. & M.S. Clemens* 26841 (BM!, HBG!, K!), 3 Nov. 1931, *J. & M.S. Clemens* 26930 (BM!, E!) & 12 Oct. 1931, *J. & M.S. Clemens* s.n. (BM!). Mount Kinabalu, Kiau, 6 Nov. 1915, *J. Clemens* 146 (AMES!, BM!, SING!) & 8 Nov. 1915, *J. Clemens* 178 (AMES, BM!, BO!, K!, SING!). Mount Kinabalu, Kilembun River, 6 or 8 June 1933, *J. & M.S. Clemens* 32432 (AMES!, BM!, E!, K!). Mount Kinabalu, Kinateki River, 25 Feb. [& ? 22 April] 1933, *J. & M.S. Clemens* 31800 (AMES!, BM!, E!, G!, HBG!, L!). Mount Kinabalu, Lubang Gorge, 23 Nov. 1915, *J. Clemens* 289 (AMES, BM!, BO!, K!, SING!). Mount Kinabalu, Marai Parai, 26 Feb. [& ? 22 March] 1933, *J. & M.S. Clemens* 32246 (AMES!, BM!), 22 March 1933, *J. & M.S. Clemens* 32247 (BO!, E!, G!, L!) & 5 April 1933, *J. & M.S. Clemens* 32617 (AMES!, BM!, E!). Mount Kinabalu, Marai Parai/Nungkek, 3 April 1933, *J. & M.S. Clemens* 32552 (BM!, E!). Mount Kinabalu, Penibukan, 30 Dec. 1932, *J. & M.S. Clemens* 30573 (AMES!, BM!, HBG!, K!), 7 Jan. 1933, *J. & M.S. Clemens* 30797 (L!, mixed collection with *Bulbophyllum flavescens* (Blume) Lindl.), 7 Feb. 1933, *J. & M.S. Clemens* 31457 (AMES!, BM!, BO!, E!, K!), Nov. 1933, *J. & M.S. Clemens* 50370 (BM!, K!), 15 Jan. 1933, *J. & M.S. Clemens* s.n. (E!) & Jan. 1933, *J. & M.S. Clemens* s.n. (BM!, mixed collection with *Bulbophyllum* sp.). Mount Kinabalu, Tahubang Falls, 4 Jan. 1933, *J. & M.S. Clemens* 30702 (BM!, BO!) & 4 Jan. 1933, *J. & M.S. Clemens* 30703 (AMES!). Mount Kinabalu, Tenompok, 14 April 1932, *J. & M.S. Clemens* 29235 (BM!, E!, K!) & *J. & M.S. Clemens* 50184 (BM!). Mount Kinabalu, Penibukan, Tahubang River, 11 Oct. 1933, *J. & M.S. Clemens* 40701 (BM!). Mount Kinabalu, Tenompok Orchid Garden, 9 Nov. 1933, *J. & M.S. Clemens* 50177 (BM!) & 9 Nov. 1933, *J. & M.S. Clemens* 50184 (BM!, E!, K!, L!). Mount Kinabalu, Tinekuk Falls, 26 Oct. 1933, *J. & M.S. Clemens* 40925 (BM!, K!, L!). Mount Kinabalu, Marai Parai Spur, Feb. 1910, *Gibbs* 4085 (holotype of *D. kinabaluense* BM, isotype K). Mount Kinabalu, Marai Parai Spur, Feb. 1910, *Gibbs* 4087 (holotype BM!, isotype K!). Crocker Range, Keningau to Kimanis road, Sept. 1987, *Lamb & Chan* in *Lamb* AL 875/87 (K!). Mount Kinabalu, Tenompok, 5 Nov. 1959, *Meijer* SAN 20326 (K!, L!, SAN). Mount Kinabalu, cult. Kinabalu Park Mountain Garden, SNP 0788 (Sabah Parks Herbarium, Kinabalu Park). Mount Kinabalu, Marai Parai Spur, 13 Feb. 1985, *SNP* 1833 & 12 March 1987, *SNP* 3038 (Sabah Parks Herbarium, Kinabalu Park). Mount Tembuyuken, 11 March 1993, *SNP* 5572 (Sabah Parks Herbarium, Kinabalu Park). Crocker Range, Keningau to Kimanis road, Dec. 1986, *Vermeulen & Duistermaat* 664 (K!, L!), Dec. 1986, *Vermeulen & Duistermaat* 688 (K!, L!) & 10 Oct. 1985, *Wood* 578 (K!). Sandakan Zone, Mount Tawai, Nov. 1986, *Vermeulen & Lamb* 705 (K!, L!). Mount Kinabalu, Park Headquarters, Silau Silau Trail, 20 Oct. 1985, *Wood* 621 (K!).

KALIMANTAN TIMUR. Apo Kayan, Kayan River near Long Sungai Barang, east of Long Nawan, Oct. 1991, *Leiden* cult. (*de Vogel*) 913520 (L!).

HABITAT. Lower montane oak-laurel forest; open montane ridge-crest forest on limestone; mossy forest on ultramafic substrate with *Gymnostoma* spp. and undergrowth of climbing bamboo; low mossy and xerophyllous scrub forest on extreme ultramafic substrate; steep open roadside banks on sandstone and shale outcrops, associated with *Melastoma* spp., *Nepenthes fusca* Danser, *Gahnia* spp., *Lycopodium* spp., ferns, etc.; limestone boulders in primary forest. Recorded as epiphytic on tree trunks and lower branches. 870–2400 m.

D. gibbsiae is the most widespread and one of the most frequently collected species of section *Cruciformia*. It is variable both vegetatively and in the dimensions and shape of the floral parts, unlike the majority of species in the genus.

Variants with narrow leaves, racemes and floral parts have been called *D. kinabaluense* Rolfe and those with strongly recurved falcate labellum mid-lobe side lobules assigned to *D. quinquelobum* Ames. Examination of the large number of collections now available shows every gradation from short and straight to strongly falcate labellum mid-lobe side lobules. *Carr* 3620 has large leaf blades up to 28 × 2 cm, but smaller flowers with rather broad sepals and petals and short, broad, straight side lobules. The small transverse septum commonly present within the horseshoe-shaped keel on the disc of the labellum is lacking in *Carr* 3134 and 3620. *Carr* SFN 28034 has leaf blades only up to 11 × 0.7 cm. Three collections from Gunung Mulu National Park in Sarawak, *viz. Argent* s.n., cult. Edinburgh Botanic Garden, *Argent & Jermy* 988 and *Leiden* cult. (*Vogel et al.*) 980299 collected from shrubbery on an exposed limestone ridge at 1500 m on Mount Api, are curious in having an elongated rhizome, dwarf stature and short leaf blades only 3.3–5.5 × 0.4–0.6 cm. The flowers, however, are typical of *D. gibbsiae* and these collections may represent an ecotype adapted to extreme exposed conditions.

A characteristic feature of *D. gibbsiae* and some related species, e.g. *D. dolichobrachium* (Schltr.) Schltr. and *D. hastilobum* J.J. Wood, are the numerous crystalline calcium oxalate bodies present in the leaf blades. These may be few in number or, more commonly, densely distributed over the entire surface and clearly visible in dried and spirit material. Their shape may be simple, or complicated and resembling an asymmetrical dentritic snowflake-like structure.

The specific epithet commemorates Lilian Suzette Gibbs (1870–1925), an English botanist interested in tropical alpine floras, who made collections on Mount Kinabalu in 1910.

10. DENDROCHILUM GRANDIFLORUM

Dendrochilum grandiflorum (Ridl.) J.J. Sm. in *Recl. Trav. bot. néerl.* 1: 66–67 (1904). Type: Malaysia, Sabah, Mount Kinabalu, 3200 m, *Haviland* 1142 (holotype K!). **Fig. 41, plate 6A–C.**

Platyclinis grandiflora Ridl. in Stapf in *Trans. Linn. Soc. Lond.* 2, 4: 233 (1894).
Acoridium grandiflorum (Ridl.) Rolfe in *Orchid Rev.* 12: 220 (1904).
Dendrochilum perspicabile auct., non Ames: Sato, *Flowers and Plants of Mount Kinabalu*: 36 (1991).

DESCRIPTION. Epiphyte, occasionally terrestrial. *Rhizome* up to 10 cm long, 0.4–0.6 cm in diameter, branches 2–4 cm long, tough, clothed in fibrous sheaths. *Roots* much branched, smooth, 0.5–1 mm in diameter. *Cataphylls* 3–4, 1.5–5 cm long, finely nerved, greyish brown to brown, finely speckled, persistent, becoming fibrous. *Pseudobulbs* caespitose, up to 1.5 cm apart, narrowly cylindrical, terete, (1.8–)2.5–6 × 0.4–0.6 cm, with narrow longitudinal furrows, usually appearing deeply sulcate when dried, surface densely minutely rugose and quite distinctive in dried material. *Leaf*: petiole sulcate below blade, 1–3 cm long, with minute brown trichomes; blade narrowly elliptic to elliptic, acute, apiculate, margin slightly

93

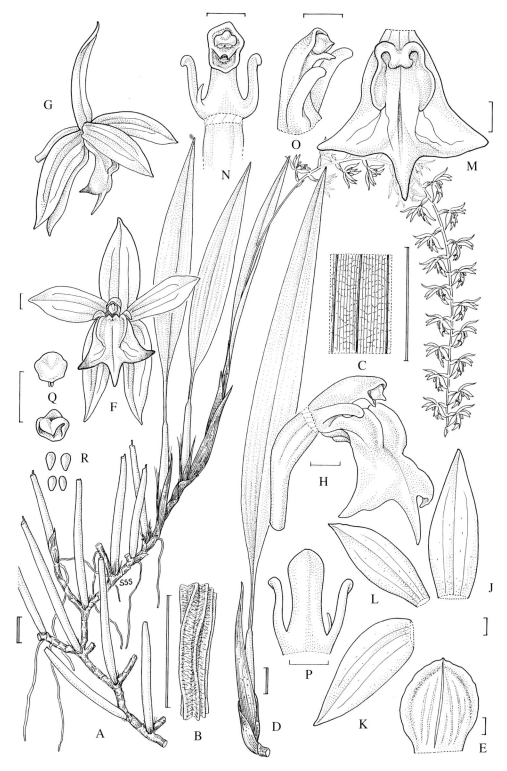

Fig. 41. *Dendrochilum grandiflorum.* **A** habit; **B** portion of old pseudobulb; **C** close-up detail of leaf venation; **D** habit; **E** floral bract; **F** flower, front view; **G** flower, side view; **H** pedicel with ovary, labellum and column, side view; **J** dorsal sepal, **K** lateral sepal; **L** petal; **M** labellum, front view; **N** column and upper portion of ovary, showing an abnormal protuberance at base of stelidium, front view; **O** column, side view; **P** column, back view; **Q** anther-cap, back and interior views; **R** pollinia. **A–C** from *J. Sinclair et al.* 9180, **D** from *Carr* C. 3476, SFN 27430, **E–R** from *Wood* 608. Scale: single bar = 1 mm; double bar = 1 cm. Drawn by Susanna Stuart-Smith.

constricted 1.5–2.5 cm below apex, 6–18 × (0.8–)1.3–2.5(–3.3) cm, main nerves 5, numerous small transverse nerves forming a reticulate pattern, most distinctive in dried material, abaxial and adaxial surface distributed with minute brown trichomes, dark green and glossy above, slightly paler beneath, coriaceous. *Inflorescences* several- to many-flowered, gently curving, subdense to lax, flowers borne 4–7 mm apart; peduncle terete, 7–16 cm long, glabrous; rachis quadrangular, sulcate, 8–15 cm long, minutely ramentaceous; non-floriferous bract solitary, 4–6 mm long; floral bracts ovate, shortly apiculate, margin minutely erose, 3–6.8 × 4 mm, nerves prominent, raised, surface ramentaceous. *Flowers* having a faintly musky, fungal scent, pedicel with ovary salmon-pink, sepals and petals flesh-coloured, salmon-pink, pinkish brown, or citron-yellow, labellum deep flesh-coloured with darker keels, column dark olive with pink base and stelidia, rostellum yellow, anther-cap cream suffused rose-pink towards apex; also variously described as light brown, old rose-pink and bright flesh-coloured. *Pedicel with ovary* clavate, curved, 4 mm long, minutely ramentaceous. *Sepals* and *petals* 3-nerved. *Dorsal sepal* oblong-elliptic to ovate-elliptic, acute, 7–7.5 × 2.5–2.6 mm, somewhat concave, minutely papillose-hairy at base inside. *Lateral sepals* obliquely ovate-elliptic, dorsally carinate at apex, acute, 6.9–7 × 2.6–2.8 mm, adaxial surface sparsely minutely papillose-hairy. *Petals* oblong-elliptic, acute, 5.5–6.5 × 2 mm, very minutely papillose at base inside. *Labellum* tridentate, 5 mm long, 5–5.1 mm wide across side lobes, 1.5–1.6 mm wide at base; nerves 3, median simple, 2 each side branching into 2–3 secondary nerves on side lobes; immobile; side lobes triangular, acute, 2–2.1 mm long, spreading; mid-lobe cuspidate, acute, 1.5–1.6 mm long; disc with an M-shaped callus composed of 2 elevated, fleshy, semi-circular wing-like keels 0.3–0.4 mm high, terminating *c.* half way along labellum and becoming low and fleshy at base where they are joined, between which is a short, fleshy, obtuse basal boss. *Column* slightly narrowed at middle, 2.8–3 mm long, *c.* 0.8 mm wide, gently curved; foot absent; apical hood ovate, rounded, entire; stelidia basal, narrowly linear, obtuse, 2 × 0.2 mm, equalling rostellum to slightly shorter than apical hood; rostellum triangular, obtuse; anther-cap ovate, cucullate, glabrous, *c.* 0.5 × 0.5 mm.

DISTRIBUTION. Borneo.

SABAH. Mount Kinabalu, near helipad, *Barkman* 10 (K!, Sabah Parks Herbarium, Kinabalu Park). Mount Kinabalu: Summit Trail, 17 Feb. 1962, *Bogle* 548 (AMES!); Kemburongoh, May 1933, *Carr* 3476, SFN 27430 (K!, SING!); Marai Parai, 31 May 1933, *Carr* SFN 27430? (SING!); Paka-paka Cave, 8 June 1933, *Carr* 3476, SFN 27430A (K!, SING!) & 16 Nov. 1931, *Carr* SFN 36564 (SING!); locality unknown, 1987, *Chan & Gunsalam* 38/87 (K!) & 43/87 (K!); Kemburongoh, 17 Nov. 1915, *J. Clemens* 209 (AMES!, BM!, E!, K!, SING!); Kemburongoh, 13 Nov. 1931, *J. & M.S. Clemens* 27142 (BM!, E!, K!, L!, SING!), 12 Nov. 1931, *J. & M.S. Clemens* 27149 (BM!, E!, K!, L!, SING!) & 8 Jan. 1932, *J. & M.S. Clemens* 27867 (BM!, E!, HBG!, K!, SING!); Kemburongoh/Paka-paka Cave, 8 Jan. 1932, *J. & M.S. Clemens* 27866 (BM!, SING!); Upper Kinabalu, 24 March 1932, *J. & M.S. Clemens* 29128 (BM!, E!, K!) & 13 Nov. 1931, *J. & M.S. Clemens* 30103 (E!, HBG!, K!, L!; *Dendrochilum dewindtianum* W.W.Sm. & *Bulbophyllum coriaceum* Ridl. also mounted on Kew sheet); Tenompok, 12 June 1932, *J. & M.S. Clemens* 30101 (AMES!, BO!, E!, HBG!, K!, L!) & *J. & M.S. Clemens* 30147 (AMES!, E!, HBG!, K!, L!, SING!); Dallas, 31 Dec. 1931, *J. & M.S. Clemens* 30102 (BO!); Marai Parai, 1 March 1933, *J. & M.S. Clemens* 31683 (BM!, E!, L!), 8 May 1933, *J. & M.S. Clemens* 33130 (E!), 19 May 1933, *J. & M.S. Clemens* 33173 (AMES!, BM!, E!, L!; sheets at BM & E mixed with *D. transversum* Carr), 19 May 1933, *J. & M.S. Clemens* 33178 (AMES!, BM!, E!, HBG!, K!, L!), & 19 May 1933, *J. & M.S. Clemens* 33180 (BM!; mixed with *D. pterogyne*); Kinateki River Head, 3 March 1933, *J. & M.S. Clemens* 31835 (AMES!, BM!, E!); Gurulau Spur, 1 Dec. 1933, *J. & M.S. Clemens* 50661

(BM!, G, K!), 6 Dec. 1933, *J. & M.S. Clemens* 50768 (BM!) & 7 Dec. 1933, *J. & M.S. Clemens* 50801 (BM!, K!). Gurulau Spur, Victoria Peak, 17 Dec. 1933, *J. & M.S. Clemens* 51534 (AMES!, BM!, K!); Kemburongoh, 16 July 1963, *Fuchs* 21067 (L!); Paka-paka Cave, Feb. 1910, *Gibbs* 4250 (BM!); locality unknown, 1931–32, *J. & M.S. Clemens* s.n. (AMES!); locality unknown, July-Aug. 1916, *Haslam* s.n. (AMES!, BM!, E!, K!, SING!; *Dendrochilum stachyodes* (Ridl.) J.J.Sm. also mounted on Kew sheet); 3200 m, *Haviland* 1142 (holotype K!); Paka-paka Cave, 15 Nov. 1931, *Holttum* SFN 36568 (AMES!, SING!); Summit Trail, 20 Feb. 1957, *Kidman Cox* 1008 (L!); Liwagu River Head, 7 March 1961, *Meijer* SAN 24131 (K!, SAN); Kemburongoh, 8 Feb. 1963 *Meijer* SAN 29165 (K!, SAN); Layang-Layang, 25 July 1981, *Sato* 699 (UKMS!); Summit Trail, 4 Aug. 1981, *Sato* 970 (UKMS!) & *Sato* 972 (UKMS!), 2 Aug. 1981, *Sato* 974 (UKMS!), 12 Sept. 1981, *Sato* 1046 (UKMS!), *Sato* 1047 (UKMS!), *Sato* 1080 (UKMS!), *Sato* 1081 (UKMS!) & *Sato* 1082 (UKMS!), 26 July 1981, *Sato* 1223 (UKMS!); Kadamaian River, just below Paka-paka Cave, 14 June 1957, *Sinclair et al.* 9180 (E!, K!, L!, SAN, SING!); Panar Laban, 1 Aug. 1978, *Smith* 543 (L!); Summit Trail, 28 Feb. 1994, *SNP* 5872 (Sabah Parks Herbarium, Kinabalu Park); Eastern Ridge, 20 Sept. 1986, *SNP* 2411 (Sabah Parks Herbarium, Kinabalu Park); Mount Tembuyuken, 5 Oct. 1990, *SNP* 4705 & *SNP* 4717 (Sabah Parks Herbarium, Kinabalu Park); Summit Trail, 18 Oct. 1985, *Wood* 608 (K!).

HABITAT. Upper montane forest, most frequently on ultramafic substrate; recorded as a terrestrial in forest composed of *Leptospermum recurvum* Hook. f.; epiphytic in thick moss cushions on the branches of dead trees and on rotting stumps; rocky wooded stream banks. 900–3800 (most commonly between 2500 and 3100) m.

D. grandiflorum is an easily recognised species with relatively large flowers, hence the specific epithet, so far only recorded on Mount Kinabalu, from where we have numerous collections from a wide altitudinal range along the well-trodden summit trail. It is curious that it has never been found at similar elevations on, for example, Mount Murud in Sarawak or Mount Trus Madi in Sabah.

Sterile herbarium specimens can easily be identified by the finely rugose pseudobulbs and reticulate nerve pattern on the leaf blades.

11. DENDROCHILUM DOLICHOBRACHIUM

Dendrochilum dolichobrachium (Schltr.) Schltr. in *Bot. Jb.* 45, Beibl. 104: 11 (1911). Type: Indonesia, Kalimantan Timur, Long Dett, Aug. 1901, *Schlechter* 13557 (holotype B, destroyed, sketch by *Carr* at K!). **Fig. 42.**

Platyclinis dolichobrachia Schltr. in *Bull. Herb. Boissier*, ser. 2, 6: 301–302 (1906).
Dendrochilum meijeri J.J. Wood in *Checklist orch. Borneo*: 185–186, fig. 23, F–H (1994).
    Type: Indonesia, Kalimantan Timur, Kutai, Mount Beratus, near Balikpapan, Terrace Sulau Mandau, 18 July 1952, *Meijer* 882 (holotype K, isotypes A, BO, L), **syn. nov.**

DESCRIPTION. Epiphyte. *Rhizome* 5–6 cm long, 1–2 mm in diameter, clothed in fibrous sheaths. *Roots* filiform, flexuous, smooth. *Cataphylls* 3–4, brown, unspotted, soon becoming fibrous, 1–1.5 cm long. *Pseudobulbs* narrowly cylindrical or fusiform, erect, 0.6–1.7 × 0.2–0.4 cm, finely furrowed and rugose when dried. *Leaf*: petiole sulcate, 0.5–1 cm long; blade linear-ligulate, attenuated below, obtuse and minutely mucronate, 3.5–7.5 × 0.2–0.6 cm, main nerves

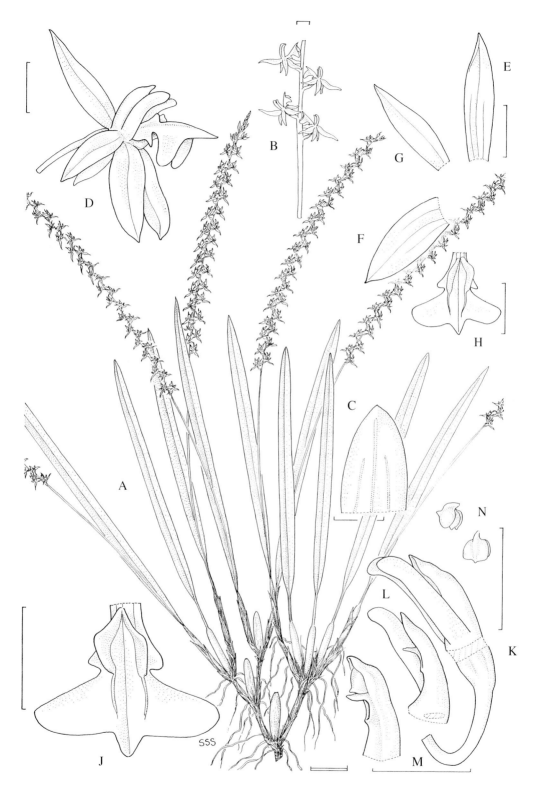

Fig. 42. *Dendrochilum dolichobrachium*. **A** habit; **B** lower portion of inflorescence; **C** floral bract; **D** flower, side view; **E** dorsal sepal; **F** lateral sepal; **G** petal; **H** labellum, front view; **J** labellum, front view; **K** pedicel with ovary and column, side view; **L** column, with stelidium and anther-cap removed, side view; **M** column, distal portion, side view; **N** anther-cap, back and side views. **A & B** from *Meijer* 867, **C, E–H, K, L & N** (back view) from *Meijer* 882, **D, J, L–N** (side view) from *Meijer* 599. Scale: single bar = 1 mm; double bar = 1 cm. Drawn by Susanna Stuart-Smith.

3–4, coriaceous, containing scattered crystalline calcium oxalate bodies similar to *D. gibbsiae*. *Inflorescences* many-flowered, curving, lax to subdense, flowers borne 1.5–2 mm apart; peduncle terete, 4–7 cm long; rachis quadrangular, 3.5–8 cm long; non-floriferous bracts absent; floral bracts ovate, acute, 1 mm long. *Flowers* yellowish green with a pale brownish red pedicel with ovary. *Pedicel with ovary* clavate, 2 mm long. *Dorsal sepal* elliptic to narrowly elliptic, ligulate, acute, concave, 2–3 × 0.9 mm, 1-nerved. *Lateral sepals* ovate-elliptic, acute, 2–3 × 0.9 mm, 3-nerved. *Petals* linear-ligulate, acuminate or somewhat obtuse, 2–2.1 × 0.6 mm, 1-nerved. *Labellum* pandurate, 1.5–1.6 mm long, 2–3 mm wide across mid-lobe; nerves 3, with 3 secondary nerves on mid-lobe side lobules; side lobes keel-like, sometimes slightly falcate, obtuse to lanceolate-falcate, subacute; mid-lobe transversely oblong to narrowly hastate, somewhat concave at centre, side lobules obtuse to subacute, mid lobule shortly acuminate to triangular-acute and tooth-like; disc with 2 prominent raised lamellate keels extending from near the base to at or just beyond the junction of the mid- and side lobes. *Column* 1.4–1.5 mm long; foot absent; apical hood ovate, obtuse, entire; stelidia basal, linear-ligulate, obtuse, 1.4–1.7 mm long, slightly shorter or longer than apical hood; anther-cap cucullate, glabrous.

DISTRIBUTION. Borneo.

KALIMANTAN TIMUR. Kutai, Mount Beratus, near Balikpapan, Terrace Sulau Mandau, 7 July 1952, *Meijer* 599 (BO, L!), 18 July 1952, *Meijer* 867 (BO!, L!), *Meijer* 874 (BO, L!), *Meijer* 877 (BO!, L!) & *Meijer* 882 (holotype of *D. meijeri* K!, isotypes A, BO, L); Mount Beratus, Terrace Berikan bulu, 18 July 1952, *Meijer* 843 (BO, K!, L!). Long Dett, Aug. 1901, *Schlechter* 13557 (holotype B, destroyed, *Carr* sketch K!).

HABITAT. Mossy forest, recorded as epiphytic on a variety of trees including *Dacrydium* spp. and *Syzygium* spp. as well as members of the *Euphorbiaceae*, *Lauraceae* and *Meliaceae* families. 800–1000 m.

A diminutive plant with tiny flowers so far known only from Kalimantan. *D. meijeri*, described in 1994, and distinguished by even smaller flowers with stelidia longer than the column is here considered conspecific.

The specific epithet is derived from the Greek *dolicho*, long, and the Latin *brachium*, an arm, and refers to the stelidia on the column.

12. DENDROCHILUM HASTILOBUM

Dendrochilum hastilobum J.J. Wood in *Orchid Rev.* 103 (1201): 5, fig. 4 (1995). Type: Malaysia, Sarawak, Limbang District, Pa Mario River, Ulu Limbang, route to Batu Lawi, Bario, 9 Aug. 1985, *Awa & Lee* S. 50754 (holotype K, isotypes L, SAR, SING). **Fig. 43.**

DESCRIPTION. Epiphyte. *Rhizome* 6–9 cm long, 0.2–0.5 cm in diameter, branches 1.5–3.5 cm long, tough, clothed in fibrous sheaths. *Roots* filiform, flexuous, smooth. *Cataphylls* 3, brown, unspotted, soon becoming fibrous, 1–1.5 cm long. *Pseudobulbs* borne 0.5– 1.5 cm apart, cylindrical or fusiform, erect, 1–1.8 × 0.3–0.5 cm, finely furrowed and rugose when dried. *Leaf*: petiole narrow, sulcate, 0.4–1 cm long; blade linear-ligulate to narrowly elliptic, obtuse and mucronate or subacute, coriaceous, 6.5–9.5 × 0.4–0.6 cm, main nerves 3–4, containing densely distributed crystalline calcium oxalate bodies similar to *D. gibbsiae*. *Inflorescences* 9- to 20-flowered, lax, curving, flowers borne 3–4 mm apart; peduncle terete, filiform, 1.8–2.8 cm long; non-floriferous bracts absent; rachis quadrangular, 3.5–7.5 cm long;

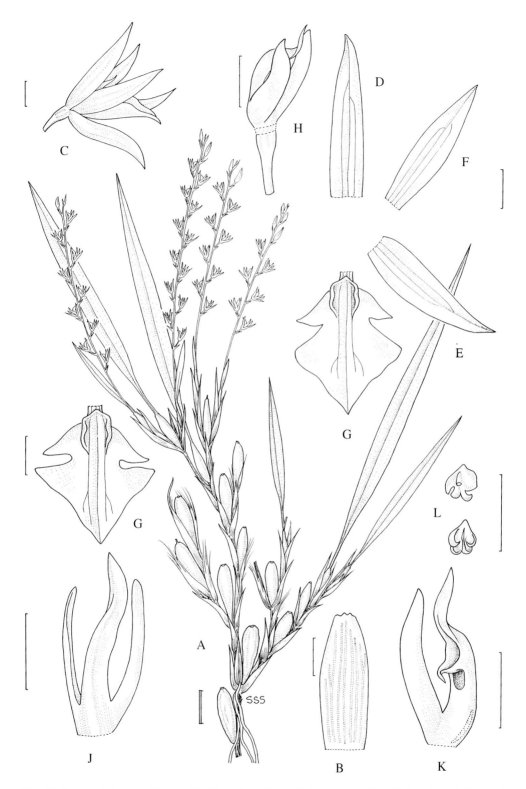

Fig. 43. *Dendrochilum hastilobum*. **A** habit; **B** floral bract; **C** flower, side view; **D** dorsal sepal; **E** lateral sepal; **F** petal; **G** labellum, front views; **H** pedicel with ovary and column, side view; **J** column, back view; **K** column with stelidium and anther-cap removed, oblique view; **L** anther-cap, back and interior views. **A–L** from *Awa & Lee* S. 50754 (holotype). Scale: single bar = 1 mm; double bar = 1 cm. Drawn by Susanna Stuart-Smith.

floral bracts ovate-elliptic, acute to acuminate, 2–3 mm long. *Flowers* greenish brown. *Pedicel with ovary* narrowly clavate, 1.2–1.3 mm long. *Sepals* and *petals* 3-nerved. *Dorsal sepal* narrowly elliptic, acute, 4 × 1 mm. *Lateral sepals* oblong-elliptic, acute, 4.1 × 1 mm. *Petals* narrowly elliptic, acute, 4 × 0.9–1 mm. *Labellum* 3.8–4 mm long, 3 mm wide across mid-lobe, 3- nerved; side lobes triangular, acute, 1–1.5 × 0.6–0.7 mm; mid-lobe broadly hastate, acute, much larger than side lobes; disc with 2 prominent lamellate keels extending from near the base to at or just below the junction of the mid- and side lobes. *Column* 1.8–1.9 mm long; foot very short; apical hood narrowly ovate-elliptic, entire, elongate, subacute, 0.8–1 mm long; stelidia basal, oblong-ligulate, obtuse, 1.6–1.7 mm long, slightly shorter than apical hood; rostellum acuminate; anther-cap cucullate, glabrous.

DISTRIBUTION. Borneo.

SARAWAK. Limbang District, Pa Mario River, Ulu Limbang, route to Batu Lawi, Bario, 9 Aug. 1985, *Awa & Lee* S. 50754 (holotype K! isotypes L!, SAR!, SING!).

HABITAT. Lower montane ridge forest. 1560 m.

D. hastilobum is closely related to *D. dolichobrachium* (Schltr.) Schltr. but is distinguished by the proportionately rather larger leaves, and larger flowers with a larger hastate labellum mid-lobe and a very short column foot.

The specific epithet is derived from the Latin *hastatus*, halbert-shaped, with equal, more or less triangular basal lobes directed outwards, and *lobus*, a lobe, in reference to the mid-lobe of the labellum.

13. DENDROCHILUM CRUCIFORME

Dendrochilum cruciforme J.J. Wood in Wood & Cribb, *Checklist orch. Borneo*: 169–170, fig. 24 A–B (1994). Type: Malaysia, Sabah, Mount Kinabalu, Penibukan, Jan. 1933, *J. & M.S. Clemens* s.n. (holotype K!, isotypes AMES!, BM!, BO!, E!, G!, HBG!, L!).

DESCRIPTION. Clump-forming epiphyte. *Rhizome* abbreviated. *Roots* flexuous, smooth, 1 mm in diameter. *Cataphylls* 3–4, ovate-elliptic, acute to acuminate, 0.3–1.5 cm long, becoming fibrous. *Pseudobulbs* cylindrical or narrowly fusiform, 0.5–2.5 × 0.2 –0.4 cm. *Leaf*: petiole 0.5–2 cm long; blade narrowly linear, subacute to acute, narrowed below, 5–11.5 × 0.1–0.3(–0.4) cm. *Inflorescences* gently curving, densely many-flowered, flowers borne 1–2 mm apart; peduncle 2.5–6.5 cm long; non-floriferous bracts absent; rachis quadrangular, 3.5–5(–6.5) cm long; floral bracts ovate, acute, 1–3.5 × 1–2 mm. *Flowers* having a musty scent according to *Bacon*, variously described as pinkish cream with purple spots, pale yellow, cream-green, cream-yellow, cream or pure white, labellum usually with a dark purple-brown blotch at base of side lobes and on apex of keels, rarely pale greenish white with a greenish labellum. *Pedicel with ovary* clavate, 0.5–1(–2) mm long. *Sepals* and *petals* 1-nerved. *Dorsal sepal* oblong-elliptic, acute, 2.2–2.6(–3) × 0.6–0.9 mm. *Lateral sepals* ovate-elliptic, acute, 2.1–2.5(–3) × 0.8–1 mm. *Petals* narrowly elliptic, acute, 2–2.6 × 0.4–0.7 mm. *Labellum* cruciform, 2–2.5 mm long, 1.5–2 mm wide across side lobes, 3-nerved, with 2 rounded basal keels joined at the base by a transverse ridge; side lobes triangular or triangular-acute, acute or subacute, often somewhat falcate; mid-lobe narrowly triangular-acuminate, cuspidate, up to 1 mm long; *Column* 1–1.2 × 0.2–0.3 mm; apical hood acute, acuminate, bifid or trifid; stelidia basal, linear-ligulate, obtuse, 0.9–2 mm long; anther-cap cucullate.

100

KEY TO THE VARIETIES OF *D. CRUCIFORME*

Mid-lobe of labellum triangular-acuminate, cuspidate; side lobes triangular or triangular-acute. Stelidia shorter than or more or less equal to apical hood of column ....
.... var. *cruciforme*
Mid-lobe of labellum long-acuminate; side lobes narrower. Stelidia longer than apical hood of column .... var. *longicuspe* J.J.Wood

a. var. **cruciforme**. **Figs. 44 & 45H–L, plate 2C.**

DISTRIBUTION. Borneo.

SABAH. Kota Kinabalu to Sinsuron road, Mile 27, from roadside stall, 1971, *Bacon* 187 (E!); Maliau Basin, Jalan Babi, 10 June 1995, *Chan & Tham Nyip Shen* s.n. (K!); Mount Kinabalu: 20 Dec. 1932, *J. & M.S. Clemens* 30471 (BM! E!) and 10 Jan. 1933, *J. & M.S. Clemens* 30826 (AMES!, BM!, E!, K!); Penibukan, 18 March 1933, *J. & M.S. Clemens* 32220 (BM!, BO!, E); Mount Kinabalu, Penataran Basin, 27 June 1933, *J. & M.S. Clemens* 34329 (AMES!, BM!, E!, HBG!, L!); Penataran Basin, 31 Aug. 1933, *J. & M.S. Clemens* 40134 (AMES!, BM!, E!, L!); Mount Kinabalu, Tinekuk Falls, 1 Nov. 1933, *J. & M.S. Clemens* 50278 (AMES!, BM!, K!); Penibukan 11 Nov. 1933, *J. & M.S. Clemens* 50322 (AMES!, BM!, K!); Penataran Basin, 6 Dec. 1933, *J. & M.S. Clemens* 50774 (BM!); Kinateki River Head, 16 Jan. 1933, *J. & M.S. Clemens* s.n. (BM!); Penibukan 12 Jan. 1933, *J. & M.S. Clemens* s.n. (BM!, E!); Penibukan, Jan. 1933, *J. & M.S. Clemens* s.n. (holotype K!, isotypes AMES!, BM!, BO!, E!, G!, HBG!, L!); Marai Parai Spur, 18 Sept. 1958, *Collenette* 41 (BM!). Crocker Range, Mount Alab, south ridge, 31 Oct. 1986, *de Vogel* 8661 (K!, L!).

HABITAT. Lower montane forest. *Leptospermum/Dacrydium* forest about 8 metres high growing on steep east facing slopes on weathered sandstone and shale. 900–2000 m.

D. cruciforme is very similar in habit to *D. graminoides* Carr, but is at once distinguished by the creamy or white, instead of yellow, flowers usually having a purplish-brown blotch on the labellum similar to *D. gibbsiae* Rolfe, cruciform labellum with a narrowly triangular-acuminate, cuspidate mid-lobe and column with longer linear-ligulate, obtuse stelidia and an elongated apical hood. The habit and floral structure also resembles *D. dolichobrachium* (Schltr.) Schltr., but the labellum of *D. cruciforme* has shorter keels, broader triangular or triangular-ovate side lobes and a much smaller narrowly triangular-acuminate, cuspidate mid-lobe. It may also be distinguished from *D. devogelii* J.J.Wood by the shorter pedicel with ovary, cruciform labellum and shorter stelidia. It differs from *D. gibbsiae* Rolfe by its much dwarfer stature, shorter, narrower grass-like leaves, shorter inflorescence with slightly smaller flowers, a three, rather than five-lobed labellum and longer column.

Clemens 40134 is a more robust specimen with pseudobulbs up to 2.5 × 0.4 cm, leaf blades up to 11.5 × 0.4 cm and pure white flowers lacking the characteristic dark blotch on the labellum. *De Vogel* 8661 is a variant with a longer filiform pedicel, greenish white flowers, again without a dark blotch on the labellum, less pronounced side lobes and a deeply bifid apical column hood.

The degree of lobing of the labellum in *D. cruciforme* appears quite variable, some specimens having more pronounced side lobes than others. This is also true of the widespread *D. gibbsiae*. The shape of the apical column hood also ranges from acute to acuminate and from entire to deeply bifid, but is nearly always longer than the stelidia. The full range of variation is difficult to ascertain given the limited material available.

Fig. 44. *Dendrochilum cruciforme* var. *cruciforme*. **A** habit; **B** flower, front view; **C** pedicel with ovary and column, back view; **D** column, oblique view; **E** habit; **F** floral bract; **G** flower, side view; **H** dorsal sepal; **J** lateral sepal; **K** petal; **L** labellum, front view; **M**, labellum, side view; **N** pedicel with ovary and column, side view; **O** pedicel with ovary and column, back view; **P** column, front view; **Q** anther-cap, interior and back views; **R** pollinia. **A–D** from *J. & M.S. Clemens* 40134, **E–R** from *J. & M.S. Clemens* s.n. (holotype). Scale: single bar = 1 mm; double bar = 1 cm. Drawn by Susanna Stuart-Smith.

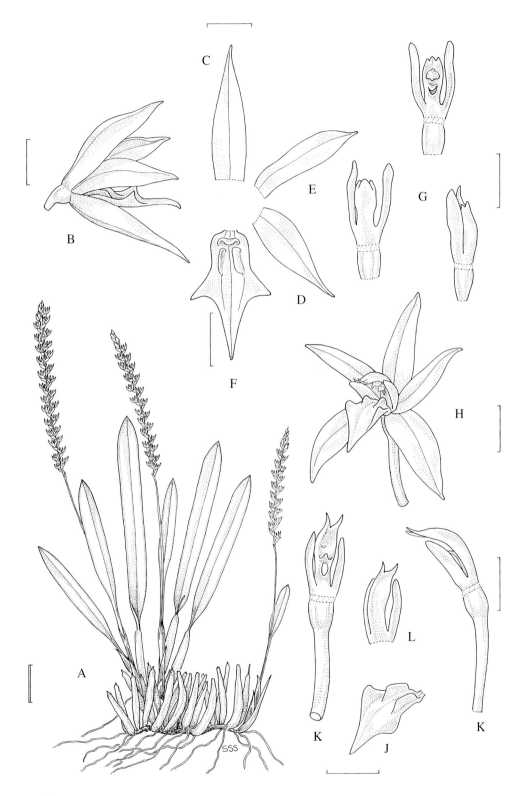

Fig. 45. *Dendrochilum cruciforme* var. *longicuspe*. **A** habit; **B** flower, side view; **C** dorsal sepal; **D** lateral sepal; **E** petal; **F** labellum, front view; **G** pedicel with ovary and column, front, back and side views. *D. cruciforme* var. *cruciforme*. **H** flower, oblique view; **J** labellum, oblique view; **K** pedicel with ovary and column, front and side views; **L** column, back view. **A–G** from *Carr* C. 3675, SFN 28004 (holotype), **H–L** from *de Vogel* 8661. Scale: single bar = 1 mm; double bar = 1 cm. Drawn by Susanna Stuart-Smith.

Chan & Tham Nyip Shen s.n., depicted in Plate 2C is unusual in having dark salmon-coloured flowers. The plant figured on the back cover of Orchids of Borneo Volume Three (Wood 1997) is possibly this species. Lack of preserved material or a close-up view of a flower has prevented positive identification (see Plate 9A).

The specific epithet is derived from the Latin *cruciformis*, cross-shaped, in reference to the labellum.

b. var. **longicuspe** J.J. Wood in Wood & Cribb, *Checklist orch. Borneo*: 170–171, fig. 24 C & D (1994), as *longicuspum*. Type: Malaysia, Sabah, Mount Kinabalu, Kadamaian River, Aug. 1933, *Carr* 3675, SFN 28004 (holotype K!, isotype SING!). **Fig. 45A–G, plate 2D.**

DISTRIBUTION. Borneo.
SABAH. Mount Kinabalu, Summit Trail, below Layang Layang, May 1995, *Barkman* TJB 194 (Sabah Parks Herbarium, Kinabalu Park); Mount Kinabalu, Kadamaian River, Aug. 1933, *Carr* 3675, SFN 28004 (holotype K!, isotype SING!).
HABITAT. Upper montane mossy forest. 2000–2500 m.

This differs from var. *cruciforme* in having a labellum with a longer, acuminate mid-lobe and narrower, somewhat falcate side lobes, and stelidia longer than the tridentate apical hood. It was incorrectly determined as *D. corrugatum* J.J. Sm. by Carr.

The varietal epithet is derived from the Latin *longus*, long and *cuspis*, a sharp, rigid point, in reference to the mid-lobe of the labellum.

14. DENDROCHILUM DEVOGELII

Dendrochilum devogelii J.J. Wood in Wood & Cribb, *Checklist orch. Borneo*: 173, fig. 24 E & F (1994). Type: Malaysia, Sabah, Sipitang District, ridge between Maga River and Pa Sia River, 18 Oct. 1986, *de Vogel* 8376 (holotype L!, isotype K!). **Fig. 46.**

DESCRIPTION. Epiphyte. *Rhizome* abbreviated. *Roots* flexuous, with a few branches, smooth, 0.5 mm in diameter. *Cataphylls* 3, ovate-elliptic, acute, 1–1.5 cm long. *Pseudobulbs* ovate-elliptic or cylindrical, 1–1.5 × 0.4–0.5 cm. *Immature leaf* only extant: petiole undeveloped; blade linear-ligulate, acute, 9.5 × 0.3 cm. *Inflorescence* laxly many-flowered, flowers borne 2 mm apart; peduncle 2.2 cm long; non-floriferous bract solitary, acute, 3.5 mm long; rachis 6.5 cm long; floral bracts ovate, acute, 1–2 mm long. *Flowers* pale green. *Pedicel with ovary* very narrowly clavate, 3 mm long. *Sepals* and *petals* 3-nerved. *Dorsal sepal* oblong, acute, 2.5 × 0.8 mm. *Lateral sepals* oblong, slightly falcate, acute, 2–2.1 × 0.8–0.9 mm. *Petals* linear, acute, margins a little uneven, 2 × 0.5–0.6 mm. *Labellum* pandurate, 1–1.1 × 0.9–1 mm, with a fleshy, U-shaped keeled basal callus; side lobes rounded to subacute; mid-lobe acute, apiculate, mucronate. *Column* 2 mm long; apical hood entire, shorter than stelidia; stelidia ligulate, obtuse, tips decurved, 2–2.1 mm long; anther-cap cucullate. **Fig. 46.**
DISTRIBUTION. Borneo.
SABAH. Sipitang District, ridge between Maga River & Pa Sia River, 18 Oct. 1986, *de Vogel* 8376 (holotype L!, isotype K!).
HABITAT. Rather dense primary forest up to 30 metres high dominated by *Agathis* spp. and *Lithocarpus* spp. on poor sandy soil over a sandstone bedrock, with much leaf litter, but little undergrowth. 1450 m.

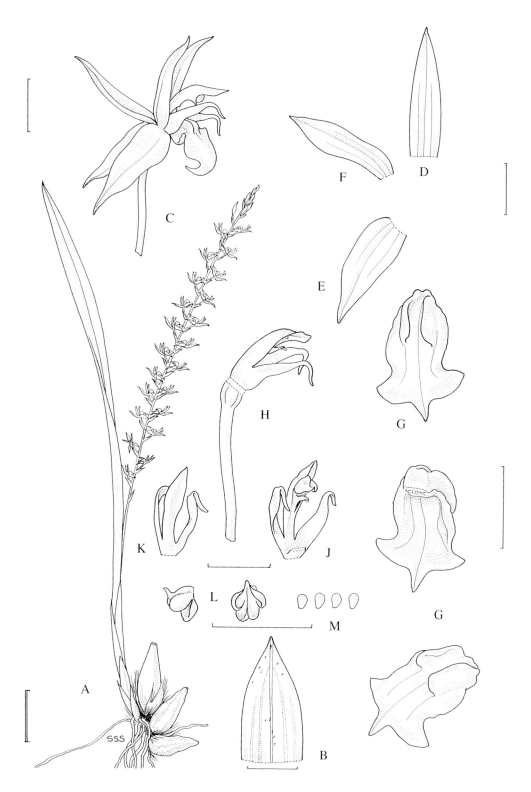

Fig. 46. *Dendrochilum devogelii*. **A** habit; **B** floral bract; **C** flower, side view; **D** dorsal sepal; **E** lateral sepal; **F** petal; **G** labellum, front, back and oblique views; **H** pedicel with ovary and column, side view; **J** column, oblique view; **K** column, back view; **L** anther-cap, side and interior views; **M** pollinia. **A–M** from *de Vogel* 8376 (holotype). Scale: single bar = 1 mm; double bar = 1 cm. Drawn by Susanna Stuart-Smith.

D. devogelii is known only from the type collection which consists of two pseudobulbs, one leaf and one inflorescence. Although related to *D. cruciforme*, it is distinguished by the pandurate labellum with a proportionately longer basal unlobed portion, stelidia longer than the apical hood and decurved at the apex, and the longer pedicel.

The specific epithet honours Dutch orchidologist Dr E.F. de Vogel of the Rijksherbarium, Leiden, The Netherlands, who collected the type.

15. DENDROCHILUM HASLAMII

Dendrochilum haslamii Ames, *Orchidaceae* 6: 53–54, pl. 85 (1920). Type: Malaysia, Sabah, Mount Kinabalu, locality unknown, July–Aug. 1916, *Haslam* s.n. (holotype AMES!).

DESCRIPTION. Epiphyte. *Rhizome* abbreviated, up to 8 cm long, usually much shorter, 2 mm in diameter, clothed in fibrous sheaths. *Roots* flexuous, produced in a large mass, 0.5–1 mm in diameter, smooth. *Cataphylls* 3–4, 0.5–1.5 cm long, reddish brown, unspotted, distinctly papillose, persistent, slowly becoming fibrous. *Pseudobulbs* caespitose, crowded on to rhizome, subfusiform, or ovoid, 0.5–2 × 0.4–0.6 cm, smooth to slightly rugose, strongly rugose in dried material, epidermal cells often clearly defined at high magnification in dried material, surface appearing almost papillose. *Leaf*: petiole sulcate, 2–7 mm long; blade linear-ligulate to linear-lanceolate, obtuse, minutely apiculate, 2.5–6 × 0.2–0.6 cm, main nerves 3, median prominent, coriaceous, containing scattered to densely distributed crystalline calcium oxalate bodies similar to *D. gibbsiae*. *Inflorescences* 15- to 30-flowered, curving, subdense, flowers borne 1.8–2.5 mm apart; peduncle terete, slender, 1.2–5(–9) cm long, dull red; non-floriferous bract solitary, ovate-elliptic, acute, 3 mm long, sometimes absent; rachis quadrangular, somewhat sulcate, becoming fractiflex, 3–7 cm long, dull red; floral bracts ovate, apiculate, 2 × 1.2–1.3 mm, median nerve prominent. *Flowers* unscented, with creamy yellow, yellowish green or yellow sepals and petals, labellum bright brown or brownish orange, apical cusp creamy yellow, column olive-brown, stelidia pale brown, tipped yellowish brown, or entirely pale creamish green. *Pedicel with ovary* clavate, gently curved, 2–3 mm long, ovary rotund, 0.5 mm long. *Sepals* and *petals* 1-nerved. *Sepals* strongly concave and reflexed. *Dorsal* and *lateral sepals* oblong-ovate, obtuse to subacute, 1.5–1.75(–1.9) × 1–1.1 mm, slightly carinate. *Petals* ovate to oblong-ovate, obtuse to subacute, 1.5(–1.8) × 1 mm, strongly concave, slightly spreading, directed forward. *Labellum* 4-lobed, 2 mm long, 1.5 mm wide below middle, usually 3 mm wide across terminal lobules (1.5–1.6 mm across in var. *quadrilobum*); nerves 3, vague; side lobes obscure, rounded, undulate or (in var. *quadrilobum*) shortly rounded; mid-lobe expanded into 2 ovate to oblong, obtuse, falcate, divaricate lobules (less divaricate in var. *quadrilobum*) between which is a short triangular, acute cusp or mucro; disc with a U-shaped fleshy callus extending from near base and terminating on middle of disc. *Column* straight, 1 mm long; foot absent; apical hood entire, obtuse; stelidia basal, linear, obtuse, 1–1.1 mm long; anther-cap cucullate, glabrous. *Capsule* ovoid.

KEY TO THE VARIETIES OF *D. HASLAMII*

Side lobes of labellum obscure, undulate; mid-lobe with divaricate, falcate lobules .... var. *haslamii*
Side lobes of labellum distinct, shortly rounded; mid-lobe with straight or only slightly falcate lobules .... var. *quadrilobum* J.J.Wood

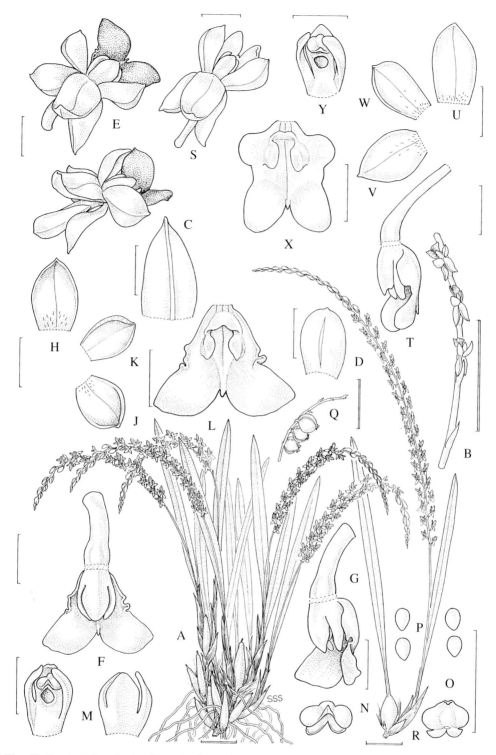

Fig. 47. *Dendrochilum haslamii* var. *haslamii*. **A** habit; **B** lower portion of inflorescence; **C & D** floral bracts; **E** flowers, viewed from above and side; **F** pedicel with ovary, labellum and column, viewed from above; **G** pedicel with ovary, labellum and column, side view; **H** dorsal sepal; **J** lateral sepal, **K** petal; **L** labellum, front view; **M** column, front and back views; **N** anther-cap, front view; **O** anther-cap, back view; **P** pollinia; **Q** infructescence. *D. haslamii* var. *quadrilobum*. **R** habit; **S** flower, viewed from above; **T** pedicel with ovary, labellum and column, side view; **U** dorsal sepal; **V** lateral sepal; **W** petal; **X** labellum, front view; **Y** column, front view. **A** from *Collenette* 21535, **B–P** from *Gunsalam* 3, **Q** from *J. & M.S. Clemens* 31663, **R–Y** from *Lewis* 369 (holotype). Scale: single bar = 1 mm; double bar = 1 cm. Drawn by Susanna Stuart-Smith.

a. var. **haslamii, Fig. 47A–Q, plate 7A–C.**

DISTRIBUTION. Borneo.

SABAH. Mount Kinabalu: Summit Trail by-pass above Layang-Layang before Pondok Villosa, 15 July 1994, *Barkman et al.* 17 (K!, Sabah Parks Herbarium, Kinabalu Park!); Kemburongoh/Paka-paka Cave, 24 July 1933, *Carr* 3528, SFN 27864 (BO!, K!, L!, SING!); Paka-paka Cave, 16 Nov. 1931, Carr SFN 36565 (SING!); Janet's Halt/Sheila's Plateau, 5 Sept. 1963, *Collenette* 21535 (K!, KSEPL, L!, SAR); Kemburongoh, above Marai Parai, 1 Mar. 1933, *J. & M.S. Clemens* 31663 (AMES!, E!); Kinateki River Head, 1–3 Mar. 1933, *J. & M.S. Clemens* 31830 (BM!, K!); Gurulau Spur, 1 Dec. 1933, *J. & M.S. Clemens* 50650 (BM!, E!, K!, L!), 6 Dec. 1933, *J. & M.S. Clemens* 50777 (AMES, mixed with *D. corrugatum* (Ridl.) J.J. Sm.!, E!, L!); Kemburongoh, 24 Mar. 1932, *J. & M.S. Clemens* s.n. (BM!); Summit Trail, 28 Oct. 1991, *Gunsalam* 3 (K!); locality unknown, July–Aug. 1916, *Haslam* s.n. (holotype AMES!); Paka-paka Cave, 16 Nov. 1931, *Holttum* s.n. (AMES!, SING); Summit Trail, Dec. 1986, *Vermeulen & Duistermaat* 544 (L!).

HABITAT. Upper montane ridge forest, frequently on ultramafic substrate; growing low down on trunks and branches, often in exposed sites; recorded as epiphytic on *Leptospermum recurvum* Hook. f. 2400–3100 m.

The specific epithet honours George A.G. Haslam who collected the type.

b. var. **quadrilobum** J.J. Wood, **var. nov.** a varietate typica lobis lateralibus labelli distinctis breviter rotundatis atque lobulis lobi medii strictioribus minus divaricatis distinguitur. Typus: Malaysia, Sarawak, Marudi District, Gunung Mulu National Park, path from sub-camp 4 to Mulu summit, 2190 m, 20 Oct. 1977, *Lewis* 369 (holotypus K). **Fig. 47 R–Y.**

Erect epiphyte. *Pseudobulbs* 0.5–0.9 × 0.4–0.6 cm. *Leaf:* petiole 2–3 mm long; blade linear-ligulate, obtuse, minutely apiculate, 5–6 × 0.2–0.25 cm, with scattered calcium oxalate bodies. *Inflorescences* erect; peduncle 3–4 cm long; rachis 5–6 cm long. *Flowers* pale creamish green. *Sepals* and *petals* 1.8–1.9 × 1 mm. *Labellum* 2 × 1.5–1.6 mm; side lobes distinct, shortly rounded; mid-lobe with straight to subfalcate lobules.

DISTRIBUTION. Borneo.

SARAWAK. Marudi District, Gunung Mulu National Park, path from sub-camp 4 to Mulu summit, 20 Oct. 1977, *Lewis* 369 (holotype K!).

HABITAT. Upper montane ridge forest. 2190 m.

Known only from the type collection, var. *quadrilobum* from Mount Mulu is distinguished from var. *haslamii*, so far only recorded from Mount Kinabalu, by the four-lobed appearance of the labellum. The side lobes are distinct and shortly rounded, while the lobules of the mid-lobe are straight or only subfalcate and not falcate-divaricate as in var. *haslamii*. The sepals and petals are also a little longer in the material from Mulu. It is not known whether intermediates exist.

The specific epithet is derived from the Latin *quadri*, four, and *lobus*, lobe, referring to the shape of the labellum.

Section **Eurybrachium**

Section **Eurybrachium** *Carr ex J.J. Wood, H. Ae. Peders. & J.B. Comber* in *Opera Bot.* 130: 31 (1997). Type species: *Dendrochilum microscopicum* J.J. Sm.

Dendrochilum subgen. *Monochlamys* Pfitzer in Engler, *Pflanzenreich* 4, 50, 2 B 7: 87 (1907).
  - Rosinski 1992: 104. - Type not designated.
Dendrochilum subgen. *Platyclinis* ser. *Edentula* Pfitzer in Engler, *Pflanzenreich* 4, 50, 2 B 7: 91, 92 (1907), p.p. - Rosinski 1992: 104 (p.p.). - Type not designated.
Dendrochilum subgen. *Platyclinis* ser. *Angustata* Pfitzer in Engler, *Pflanzenreich* 4, 50, 2 B 7: 92 105 (1907), p.p. - Rosinski 1992: 104 (p.p.). - Type not designated.
Basigyne J.J. Sm., *Bull. Jard. Bot. Buitenz.* 2, 25: 4–5 (1917). - Smith 1934: 201, 211; Farr *et al.* 1979: 182; Dressler 1981: 215; Schlechter 1986, II: iv, vii; Burns-Balogh 1989: 35, 40; Webster 1992: 1.4, 4.15; Dressler 1993: 190, 274; Greuter *et al.* 1993: 118. - Type: *Basigyne muriculata* J.J. Sm.
Dendrochilum sect. *Eurybrachion* Carr, *Gdns. Bull. Straits Settl.* 8: 232 (1935), *nom. prov.* - Type not designated.

Labellum firmly attached to the column, making a right to obtuse angle to the latter; entire to indistinctly (rarely distinctly) 3-lobed, never cruciform, neither replicate nor coiled-up, not divided into a hypochile and epichile by two prominent calli. *Column* shorter than the dorsal sepal, comparatively stout and straight (somewhat incurved in the Philippine *D. rotundilabium*), not tapering from the base upwards; foot absent; stelidia usually present.

DISTRIBUTION. Sumatra, Java, Borneo, Lesser Sunda Islands, Sulawesi and the Philippines.

In his treatment of *Dendrochilum* for *The Orchids of Mount Kinabalu, British North Borneo* (1920: 44), Oakes Ames referred to several new species he described as being of unusual interest in that, "…they seem to indicate the probable existence of alliances that may reward intensive explorations of the higher altitudes of Borneo". One of these alliances he referred to is now recognised as section *Eurybrachium*. Indeed, intensive exploration since his treatment in 1920 has led to the discovery of seven species belonging to this section at elevations near or above 2000 m on Mount Kinabalu, Mount Trus Madi and the Crocker Range in Sabah, and Mount Mulu in Sarawak.

KEY TO THE SPECIES OF SUBGENUS *PLATYCLINIS* SECTION *EURYBRACHIUM* IN BORNEO

1. Labellum broader than long 2
   Labellum as long as or longer than broad 5

2. Labellum strongly concave, strongly cupulate, with 2 small rounded and separate central calli, and a separate basal ridge. Pseudobulbs borne up to 2.2 cm apart on the rhizome. Leaves generally 4–10 × 0.4–0.5 cm; petioles 0.1–0.5 cm long ....
   .... 16. *D. cupulatum* J.J. Wood
   Labellum flatter, never strongly cupulate, sometimes concave at the base, calli much larger and keel-like, often united and forming a horseshoe-shaped structure. Pseudobulbs borne closer together on the rhizome, often caespitose. Leaves larger, generally 8–23 × 0.5–1.2 cm; petioles 0.8–4 cm long 3

3. Margin of labellum shortly and irregularly fimbriate. Labellum sometimes shallowly three-lobed. Stelidia shorter than the column. Petals denticulate .... 19. *D. corrugatum* (Ridl.) J.J. Sm.
    Margin of labellum entire. Labellum never shallowly lobed. Stelidia as long as or slightly longer than the column. Petals entire 4

4. Labellum shallowly retuse, *c.* 1.8 × 2.7 mm. Pseudobulbs 1.5–2.5 cm long. Petals porrect, never twisted and aligned 90° from vertical. Stelidia glabrous, dark reddish-brown .... 25. *D. scriptum* Carr
    Labellum subapiculate, *c.* 3 × 4.4 mm. Pseudobulbs generally longer, up to 4.5 cm long. Petals sometimes twisted and aligned up to 90° from vertical. Stelidia minutely papillose, ochre, tipped with dark ochre-brown .... 26. *D. transversum* Carr

5. Flowers minute, the sepals 1–1.5 × 0.75–0.9 mm. Labellum concave, *c.* 0.5–0.7 × 0.5 mm, cuneate-cucullate, entire or obscurely lobed 6
    Flowers larger, the sepals 2.5–8 × 1–4.4 mm. Labellum flat, 2–6.4 × 0.8–5.8 mm, ovate to oblong–obovate and entire, oblong to oblong-elliptic or narrowly elliptic and shallowly lobed, or sharply deflexed near base, oblong-lanceolate, acuminate, obscurely auriculate at base, or shallowly cymbiform and obscurely lobed 7

6. Labellum entire, apex retuse .... 17. *D. microscopicum* J.J. Sm.
    Labellum obscurely lobed, apex acutely apiculate .... 18. *D. minimiflorum* Carr

7. Petals twisted and aligned 90° from vertical. Labellum entire, longer than broad 8
    Petals porrect to ascending, not twisted and aligned 90° from vertical. Labellum entire, auriculate at base, or obscurely lobed 11

8. Flowers relatively large, the sepals 7.8 × 4–4.4 mm, the labellum 6.4 × 5.8 mm .... 23. *D. alpinum* Carr
    Flowers smaller, the sepals 3–6 × 1–3 mm, the labellum 2.5–3.5 × 1.5–3.5 mm 9

9. Labellum as long as broad, nearly circular or cordate in outline, sometimes cymbiform. Callus consisting of two broader than high, very fleshy keels which touch but are not united at the base and are distinctly horseshoe-shaped .... 24. *D. pseudoscriptum* T.J. Barkman & J.J. Wood
    Labellum generally somewhat longer than broad, oblong-obovate to obcuneate or ovate to oblong. Callus keels higher and united at the base by a transverse ridge, less obviously horseshoe-shaped 10

10. Labellum oblong-obovate to obcuneate, shortly apiculate, margins not recurved, *c.* 3 × 1.5 mm .... 20. *D. alatum* Ames
    Labellum ovate or oblong, minutely apiculate, margins sometimes strongly recurved in the upper half, *c.* 2.6–3.5 × 2.7–3 mm .... 22. *D. pterogyne* Carr

11. Stelidia normally absent. Exclusively lithophytic. Flowers creamy-white with a pink pedicel with ovary and column .... 28. *D. stachyodes* (Ridl.) J.J. Sm.
    Stelidia present. Not this combination of characters 12

12. Labellum sharply deflexed near base, otherwise flat, oblong-lanceolate, acute to acuminate, obscurely auriculate and somewhat erose at base ..... 27. *D. lewisii* J.J. Wood
Labellum not sharply deflexed near base. Not this combination of characters 13

13. Stelidia very slender, linear, subacute, reaching only as far as rostellum. Labellum flat, mid-lobe ovate or transversely elliptic, side lobes short, but distinct, erose ....
.... 29. *D. acuiferum* Carr

Stelidia oblong, rounded to truncate, as long as or slightly shorter than column. Labellum shallowly cymbiform, entire or obscurely lobed, side lobes, when present, not erose ......
.... 14

14. Plant small, up to *c.* 10 cm high. Leaves 0.3–0.5 cm wide. Flowers salmon-pink, sepals 3 × 1 mm. Labellum elongate-acuminate .... 21. *D. joclemensii* Ames
Plant large, up to 28 cm high. Leaves 0.8–2 cm wide. Flowers translucent pale lemon-yellow, labellum chocolate brown, sepals 5.5–6 × 2–2.5 mm. Labellum obtuse to subacute .... 30. *D. trusmadiense* J.J. Wood

16. DENDROCHILUM CUPULATUM

Dendrochilum cupulatum J.J. Wood in Wood & Cribb, *Checklist orch. Borneo*: 171, 173, fig. 23 J–L (1994). Type: Malaysia, Sarawak, Marudi District, Gunung Mulu National Park, above Camp 4, *c.* 1700 m, 21 Mar. 1981, *Lamb* MAL 12 (holotype K!). **Fig. 48, plate 2E.**

DESCRIPTION. Creeping, clump-forming epiphyte. *Rhizome* tough, branching profusely, 2 mm in diameter. *Roots* wiry, flexuous, branching, 0.5–1 mm in diameter, smooth. *Cataphylls* 3, ovate-elliptic, acute to acuminate, 1–2.5 cm long, reddish brown, unspotted, eventually becoming fibrous. *Pseudobulbs* cylindrical or elliptic, borne 0.3–2.4 cm apart on rhizome, 0.9–2.3 × 0.4–0.8 cm. *Leaf*: petiole 0.1–0.5 cm long; blade erect, linear-ligulate, conduplicate at base, apex slightly carinate, apiculate, 4–10 × 0.4–0.5(–0.8) cm. *Inflorescences* erect, densely many-flowered, flowers borne 1 mm apart; peduncle 1.5–3 cm long, greenish-yellow; non-floriferous bract solitary, acuminate, 2–3 mm long; rachis quadrangular, 6–14 cm long, greenish-yellow; floral bracts ovate, apiculate, carinate, 2–2.5 × 1.1–1.2 mm. *Flowers* with a slightly spicy scent or strongly honey-scented, 3 mm across, borne in 2 ranks, greenish yellow, very pale greenish with a brighter green labellum or white with a greenish white labellum, calli often bright green, anther-cap pale brown. *Pedicel with ovary* slightly curved, 1.8 mm long. *Sepals* and *petals* spreading. *Dorsal sepal* ovate-elliptic, acute, 2 × 1 mm. *Lateral sepals* ovate-elliptic, acute, 2 × 1 mm. *Petals* ovate-elliptic, acute, *c.* 1.6–1.7 × 0.8–0.9 mm. *Labellum* broadly ovate, concave, cupulate, apex apiculate, 1.1 × 1.5–1.6 mm; disc with a basal ridge and 2 small rounded central calli. *Column* cuneate, truncate, 0.2 × 0.2–0.3 mm; stelidia basal, oblong, somewhat truncate, 0.5 mm long; anther-cap minute, cucullate.

DISTRIBUTION. Borneo.

SARAWAK. Marudi District, Gunung Mulu National Park, southern summit, 28 April 1978, *Argent & Coppins* 1126 (E!, K!); Marudi District, Gunung Mulu National Park, Camp 4, 18 March 1978, *Hansen* 498 (C!, K!); Marudi District, Gunung Mulu National Park, above Camp 4, 21 Mar. 1981, *Lamb* MAL 12 (holotype K!); Marudi District, Gunung Mulu National Park, west ridge near Camp 4, 22 Mar. 1978, *Nielsen* 806 (AAU!, K!). Kapit District, northern Hose Mountains, ridge leading to Batu Hill, 3 Dec. 1991, *de Vogel* 1013 (K!, L!).

111

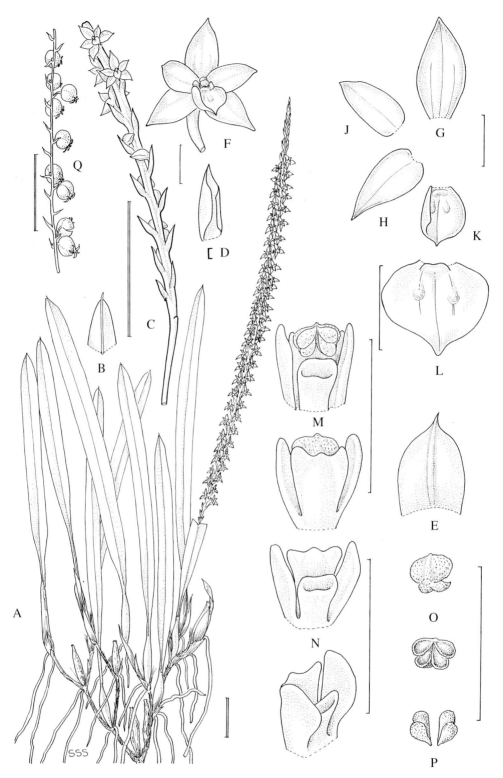

Fig. 48. *Dendrochilum cupulatum*. **A** habit; **B** leaf apex, abaxial surface; **C** lower portion of inflorescence; **D** floral bract, natural position; **E** floral bract, flattened; **F** flower, front view; **G** dorsal sepal; **H** lateral sepal; **J** petal; **K & L** labellum, front views; **M** column with anther-cap, front and back views; **N** column with anther-cap removed, front and oblique views; **O** anther-cap, back and interior views; **P** pollinia; **Q** infructescence. **A & B** from *Hansen* 498, **C–P** from *de Vogel* 1013, **Q** from *Kitayama* 893. Scale: single bar = 1 mm; double bar = 1 cm. Drawn by Susanna Stuart-Smith.

SABAH. Crocker Range, Kimanis to Keningau road, *Barkman* 228 (K!, Sabah Parks Herbarium, Kinabalu Park). Crocker Range, Kimanis to Keningau road, 6 April 1985, *Kitayama* 893 (UKMS!) & Dec. 1986, *Vermeulen & Duistermaat* 667 (K!, L!).

HABITAT. Upper montane mossy ridge forest up to 25 metres high on sandstone; ridge-top scrub on sandstone, often in exposed places. 1400–2100 m.

D. cupulatum is probably most closely related to *D. corrugatum* (Ridl.) J.J. Sm. from Sabah but is distinguished by its well-spaced pseudobulbs, linear-ligulate leaves and slightly smaller flowers with a shorter dorsal sepal, non falcate lateral sepals, entire petals, an entire, strongly concave, cupulate labellum with two tiny rounded central calli and a shorter column. It may also be distinguished from *D. alatum* Ames, also from Sabah, by the well-spaced pseudobulbs, denser inflorescence with slightly smaller flowers with entire petals, shorter, strongly concave, cupulate labellum with an apiculate apex and tiny calli, and a shorter, truncate column. The spikes of tiny crowded flowers look remarkably like those of *D. sublobatum* Carr, a species belonging to section *Platyclinis* so far only recorded from Sarawak. *D. sublobatum*, however, has crowded pseudobulbs and longer leaves. On closer inspection, the flowers of *D. sublobatum* have a decurved, slightly lobed ovate-acuminate labellum with a large horseshoe-shaped callus and a column with a conspicuous hood and long linear stelidia.

The specific epithet is derived from the Latin *cupula*, a cup, in reference to the concave, cup-like labellum.

17. DENDROCHILUM MICROSCOPICUM

Dendrochilum microscopicum J.J. Sm. in *Bull. Jard. Bot. Buitenz.*, ser. 3, 11: 110–111 (1931). Type Indonesia, Kalimantan Timur, Mount Kemul (Kemoel), 1500 m, 13 Oct. 1925, *Endert* 4103 (holotype L!, isotypes BO!, K!). **Fig. 49.**

DESCRIPTION. Epiphyte. *Rhizome* abbreviated or up to *c.* 5 cm long, branching. *Roots* flexuous, filiform, branching, glabrous. *Cataphylls* 3, ovate-elliptic, acute to acuminate, 0.8–1.9 cm long. *Pseudobulbs* caespitose, borne 2–3 mm apart on rhizome, elongate-ovoid (in dried state), 0.8–1 × 0.2–0.3 cm, rugulose, brownish yellow. *Leaf:* petiole filiform, sulcate, 1–1.5 cm long; blade narrowly linear, apex canaliculate-conduplicate, shortly acute, 4.7–7.5 × 0.16–0.2 cm, rigid, coriaceous, main nerves 3. *Inflorescences* erect to gently curving, laxly many-flowered, flowers borne 0.8–1 mm apart; peduncle filiform, 4.3–5.5 cm long; non-floriferous bract solitary, ovate-elliptic, acute, 1 mm long; rachis slender, slightly flexuous, flattened and narrowly sulcate, 4.7–5.7 cm long; floral bracts oblong, obtuse, irregularly erose, 1 × 1 mm, 3-nerved. *Flowers* minute, described as yellowish with a brownish red centre. *Pedicel with ovary* obscurely clavate, 1 mm long. *Sepals* and *petals* spreading, 1-nerved. *Dorsal sepal* ovate or triangular-ovate, shortly acute, 1 × 0.75 mm. *Lateral sepals* obliquely ovate, shortly acute, dorsally carinate, 1 × 0.75 mm. *Petals* obliquely narrowly elliptic, shortly acute, somewhat concave, 1 × 0.75–0.8 mm. *Labellum* entire, shallowly reniform from a cuneate base, strongly concave, lateral margins incurved, borne at an obtuse angle to the ovary, 0.5 × 0.5 mm; disc with a small rounded callus each side extending from just above the base to about the middle. *Column* fleshy, oblong, 0.5 × 0.5 mm; apical hood truncate; stelidia basal, oblong, obtuse, truncate, shorter than column; rostellum recurved; stigma transversely triangular, apex truncate, margin elevate below; anther-cap transversely cucullate, obtusely conical. *Capsule* globose, 2 × 2.5–2.7 mm.

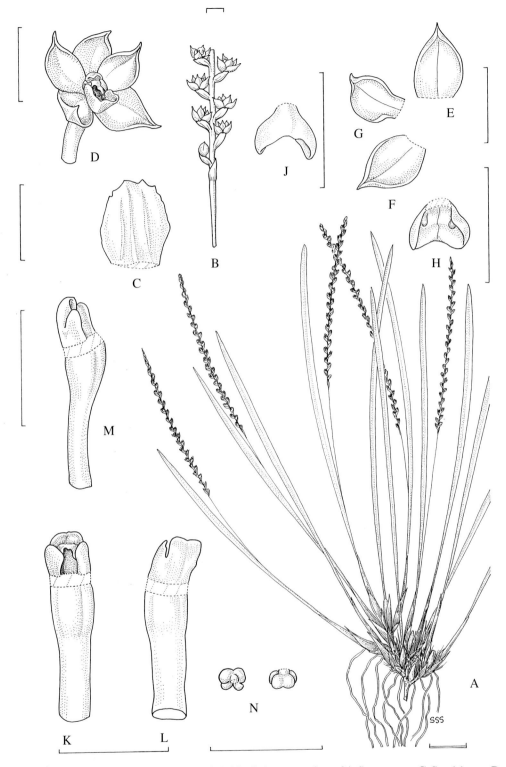

Fig. 49. *Dendrochilum microscopicum.* **A** habit; **B** lower portion of inflorescence; **C** floral bract; **D** flower, oblique view; **E** dorsal sepal; **F** lateral sepal, **G** petal; **H** labellum, front view; **J** labellum, back view; **K** pedicel with ovary and column, anther-cap removed, front view; **L** pedicel with ovary and column, anther-cap removed, back view; **M** pedicel with ovary and column, anther-cap removed, side view; **N** anther-cap, back and side views. **A–N** from *Endert* 4103 (holotype). Scale: single bar = 1 mm; double bar = 1 cm. Drawn by Susanna Stuart-Smith.

DISTRIBUTION. Borneo.

KALIMANTAN TIMUR. Mount Kemul, 13 Oct. 1925, *Endert* 4103 (holotype BO!, isotypes K!, L!).

HABITAT. Lower montane ridge forest. 1500 m.

D. microscopicum is known only from the type specimen and has the distinction of having the smallest flowers of any Bornean orchid, hence the specific epithet. It would be interesting to discover which insects effect pollination, possibly tiny flies or fungus gnats.

18. DENDROCHILUM MINIMIFLORUM

Dendrochilum minimiflorum Carr in *Gdns. Bull. Straits Settl.* 8: 90–91 (1935). Type: Malaysia, Sarawak, Marudi District, Mount Dulit, Dulit Ridge, 1300 m, 13 Sept. 1932, *Richards* S.476 (holotype SING!, isotypes K!, L!). **Fig. 50, plate 11D.**

DESCRIPTION. Epiphyte. *Rhizome* up to 2 cm long, branching. *Roots* flexuous, filiform, branching, smooth, 0.1–0.8 mm in diameter. *Cataphylls* 3(–4), ovate-elliptic, acute, 0.4–0.5 cm long. *Pseudobulbs* caespitose, borne 2–5 mm apart on rhizome, ovoid, rugulose, sometimes gently curved, 0.5–0.9 × 0.4–0.5 cm, chestnut-brown. *Leaf*: petiole very narrow, sulcate, 0.3–1.5 cm long; blade erect, very narrowly linear to linear-oblanceolate, acute, texture thinly coriaceous, margins becoming revolute in dried material, 2.5–7.5 × 0.15–0.2 cm, 1-nerved. *Inflorescences* erect to gently curving, subdensely to densely many-flowered, flowers borne 0.5–1 mm apart; peduncle filiform, 3–8.5 cm long; non-floriferous bract solitary, ovate, acute, 1.5–1.8 mm long; rachis quadrangular, narrowly sulcate, 2.4–5 cm long; floral bracts broadly ovate, oblong-ovate or suborbicular, obtuse, margins erose-crenate, 1–1.1 × 0.8 mm, 1- to 2-nerved. *Flowers* unscented or with a slight sweet scent, minute, described as pale yellowish green with a chestnut column or pale brown with a darker brown tip to the column, labellum reddish brown with a red keel. *Pedicel with ovary* clavate, 0.6–0.8 mm long. *Sepals* and *petals* spreading, 1-nerved. *Dorsal sepal* oblong-ovate or ovate-elliptic, abruptly and shortly acuminate or acute, 1.5 × 0.9 mm. *Lateral sepals* oblong-ovate or ovate-elliptic, very shortly acuminate, acute or obtuse, 1.3 × 0.8 mm. *Petals* broadly ovate to ovate-elliptic, shortly acuminate or obtuse, somewhat falcate, 1.3 × 0.8 mm. *Labellum* very obscurely lobed, cymbiform-cupulate, broadly ovate to ovate-elliptic from a cuneate base, side lobes tiny, triangular, obtuse, apex of mid-lobe acutely apiculate, fleshy, concave, minutely papillose, 0.7 × 0.5 mm when flattened; disc with a basal transverse keel, and with a thickening on inner surface of side lobes. *Column* oblong, stout, *c.* 0.4 mm long; apical hood truncate or very obscurely 3-lobed; stelidia basal, oblong to triangular, obtuse, *c.* 0.3 mm long; anther-cap cucullate.

DISTRIBUTION. Borneo.

SARAWAK. Limbang District, Batu Buli, trail from Bario to Batu Lawi, 14 March 1998, *Leiden* cult. (*Vogel et al.*) 980074 & 980087 (K!, L!). Marudi District, Mount Dulit, Dulit Ridge, 13 Sept. 1932, *Richards* S.476 (holotype SING!, isotypes K!, L!). Marudi District, Dulit Ridge, 1 Oct. 1932, *Richards* S.548 (K!). Marudi District, Dulit Ridge, 7 Sept. 1932, *Synge* S.446 (K!).

HABITAT. Sand forest and lower montane mossy ridge forest. 1000–1750 m.

This species has minute flowers only slightly larger than those of *D. microscopicum*. No further collections appear to have been made since the Oxford University Expedition to

115

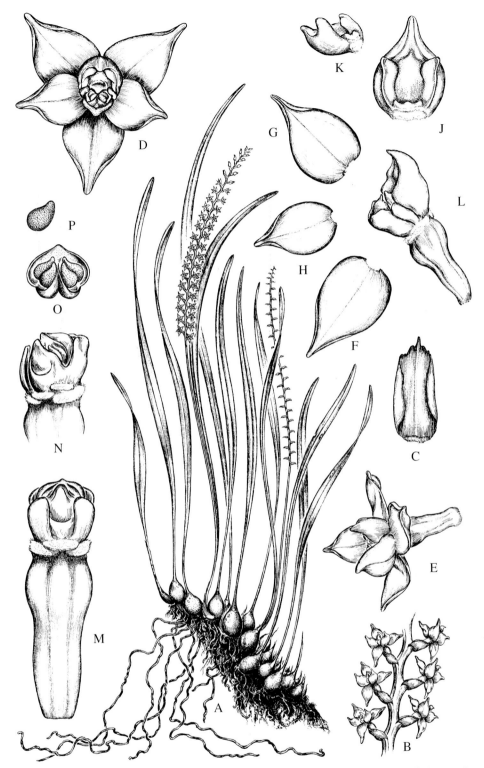

Fig. 50. *Dendrochilum minimiflorum*. **A** habit; **B** portion of inflorescence; **C** floral bract; **D** flower, front view; **E** flower, side view; **F** dorsal sepal; **G** lateral sepal; **H** petal; **J** labellum, front view; **K** labellum, side view; **L** pedicel with ovary, labellum and column, side view; **M** pedicel with ovary and column, front view; **N** upper portion of ovary and column, stelidium removed, side view; **O** anther cap with pollinia, inner view; **P** pollinium. **A** × 2, **B** × 9, **C–E** × 30, **F–H** × 40, **J** × 60, **K & L** × 40, **M & N** × 60, **O & P** × 75, **A** from *Leiden* cult. (*Vogel et al.*) 980087, **B–P** from *Leiden* cult. (*Vogel et al.*) 980074. Drawn by Oliver Whalley.

Sarawak in 1932 until recently. Living material collected on Mount Batu Buli in 1998 by Vogel, Schuiteman and Roelfsema has flowered in cultivation in the Hortus Botanicus at Leiden University, The Netherlands. Close examination of fresh flowers reveals Carr's description of the labellum morphology to be inaccurate. Carr describes the labellum as entire, having a horseshoe-shaped keel and a tiny rounded callus each side at the base. The labellum is actually obscurely three lobed, with a low transverse basal callus. What Carr describes as tiny rounded basal calli are in fact small thickenings on the inner surface of the side lobes.

19. DENDROCHILUM CORRUGATUM

Dendrochilum corrugatum (Ridl.) J.J. Sm. in *Recl. Trav. bot. néerl.* 1: 65 (1904). Type: Malaysia, Sabah, Mount Kinabalu, Marai Parai Spur, 1700 m, *Haviland* 1814 (holotype SAR!). **Figs. 51 & 52, plate 1E.**

Platyclinis corrugata Ridl. in Stapf, *Trans. Linn. Soc. Lond.* 2, 5: 233 (1894).
Acoridium corrugatum (Ridl.) Rolfe in *Orchid Rev.* 12: 220 (1904).
Dendrochilum fimbriatum Ames, *Orchidaceae* 6: 51–52 (1920). Type: Malaysia, Sabah, Mount Kinabalu, Marai Parai Spur, 23 Nov. 1915, *J. Clemens* 248 (holotype AMES!).

DESCRIPTION. Epiphyte, sometimes lithophytic. *Rhizome* abbreviated, or up to 6 cm long, 0.2–0.3 cm in diameter. *Roots* flexuous, simple or sometimes with a few short branches, smooth, often up to 25(–45) cm long, 0.5–1 mm in diameter. *Cataphylls* 2–3, acute to acuminate, 0.5–4 cm long, pale brown, with darker brown speckles, becoming fibrous. *Pseudobulbs* aggregated, often into large clumps, ovoid-globose or broadly ellipsoid, rugulose, 0.5–1.6 × 0.5–0.7(–1.2) cm, green suffused red, or entirely red, yellowish when dry. *Leaf*: petiole sulcate, 0.8–3 cm long; blade linear to linear-lanceolate, acute, minutely apiculate, rigid, sulcate along the middle, 5.5–13.5 × 0.5–1 cm, dark olive-green, main nerves 3–4. *Inflorescences* many-flowered, subdense to dense, curving, flowers borne 1.8–2 mm apart in 2 regular ranks; peduncle terete, filiform, 5.5–10 cm long; non-floriferous bracts (1–)3–4, ovate-elliptic, acute, 1–3.5 mm long; rachis quadrangular, 5–16 cm long, reddish; floral bracts broadly ovate, subacute, (1.8–)2.5–3 × 1.5–2.5 mm, rigid, reddish. *Flowers* unscented, sepals and petals creamy white, labellum yellowish green, column deep pink or orange, anther-cap yellow. *Pedicel with ovary* narrowly clavate, geniculate, 1.8–2 mm long. *Sepals* 3-nerved. *Dorsal sepal* lanceolate, acute, somewhat reflexed distally, 4–4.5 × 1.5 mm. *Lateral sepals* ovate-lanceolate, falcate, apex acute and strongly reflexed, (3–)3.5 × 1.9–2 mm. *Petals* lanceolate, acute to acuminate, margin minutely erose-denticulate, 3.5–4 × 1 mm, 1-nerved. *Labellum* broadly ovate-rotundate, sometimes shallowly 3-lobed, somewhat concave at base, triangular-apiculate, margin shortly and irregularly fimbriate, 2 × 3–3.1 mm, 3-nerved; disc bicallose, calli conspicuous, semi-orbicular, erect, fleshy, about half as long as labellum, one on each lateral nerve, confluent at base by an obscure transverse ridge. *Column* fleshy, (0.5–)0.7–0.8 mm long; apical hood rounded, entire; stelidia basal, oblong, obtuse, 0.3–0.4 mm; rostellum prominent, porrect, narrowly triangular, linguiform, acute; anther-cap cordate, cucullate, glabrous.

DISTRIBUTION. Borneo.

SABAH. Mount Kinabalu: Marai Parai Spur, 25 April 1983, *Bailes & Cribb* 841 (K!); Marai Parai Spur, *Barkman* 11 & 12 (K!, Sabah Parks Herbarium, Kinabalu Park); Marai Parai

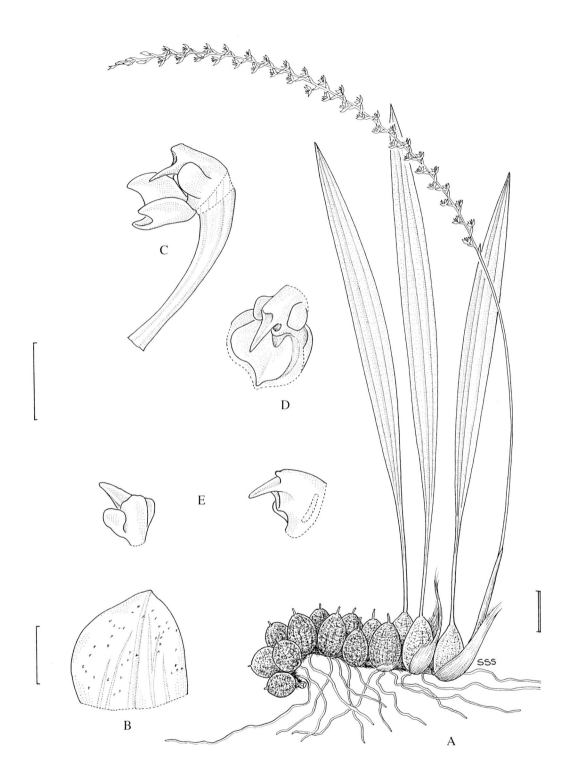

Fig. 51. *Dendrochilum corrugatum*. **A** habit; **B** floral bract; **C** pedicel with ovary, callus portion of labellum and column; **D** base of labellum showing callus, and column; **E** column, oblique and side views showing elongate rostellum. **A–E** from *Haviland* 1814 (holotype). Scale: single bar = 1 mm; double bar = 1 cm. Drawn by Susanna Stuart-Smith.

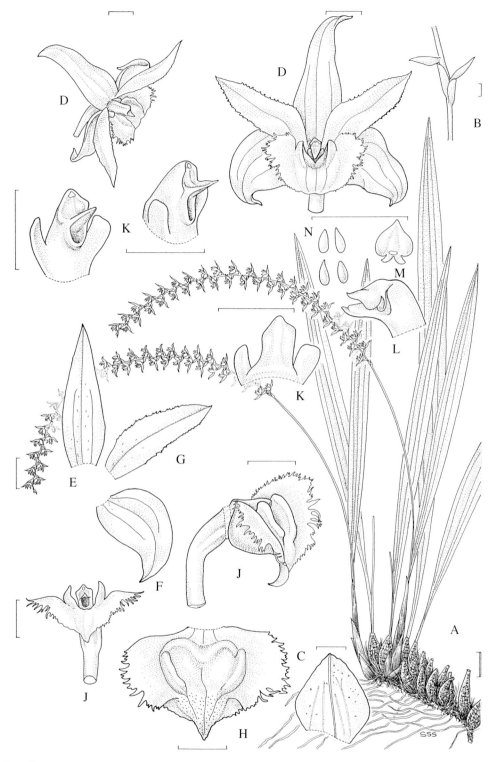

Fig. 52. *Dendrochilum corrugatum*. **A** habit; **B** junction of rachis with peduncle showing two non-floriferous bracts and two floral bracts; **C** floral bract; **D** flower, front and top views; **E** dorsal sepal; **F** lateral sepal; **G** petal; **H** labellum, front view; **J** pedicel with ovary, labellum and column, front and side views; **K** column with anther-cap removed, front, oblique and back views; **L** column and anther-cap, side view; **M** anther-cap, back view; **N** pollinia. **A** from *Carr* C. 3128, SFN 24728, **B** from *J. & M.S. Clemens* 32244, **C–N** from *Bailes & Cribb* 841. Scale: single bar = 1 mm; double bar = 1 cm. Drawn by Susanna Stuart-Smith.

Spur, 30 May 1933, *Carr* 3128, SFN 27428 (BM!, BO!, K!, L!, SING!); Marai Parai Spur, 23 Nov. 1915, *J. Clemens* 248 (holotype of *D. fimbriatum*, AMES!); Marai Parai Spur, 22 March 1933, *J. & M.S. Clemens* 32244 (AMES!, BM!, BO!, E!, L!); Marai Parai Spur, 12 Oct. 1985, *SNP* 2712 (Sabah Parks Herbarium, Kinabalu Park); Tahubang River Head, 3 April 1933, *J. & M.S. Clemens* 32548 (BM!, E!, HBG!, L!); Numeruk Ridge, 18 Aug. 1933, *J. & M.S. Clemens* 40063 (AMES!); Penataran Basin, 31 Aug. 1933, *J. & M.S. Clemens* 40135 (BM!); Marai Parai Spur, 15 Sept. 1958, *Collenette* 31 (BM!); Marai Parai Spur, *Haviland* 1814 (holotype of *Platyclinis corrugata*, SAR!); Marai Parai Spur, 16 Sept. 1965, *Meijer* SAN 54020 (SAN!).

HABITAT. Mostly a twig epiphyte in lower montane forest and *Leptospermum* ridge scrub on ultramafic substrate. Preferring open sites on extreme ultramafic substrate where it avoids the shade of a well-developed canopy. 1500–2100 m.

D. corrugatum was, until recently, known only from the type collection (*Haviland* 1814) from which Ridley prepared one of his characteristically brief original descriptions. The extant flowers on the type have, unfortunately, suffered damage since Ridley's time (see Fig. 51). The sepals and petals and much of the labellum are missing, although the column, minus the anther-cap, is intact. Comparison of *Haviland* 1814 with the type and several other excellent collections of *D. fimbriatum* however, leaves me in no doubt that they are conspecific. The types of both species were collected from Marai Parai Spur, an area of ultramafic rocks on the western slopes of Mount Kinabalu. Both have unusually long roots for the size of the plant, strongly wrinkled pseudobulbs, similarly textured leaves, identical floral bracts, a similarly shaped denticulate-fimbriate labellum and an identical column with oblong, obtuse stelidia and a distinctive, prominent rostellum. An expanded description prepared from an examination of the type of *Platyclinis corrugata* and incorporating Ridley's points is given below.

The epithet *corrugatum,* which refers to the strongly wrinkled pseudobulbs, may, to some extent, be a misnomer since examination of dried material of related species shows that extensive shrinkage, distortion and wrinkling results in part from the process of pressing and drying. However, mature pseudobulbs of many orchids do become wrinkled to varying degrees and it is certainly not a character confined to this species.

A Mount Kinabalu endemic, *D. corrugatum* bears a close resemblance and is closely related to *D. alatum* Ames. It can be distinguished primarily by the broader, fimbriate labellum, falcate lateral sepals, and petals which are not twisted and aligned 90° from vertical.

Revised description of the type of *Platyclinis corrugata* Ridl.:

Epiphyte. *Rhizome* slender, 5 cm long. *Roots* with a few branches, wiry, flexuous, smooth, up to 45 cm long, 0.5–1 mm in diameter. *Cataphylls* 2(–3?), acute, to 3 cm long. *Pseudobulbs* crowded, ovoid-globose, deeply wrinkled when dried, 0.9–1.3 × 0.9–1.2 cm, yellow. *Leaf:* petiole sulcate, 3 cm long; blade linear-lanceolate, acute, rigid, 8–9 × 0.7 cm, main nerves 3. *Inflorescences* subdensely many-flowered, flowers borne 1.8–2 mm apart; peduncle slender, wiry, 10 cm long; non-floriferous bracts solitary?; rachis 12 cm long; floral bracts broadly ovate, obtuse to subacute, 1.8–2 × 1.5 mm. *Flowers* minute, colour unknown. *Pedicel with ovary* slender, *c.* 1.8 mm long. *Sepals* ovate, caudate, 3 mm long. *Petals* lanceolate, acuminate. *Labellum* cymbiform, short, ovate and acuminate (when flattened); side lobes short, truncate; margin of mid-lobe denticulate; disc with a large, fleshy, somewhat elevated horseshoe-shaped basal callus. *Column c.* 0.5 mm long; foot absent; apical hood undeveloped; stelidia basal, oblong, obtuse, truncate; rostellum elongate, linguiform, acute, porrect; area below stigma prominent.

20. DENDROCHILUM ALATUM

Dendrochilum alatum Ames, *Orchidaceae* 6: 45–47, Plate 82, fig. 3 (1920). Type: Malaysia, Sabah, Mount Kinabalu, Marai Parai Spur, 2 Dec. 1915, *J. Clemens* 383 (holotype AMES!, isotypes BM!, K!, SING!). **Fig. 53, plate 1G.**

DESCRIPTION. Epiphyte. *Rhizome* abbreviated or up to 3 cm long. *Roots* flexuous, branching, smooth to very minutely papillose, 0.8–1 mm in diameter. *Cataphylls* 3, ovate-elliptic, acute, soon becoming fibrous, 1–2.5 cm long, finely speckled. *Pseudobulbs* crowded, ovoid or obpyriform, rugose, 0.5–1.6 × 0.5–0.9 cm, orange-yellow to red. *Leaf:* dark olive-green; petiole sulcate, 1–2.5 cm long; blade linear-lanceolate, obtuse or subacute, apiculate, attenuated at each end, 6–15 × (0.3–)0.5–0.6(–0.9) cm, main nerves 3. *Inflorescences* gently curving, to pendent, subdensely many-flowered, flowers borne 2 mm apart; peduncle slender, terete, 5.5–8.5 cm long; non-floriferous bracts 1–3, ovate-elliptic, acute, 2–2.5 mm long; rachis quadrangular, 5–15 cm long; floral bracts ovate, acute, 2–3 × 1.8–1.9 mm, 3-nerved. *Flowers* yellow or greenish yellow with an orange or brownish red column; *J. & M.S. Clemens* also noted the colour as: cream green petals, centre brown with cream spot in throat and cream or dull flesh, interior pinkish brick colour with a touch of greenish. *Pedicel with ovary* narrowly clavate, curved, 1.3 mm long. *Sepals* and *petals* with 1 main nerve and 1 or 2 secondary nerves. *Dorsal sepal* curving forward, narrowly elliptic, acute, 3 × 1.1–1.2 mm. *Lateral sepals* narrowly ovate-elliptic, acute, slightly concave, apex conduplicate, 3–3.75 × 1.5 mm. *Petals* twisted at 90° from vertical, narrowly oblong-elliptic, acute or obtuse, margin minutely denticulate, 3 × 1.1 mm. *Labellum* horizontal, entire, obcuneate, apiculate, distal margins minutely denticulate to erose, 2–2.1(–3) × 1.9 mm, 3-nerved, lateral nerves branching; disc concave between the keels, keels 2, tall, fleshy, confluent at base by a transverse ridge, terminating just beyond middle of labellum. *Column* straight, dorsally somewhat carinate, 1–1.1 mm long; apical hood tridentate, acute; stelidia basal, adpressed to labellum keels, elliptic to oblong, obtuse, 0.8–0.9 mm long; rostellum prominent, acuminate; anther-cap cucullate.

DISTRIBUTION. Borneo.

SABAH. Mount Kinabalu: Pig Hill, 25 Feb. 1995, *Barkman et al.* 137 (K!, Sabah Parks Herbarium, Kinabalu Park!); Mesilau Cave Trail, 20 Dec. 1983, *Beaman* 8005 (K!); Kemburongoh, Oct. 1933, *Carr* 3715 (SING!); Marai Parai Spur, 2 Dec. 1915, *J. Clemens* 383 (holotypes AMES!, isotypes BM!, K!, SING!); Tenompok, 27 Jan. 1932, *J. & M.S. Clemens* 04959 (BM!); Gurulau Spur, 1 Dec. 1933, *J. & M.S. Clemens* 50652 (AMES!, BM!, E!, K!, L!); Gurulau Spur, 6 Dec. 1933, *J. & M.S. Clemens* 50777 (BM!, E!, K!); Paka-paka Cave, 16 Nov. 1931, *Holttum* SFN 36569 (AMES!); Pig Hill, 1987, *Lamb* AL 735/87 (K!); Panar Laban, *Smith & Everard* 155 (K!).

HABITAT. Lower and upper montane forest; in ridge scrub; *Dacrydium/Leptospermum* scrub; on ultramafic and granite substrates. 1700–3200 m.

Resembling the Philippine *D. graciliscapum* (Ames) Pfitzer but easily distinguished by the large elliptical stelidia which superficially resemble side lobes of the labellum. One of its nearest Bornean allies is *D. corrugatum* (Ridl.) J.J. Sm., from which it differs by the narrower, non-fimbriate labellum. *D. pterogyne* Carr is also closely related but can be distinguished by its larger flowers with a labellum usually broadest at the middle and a longer column with narrower stelidia.

The specific epithet is from the Latin *alatus*, winged, and refers to the large basal stelidia on the column.

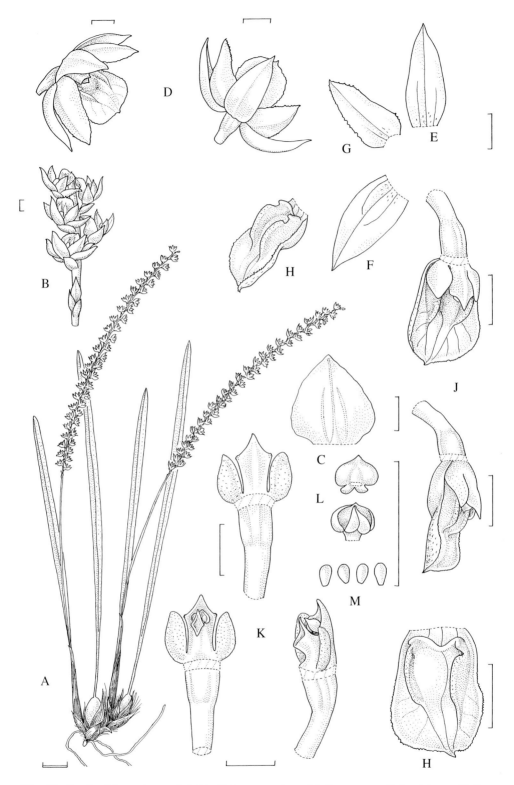

Fig. 53. *Dendrochilum alatum*. **A** habit; **B** lower portion of inflorescence; **C** floral bract; **D** flower, oblique and top views; **E** dorsal sepal; **F** lateral sepal; **G** petal; **H** labellum, front and oblique views; **J** pedicel with ovary, labellum and column, oblique and side views; **K** pedicel with ovary and column, front, back and side views; **L** anther-cap, interior and back views; **M** pollinia. **A** from *J. & M.S. Clemens* 50652, **B–M** from *Lamb* AL 735/87. Scale: single bar = 1 mm; double bar = 1 cm. Drawn by Susànna Stuart-Smith.

21. DENDROCHILUM JOCLEMENSII

Dendrochilum joclemensii Ames, *Orchidaceae* 6: 55, pl. 83, top (1920). Type: Malaysia, Sabah, Mount Kinabalu, Marai Parai Spur, 22 Nov. 1915, *J. Clemens* 247 (holotype AMES!). **Fig. 54, plate 9B & C.**

DESCRIPTION. Epiphyte, rarely terrestrial. *Rhizome* shortly creeping, forming a clump, up to 6 cm long. *Roots* branching, minutely papillose, flexuous, 1 mm in diameter. *Cataphylls* 3, ovate-elliptic, acute to acuminate, 0.5–1.6 cm long. *Pseudobulbs* crowded, subfusiform, smooth, becoming rugulose, 1–1.4 × 0.3–0.5 mm, green. *Leaf:* petiole slender, sulcate, 3–9 mm long; blade linear, obtuse and apiculate to shortly acute, 4.5–10 × 0.3–0.4 cm, erect, main nerves 3. *Inflorescences* erect to gently curving, laxly few- to about 24-flowered, flowers opening from top of inflorescence (always ?), and borne 2–3 mm apart; peduncle slender, 2.2–4 cm long, greenish-orange; non-floriferous bracts absent; rachis quadrangular, 4–5.5 cm long, salmon-pink; floral bracts narrowly elliptic, acute, 2 × 0.5–0.6 mm, 3-nerved, pale orange. *Flowers* wide opening, about 5.5 mm across, unscented, translucent salmon-pink, pale orange or yellow with pale salmon-pink at base of keels on labellum, or very pale reddish ochre to brownish salmon-pink with very pale green petals, column cream, stelidia translucent white. *Pedicel with ovary* narrowly clavate, curved, 2 mm long. *Sepals* and *petals* spreading to reflexed. *Sepals* narrowly ovate-elliptic, acute to acuminate, 3-nerved. *Dorsal sepal* 2.5–3 × 1 mm. *Lateral sepals* slightly carinate, 2.5–3 × 1.1 mm. *Petals* narrowly elliptic, acute, (2–)2.5–2.8 × (0.5–)0.9–1 mm, 1-nerved. *Labellum* very obscurely auriculately lobed, lobes hidden by keels on disc, oblong, abruptly contracted near the apex into an elongated acuminate tip, 2–2.2 × 0.8–1(–1.3) mm, 3-nerved; disc with 2 suborbicular, laterally flattened, spreading keels confluent with outer nerves, each thickened and joined just above base of labellum and terminating midway along, median nerve prominent and slightly raised. *Column* 1–1.5 mm long; apical hood strongly concave, almost orbicular, entire; stelidia basal, oblong, rounded to truncate, rather fleshy, *c.* 0.8(–1) × 0.5 mm, erect; anther-cap minute, cucullate.

DISTRIBUTION. Borneo.

SABAH. Mount Kinabalu: Road to Power Station, *Barkman* 261 (Sabah Parks Herbarium, Kinabalu Park); Kadamaian River, 9 Oct. 1933, *Carr* 3710 (SING!); Marai Parai Spur, 22 Nov. 1915, *J. Clemens* 247 (holotype AMES!); Kiau View Trail, 28 Oct. 1991, *Gunsalam* 10 (K!). Crocker Range: Kimanis road, 12 July 1987, *Lamb* AL 856/87 (K, sketch only!); Keningau to Kimanis road, Dec. 1986, *Vermeulen & Duistermaat* 671 (L!).

HABITAT. A twig epiphyte in lower montane ridge forest on ultramafic and sandstone substrates. Also recorded as a terrestrial on a mossy roadside bank by *Lamb*. 1050–2000 m.

Barkman (PhD dissertation, 1998) has carried out a molecular analysis of *D. joclemensii* and found similarities between it and *D. gibbsiae* (section *Cruciformia*) implying that they are sister taxa. He notes, however, that the level of ITS divergence was "relatively high between the two suggesting that, although sister taxa in our analysis, they may have closer relatives yet unsampled." He also postulates that *D. pandurichilum* (section *Falsiloba*) may represent a possible link between the species "which has a lobed labellum somewhat similar to *D. gibbsiae* but has enlarged labellum keels like that of *D. joclemensii.*"

The specific epithet commemorates Joseph Clemens (1862–1936), a U.S. Army chaplain who, together with his wife Mary Strong Clemens and others, made extensive collections on Mount Kinabalu.

123

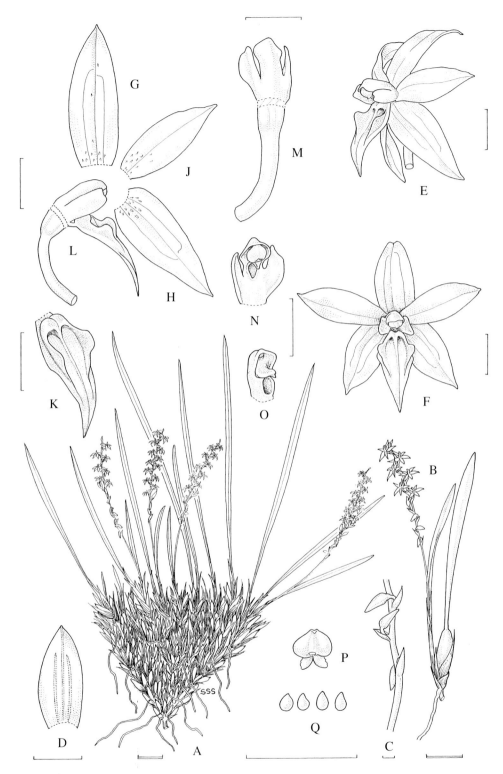

Fig. 54. *Dendrochilum joclemensii*. **A & B** habits; **C** lower portion of inflorescence showing flower buds; **D** floral bract; **E** flower, oblique view; **F** flower, front view; **G** dorsal sepal; **H** lateral sepal; **J** petal; **K** labellum, oblique view; **L** pedicel with ovary, labellum and column, side view; **M** pedicel with ovary and column, back view; **N** column, oblique view; **O** column apex, anther-cap removed, oblique view; **P** anther-cap, back view; **Q** pollinia. **A** from *Carr* C. 3710, **B–Q** from *Gunsalam* 10. Scale: single bar = 1 mm; double bar = 1 cm. Drawn by Susanna Stuart-Smith.

22. DENDROCHILUM PTEROGYNE

Dendrochilum pterogyne *Carr* in *Gdns. Bull. Straits Settl.* 8: 236 (1935). Type: Malaysia, Sabah, Mount Kinabalu, Paka-paka Cave, 3100 m, June 1933, *Carr* 3541, SFN 27597 (holotype SING!, isotypes AMES!, C!, K!). **Fig. 55, plate 14A & B.**

D. grandiflorum auct., non (Ridl.) J.J. Sm.: Sato, *Flowers and Plants of Mount Kinabalu*: 35 (1991).

D. sp.: Sato, *Flowers and Plants of Mount Kinabalu*: 37, top (1991).

D. alatum auct., non Carr: Wood, Beaman & Beaman, *Plants of Mount Kinabalu 2, Orchids*, pl. 40 C & D (1993).

DESCRIPTION. Epiphyte or lithophyte. *Rhizome* shortly creeping, much branched, up to 5 cm long. *Roots* much branched, smooth, 1 mm in diameter. *Cataphylls* 3, ovate-elliptic, acute, becoming fibrous, 0.8–2 cm long. *Pseudobulbs* crowded or up to 1 cm apart, ovoid, rugulose, 0.8–2.2 × 0.5–1.2 cm, orange, red or green. *Leaf*: petiole sulcate, 0.5–2.5 cm long; blade linear-lanceolate, apex conduplicate, acute or apiculate, rigid, (2.8–)3.5–11.5 × (0.3–)0.6–0.8(–1) cm, main nerves 5, dark olive-green. *Inflorescences* gently curving to pendent, laxly 8–15(–18)-flowered, flowers borne 3–4 mm apart; peduncle filiform, 1.5–8 cm long; non-floriferous bracts 1–2, sometimes absent, ovate-elliptic, acute, 3–5 mm long; rachis flexuous, forming an obtuse angle with peduncle, quadrangular, 3–7 cm long; floral bracts ovate, erose, acute, 3–3.3 mm long, many-nerved. *Flowers* unscented (according to *Vermeulen & Duistermaat*), sepals and petals pale yellow tinted pale salmon-pink, with darker apex and median line, labellum yellow with bright salmon-pink base and keels, or entirely salmon-pink with darker base and keels, column salmon-pink, anther-cap cream, or entire flower lemon-yellow with darker olive-green to ochre column apex and stelidia; *Vermeulen & Duistermaat* also record the flower colour as sometimes pale greenish. *Pedicel with ovary* narrowly clavate, curved, 3 mm long. *Sepals* and *petals* 3-nerved, concave, a little incurved towards apex, sparsely papillose on inner surface. *Dorsal sepal* ovate-elliptic, acute, 5 × 2.5 mm. *Lateral sepals* ovate, sometimes falcate, acute or shortly apiculate, apex dorsally carinate, 4.8 × 2.7 mm. *Petals* narrowly elliptic, acute, 4.7 × 2.5 mm. *Labellum* entire, ovate, minutely apiculate, margins sometimes strongly recurved in the upper half, 2.6–3.5 × 2.7–3 mm, 3-nerved; disc with 2 tall, curved, extrorse keels incurved and united at base forming a horseshoe often produced inside above base as 2 short adnate extrorse keels, nerves prominent from apex of keels almost to apex of labellum. *Column* narrowly oblong, 2 mm long; apical hood slightly dilated and obscurely 3-lobed, if at all; stelidia basal, oblong-ovate, obtuse, as long as or slightly shorter than column; margins of stigmatic cavity elevate below; rostellum conspicuous, triangular, acute; anther-cap cucullate. *Capsule* oblong-obovoid, 1–1.1 cm long.

DISTRIBUTION. Borneo.

SABAH. Mount Kinabalu: Panar Laban, *Barkman* 2, 3, 19 & 20 (K!, Sabah Parks Herbarium, Kinabalu Park); Paka-paka Cave, 13 June 1933, *Carr* 3541, SFN 27597 (holotype SING, isotypes AMES!, C!, K!); near Paka-paka Cave, 15 June 1933, *Carr* 3548, SFN 27635 (K!, SING!); Paka-paka Cave, 8 Jan. 1932, *J. & M.S. Clemens* 27864 (BM!, E!, K!, L!); Dallas, 31 Dec. 1931, *J. & M.S. Clemens* 30102 (AMES!, K!, L!); Kinateki River Head, 3 March 1933, *J. & M.S. Clemens* 31831 (BM!, BO!); Marai Parai, 24 May 1933, *J. & M.S. Clemens* 32322 (AMES!, BM!, E!,) & 32324 (AMES!, BM!, E!, L!), 19 May 1933, *J. & M.S. Clemens* 33180 (AMES!, BM!, E!; sheet at BM mixed with *D. grandiflorum*); Gurulau Spur, Dec. 1933, *J. & M.S. Clemens* 51079 (AMES!, BM!, E!, K!, L!); summit area, 13 Dec. 1960,

125

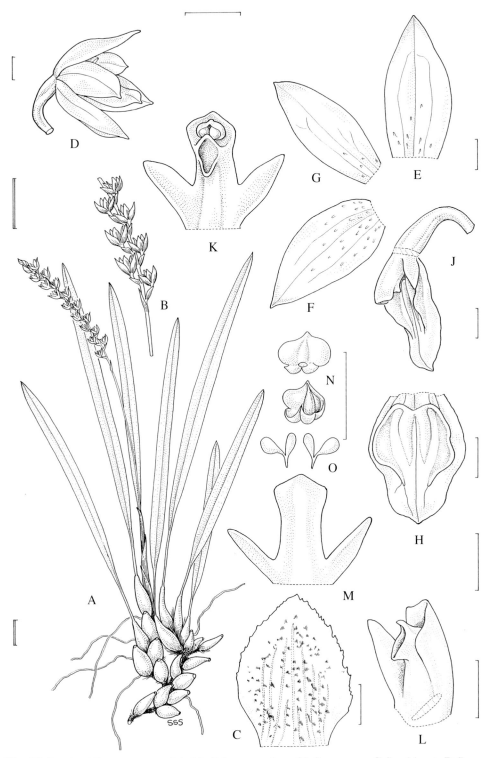

Fig. 55. *Dendrochilum pterogyne.* **A** habit; **B** lower portion of inflorescence; **C** floral bract; **D** flower, side view; **E** dorsal sepal; **F** lateral sepal; **G** petal; **H** labellum, front view; **J** pedicel with ovary, labellum and column, side view; **K** column, front view; **L** column, anther-cap and stelidium removed, side view; **M** column, back view; **N** anther-cap, back and interior view showing two pollinia; **O** pollinia. **A & B** from *Carr* C. 3541, SFN 27597, **C–O** from *J. & M.S. Clemens* 33180. Scale: single bar = 1 mm; double bar = 1 cm. Drawn by Susanna Stuart-Smith.

Collenette 616 (K!); Kemburongoh, 16 Nov. 1931, *Holttum* s.n. (SING!); between Kemburongoh and waterfalls, 17 May 1963, *Iwatsuki* s.n. (TI!); Layang-Layang, 15 April 1984, *Sands* 3880 (K!); summit area, 22 Feb. 1981, *Sato* 162 (UKMS!); Kadamaian River, just below Paka-paka Cave, 14 June 1957, *J. Sinclair et al.* 9181 (E!, K!, L!, SING!); Summit Trail, Dec. 1986, *Vermeulen & Duistermaat* 545 (L!).

HABITAT. Upper montane forest on ultramafic or non-ultramafic substrates. Able to tolerate deep shade or full sunlight. Usually epiphytic, less often lithophytic. 900–3800 m (usually above 2600 m).

D. pterogyne may be found in bloom throughout the year, often alongside *D. alpinum*, but no hybrids have been reported.

The specific epithet is derived from the Greek *pter*, a wing, and *gyn-* or *gyno-*, female or pertaining to female organs, and refers to the wing-like stelidia on the column.

23. DENDROCHILUM ALPINUM

Dendrochilum alpinum Carr in *Gdns. Bull. Straits Settl.* 8: 235 (1935). Type: Malaysia, Sabah, Mount Kinabalu, below Sayat Sayat, 3450 m, June 1933, *Carr* 3545, SFN 27624 (holotype SING!, isotypes AMES!, K!). **Fig. 56, plate 1F.**

DESCRIPTION. Lithophyte or epiphyte. *Rhizome* shortly creeping, to 5 cm long, branching. *Roots* stout, flexuous, with a few branches, minutely papillose, 1–1.5 mm in diameter. *Cataphylls* 3, ovate-elliptic, acute to acuminate, 1.5–4 cm long. *Pseudobulbs* crowded or borne up to 1 cm apart on rhizome, ovoid, grooved and minutely rugulose, sometimes somewhat curved, 1.8–2.7 × 0.8–1 cm, golden yellow, orange or red. *Leaf*: petiole sulcate, 0.8–3 cm long; blade narrowly elliptic, narrowed above the middle, apex conduplicate, acute, texture rigid, (5–)7–14.5 × 0.7–1.5 cm, main nerves 5. *Inflorescences* pendent, laxly 11- to 20-flowered, flowers borne 5–6 mm apart; peduncle slender, terete, slightly dilated distally, 3–10 cm long; non-floriferous bracts 1 or 2, ovate, acute, 4–9 mm long; rachis quadrangular, 7.5–12 cm long; floral bracts ovate, acute, margins minutely erose towards apex, 5 × 4.5 mm, nerves several, prominent. *Flowers* up to *c.* 1 cm across, sepals and petals either entirely salmon-pink or yellow, suffused salmon-pink, labellum either brownish or salmon-pink with a brownish salmon-pink median line, keels and apex, column deep salmon-pink, stelidia brown, anther-cap cream suffused salmon-pink at base. *Pedicel with ovary* narrowly clavate, 2–5 mm long. *Sepals* and *petals* with 3 main nerves. *Dorsal sepal* curving forward, oblong to oblong-elliptic, obtuse, sometimes minutely apiculate, 7.8 × 4.4 mm. *Lateral sepals* spreading, ovate-oblong, obtuse to acute, apex somewhat dorsally carinate on reverse, 7.8–7.9 × 4 mm. *Petals* oblong or obliquely-oblong, obtuse, 7.8 × 3.8 mm. *Labellum* entire, broadly ovate, shallowly retuse or subacute, rather incurved above the middle, elevate along median nerve, 6.4 × 5.8 mm, disc with a thickened central area and provided with 2 tall curved, suberect or rather extrorse fleshy keels which form, with the central thickened area, 2 grooves with a low cushion at the base. *Column* 1.2–2 mm long; apical hood dilated, obscurely 3-lobed to rounded; stelidia basal, oblong-oblanceolate, obtuse, 1.7–1.8 mm long; stigma transversely oblong; rostellum conspicuous, triangular, acute; anther-cap cucullate. *Capsule* obovoid, with 3 rounded keels along suture lines, 1.6 × 0.7 cm.

DISTRIBUTION. Borneo.

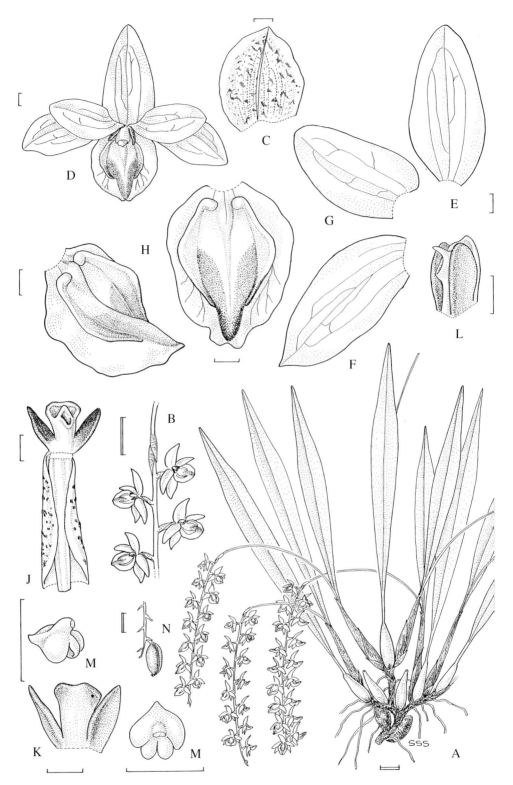

Fig. 56. *Dendrochilum alpinum*. **A** habit; **B** lower portion of inflorescence; **C** floral bract; **D** flower, front view; **E** dorsal sepal; **F** lateral sepal; **G** petal; **H** labellum, front and oblique views; **J** floral bract, pedicel with ovary and column, anther-cap removed, front view; **K** column, back view; **L** column, excluding stelidia, oblique view; **M** anther-cap, side and back views; **N** fruit capsule. **A–N** from *Carr* C. 3545, SFN 27624 (holotype). Scale: single bar = 1 mm; double bar = 1 cm. Drawn by Susanna Stuart-Smith.

SABAH. Mount Kinabalu: above Panar Laban, *Barkman* 4 & 6 (K!, Sabah Parks Herbarium, Kinabalu Park); 142 (Sabah Parks Herbarium, Kinabalu Park); below Sayat Sayat, June 1933, *Carr* 3545, SFN 27624 (holotype SING!, isotypes AMES!, K!); Gurulau Spur, 1 Dec. 1933, *J. & M.S. Clemens* 50669 (BM!); Panar Laban/Sayat Sayat, 26 July 1981, *Sato* 759 & *Sato* 762 (UKMS!); summit area, Oct. 1981, *Sato et al.* s.n. (UKMS!).

HABITAT. Upper montane forest and on the sides of granitic rocks sheltered by stunted trees. 2400–3700 m, abundant above 3200 m.

D. alpinum has the largest pseudobulbs and flowers of any species in the section represented on Mount Kinabalu. It is easily identified by the petals which are twisted 90° from vertical and the darkly pigmented labellum contrasting with the pale sepals and petals.

24. DENDROCHILUM PSEUDOSCRIPTUM

Dendrochilum pseudoscriptum T.J. Barkman & J.J. Wood in *Orchid Rev.* 104 (1209): 179–182, figs. 91–93 (1996). Type: Malaysia, Sabah, Mount Kinabalu, near Kinabalu Lipson, 2800 m, 20 May 1995, *Barkman* 198 (holotype Sabah Parks Herbarium, Kinabalu Park!, isotype K!). **Fig. 57, plate 13B–D.**

DESCRIPTION. Epiphyte. *Rhizome* abbreviated. *Roots* smooth, 1–2 mm in diameter. *Cataphylls* 3, finely spotted, persistent, becoming fibrous, 1–4 cm long. *Pseudobulbs* ovoid, 1–2 × 0.6–1 cm, red. *Leaf*: petiole sulcate, 1–1.5 × 0.1–0.2 cm; blade linear-lanceolate, obtuse and mucronate, 4–9.5 × 0.8–1.2 cm, main nerves 5. *Inflorescences* laxly 14- to 17-flowered, flowers borne 5 mm apart; peduncle erect to pendent, 3.5–6 cm long; non-floriferous bract solitary, acute, 3–4 mm long; rachis 6–6.5 cm long, less than 1 mm wide; floral bracts ovate to circular, obtuse or retuse, bearing a small apical mucro, margins erose, 3 × 2.5 mm, nerves 5. *Flowers* lemon-yellow, column bright orange, lip chestnut-brown or orange with a yellow callus, or whole flower entirely salmon-pink. *Pedicel with ovary* 4 mm long, less than 1.5 mm wide. *Sepals* and *petals* 3-nerved. *Dorsal sepal* ovate, acute, slightly concave, 4–6 × 1.5–3 mm. *Lateral sepals* lanceolate to ovate, acute, slightly concave, 3.5–6 × 1.5–3 mm. *Petals* oblanceolate to ligulate, obtuse, twisted 90° from vertical and aligned perpendicular to plane formed by sepals, 3–5.5 × 1–2.5 mm. *Labellum* entire, nearly circular or cordate in outline, shortly apiculate, sometimes cymbiform, margins slightly recurved, convex in basal half between keels, 2.5–3.5 × 2.5–3.5 mm, 3-nerved; disc with a callus consisting of two shortly raised keels *c.* 1 mm long or nearly half the length of the labellum, touching but not joined at the base, horseshoe-shaped or somewhat M-shaped, much broader than high, very fleshy. *Column* 1–1.5 mm long, less than 1 mm wide; apical hood slightly longer than anther-cap, not conspicuously lobed, but tending towards three, margins entire; stelidia shorter than apical hood, *c.* 1 × 0.3 mm; anther-cap cucullate. *Capsule* cylindrical, oblong, mostly without ridges or wings, 1.1–1.4 × 0.7 cm, yellow with orange ridges or completely salmon-pink, column persistent.

DISTRIBUTION. Borneo.

SABAH. Mount Kinabalu: Summit Trail, Kinabalu Lipson, above Layang Layang, 25 April 1994, *Barkman et al.* 16 & *Barkman et al.* 18 (K!, Sabah Parks Herbarium, Kinabalu Park!); northern ridge adjacent to Low's Gully, along trail leading to Melangkap Tomis, 20 April 1995, *Barkman et al.* 183 (Sabah Parks Herbarium, Kinabalu Park!); near Kinabalu Lipson, 20

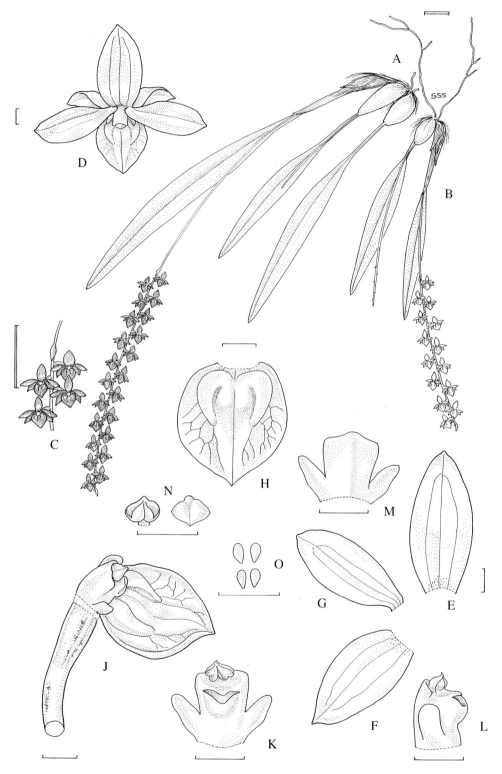

Fig. 57. *Dendrochilum pseudoscriptum*. **A & B** habits; **C** basal portion of inflorescence; **D** flower, front view; **E** dorsal sepal; **F** lateral sepal; **G** petal; **H** labellum, front view; **J** pedicel with ovary, labellum and column, oblique view; **K** column, front view; **L** column, side view; **M** column, back view; **N** anther-cap, interior and back views; **O** pollinia. **A & C** from *Barkman et al.* 183, **B & D–O** from *Barkman* 198 (holotype). Scale: single bar = 1 mm; double bar = 1 cm. Drawn by Susanna Stuart-Smith.

May 1995, *Barkman* 198 (holotype Sabah Parks Herbarium, Kinabalu Park!, isotype K!).

HABITAT. Upper montane scrub forest composed of *Dacrydium gibbsiae* Stapf and *Leptospermum recurvum* Hook.f. on ultramafic substrate. 2700–2800 m.

D. pseudoscriptum can be found growing sympatrically with five other *Dendrochilum* species on Mount Kinabalu and may even bloom at the same time. The plant is similar to *D. scriptum* Carr in appearance, as the specific epithet suggests, although the callus on the labellum is reminiscent of *D. transversum* Carr and the angled petals and sometimes cymbiform labellum recall *D. alpinum* Carr. The short stelidia and twisted petals are characters shared with *D. alatum* Ames. Two distinct flower colours have been observed, one with lemon-yellow sepals and petals and a chestnut-brown or orange labellum, the other entirely salmon-pink.

25. DENDROCHILUM SCRIPTUM

Dendrochilum scriptum Carr in *Gdns. Bull. Straits Settl.* 8: 234–235 (1935). Type: Malaysia, Sabah, Mount Kinabalu, above Kemburongoh, July 1933, *Carr* 3597 (holotype SING!). **Fig. 59, plate 15B & C.**

DESCRIPTION. Epiphyte, occasionally lithophytic. *Rhizome c.* 4 cm long, shortly creeping, densely covered with dry sheaths, which become fibrous with age. *Roots* flexuous, much branched, smooth, 1 mm in diameter. *Cataphylls* 3, ovate-elliptic, acute to acuminate, 0.5–2.5 cm long, pale brown with darker brown speckles, becoming fibrous. *Pseudobulbs* narrowly ovoid, minutely wrinkled, 1.5–2.5 × 0.4–0.7 cm, borne up to 1 cm apart on rhizome, but usually much less, forming an acute angle with rhizome, red. *Leaf:* petiole sulcate, 1.6–3 cm long; blade rigid, narrowly ligulate-elliptic, shortly acute, obtuse or minutely cuspidate, 5.5–13 × 0.7–1.1 cm, main nerves 5 or 6. *Inflorescences* erect, subdensely many-flowered, flowers borne 2–2.5 mm apart; peduncle terete, to 6 cm long; non-floriferous bracts absent; rachis quadrangular, to 8.5 cm long; floral bracts ovate, obtuse, 2–3.5 mm long. *Flowers* non-resupinate, *c.* 6 mm across, with a rather musty odour of over-ripe fruit, sepals and petals salmon-pink, ochre, orange or yellow, sepals often tipped brownish salmon with a salmon-pink median line outside, labellum salmon-pink or reddish brown with paler keels, margins ochre, or orange, column dark reddish brown, anther-cap cream with a dark salmon-pink median streak. *Pedicel with ovary* narrowly clavate, 1.8–2 mm long. *Sepals* 3-nerved. *Dorsal sepal* oblong-elliptic, obtuse or subacute, 3 × 2 mm. *Lateral sepals* falcate-ovate, obtuse or acute, 3 × 2.5–3 mm. *Petals* porrect, oblong, obtuse, 3 × 1.7 mm, 1- to 3-nerved. *Labellum* transversely oblong, shallowly retuse, *c.* 1.8 × 2.7 mm; disc with a large M-shaped keel in the lower two thirds, provided below the apex with a low fleshy keeled cushion. *Column* 1 mm long; apical hood very shortly triangular, obtuse; stelidia basal, broadly elliptic, obtuse, as long as apex of hood; stigmatic cavity with a swollen lower margin; anther-cap ovate, cucullate.

DISTRIBUTION. Borneo.

SABAH. Mount Kinabalu: Kadamaian River, *Barkman* 63 (Sabah Parks Herbarium, Kinabalu Park); above Kemburongoh, July 1933, *Carr* 3597 (holotype SING!).

HABITAT. A twig epiphyte in upper montane forest and scrub on extreme ultramafic substrate, preferring an open canopy. Also found abundantly as a riverine species on mossy limbs overhanging rivers and sometimes on boulders in and adjacent to rivers. 2600 m and above, not observed above 3200 m.

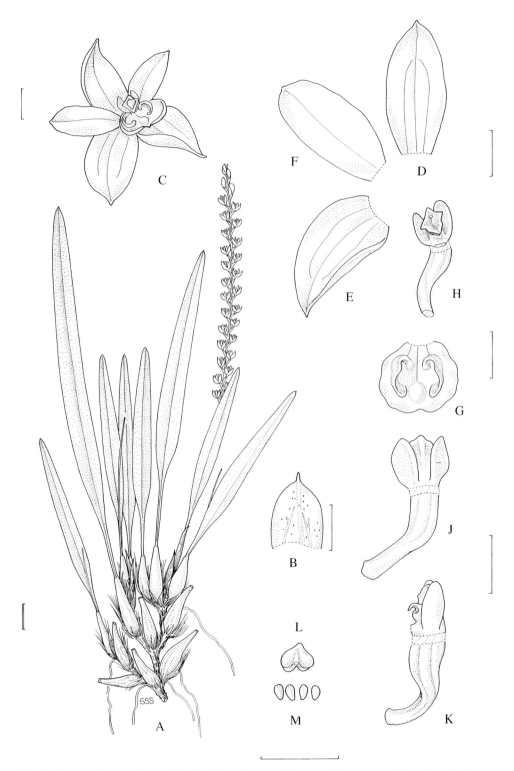

Fig. 58. *Dendrochilum scriptum*. **A** habit; **B** floral bract; **C** flower, oblique view; **D** dorsal sepal; **E** lateral sepal; **F** petal; **G** labellum, front view; **H** pedicel with ovary and column, anther-cap removed, front view; **J** pedicel with ovary and column, back view; **K** pedicel with ovary and column, anther-cap removed, side view; **L** anther-cap, back view; **M** pollinia. **A–M** from *Carr* C. 3597 (holotype). Scale: single bar = 1 mm; double bar = 1 cm. Drawn by Susanna Stuart-Smith.

This species was, until recently, known only from the type collection. Barkman (pers. comm.), however, reports that it is almost a 'weed' in certain places on Mount Kinabalu where one specimen seen growing on a boulder in a river had over 25 flower spikes. There are two colour forms, the more common being salmon-pink or orange, the other having yellow sepals and petals and an orange labellum.

The specific epithet *scriptum* is Latin, meaning written matter, and refers to the letter M-shaped keel on the labellum.

26. DENDROCHILUM TRANSVERSUM

Dendrochilum transversum Carr in *Gdns. Bull. Straits Settl.* 8: 233–234 (1935). Type: Malaysia, Sabah, Mount Kinabalu, Marai Parai Spur, May 1933, *Carr* 3477, SFN 27431 (holotype SING!, isotypes AMES!, K!, LAE). **Fig. 59, plate 17D.**

DESCRIPTION. Epiphyte. *Rhizome* shortly creeping, branching, up to 6 cm long, 0.5 cm in diameter. *Roots* flexuous, branching, smooth, 1 mm in diameter. *Cataphylls* 3–4, ovate-elliptic, acute, 0.5–5.5 cm long, pale brown, speckled darker brown. *Pseudobulbs* caespitose, narrowly cylindric, 2.5–4.5 × 0.5–0.7 cm, minutely rugulose in dried material (also in living state?), green, suffused red or entirely red. *Leaf*: petiole sulcate, 1.5–4 cm long; blade linear-lanceolate or linear-oblanceolate, abruptly narrowed below the acute, shortly cuspidate apex, 9–23 × 0.6–1.2 cm, thin-textured, main nerves 5, mid-nerve prominent on abaxial surface, deep olive-green. *Inflorescences* erect to curving, laxly to subdensely many-flowered, flowers borne 2–3 mm apart; peduncle terete, very slender, (9–)13–20 cm long; non-floriferous bracts 1–2, ovate-elliptic, acute, 3.5–5 mm long; rachis quadrangular, 7–9 cm long; floral bracts ovate, apiculate, 2.8–3.8 mm long. *Flowers* 7–8 mm across, sepals and petals rose-pink or salmon-pink, labellum deep ochre with darker apex and margins, column pinkish brown, stelidia ochre tipped dark ochre-brown, occasionally entire flower may be orange-yellow. *Pedicel with ovary* narrowly clavate, slightly curved, 2.8–4.5 mm long. *Sepals* and *petals* 3-nerved. *Dorsal sepal* ovate-elliptic, acute, 4.7–5 × 2.1 mm. *Lateral sepals* obliquely ovate-elliptic, acute, 4.5–4.7 × 2–2.5 mm. *Petals* oblong-elliptic, acute, 4.5–5 × 1.6–1.7 mm, sometimes twisted and aligned up to 90° from vertical. *Labellum* entire, transversely oblong, subapiculate, recurved distally, margins recurved in the upper half, minutely erose, 2.6–3 × 4–4.4 mm; disc provided with a tall, broadly horseshoe-shaped keel enclosing a concave area, extending ³/4 the length of the labellum; surface minutely papillose, especially distally. *Column* 1(–1.7) mm long; apical hood slightly dilated and shallowly 3-lobed or entire and transversely elliptic; stelidia basal, oblong, obtuse, acutate, 1.1–1.2 mm long, minutely papillose distally; rostellum ovate, obtuse; stigmatic cavity excavate, suborbicular, margins prominent, elevated; anther-cap lost.

DISTRIBUTION. Borneo.

SABAH. Mount Kinabalu: Marai Parai Spur, 15 Feb. 1995, *Barkman et al.* 95 (K!, Sabah Parks Herbarium, Kinabalu Park!); Marai Parai Spur, May 1933, *Carr* 3477, SFN 27431 (holotype SING!, isotypes AMES!, K!, LAE); Marai Parai Spur, 8 May 1933, *J. & M.S. Clemens* 33130 (AMES!, BM!, HBG!, K!, L!) & 19 May 1933, *J. & M.S. Clemens* 33173 (BM!, mixed with *D. grandiflorum*).

HABITAT. A twig epiphyte in upper montane mossy forest on ultramafic and non-ultramafic substrates. 2400–2800 m.

The specific epithet refers to the transversely oblong labellum.

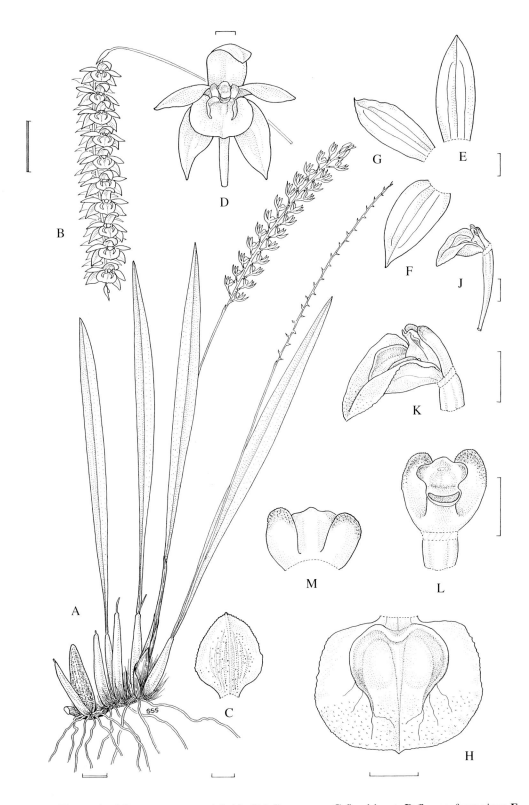

Fig. 59. *Dendrochilum transversum*. **A** habit; **B** inflorescence; **C** floral bract; **D** flower, front view; **E** dorsal sepal; **F** lateral sepal; **G** petal; **H** labellum, front view; **J** pedicel with ovary, labellum and column, side view; **K** upper portion of ovary, labellum and column, side view; **L** upper portion of ovary and column, anther-cap lost, front view; **M** column, back view. **A, C & E–M** from *Carr* C. 3477, SFN 27431 (holotype), **B & D** from *Barkman et al.* 95. Scale: single bar = 1 mm; double bar = 1 cm. Drawn by Susanna Stuart-Smith.

27. DENDROCHILUM LEWISII

Dendrochilum lewisii J.J. Wood in *Kew Bull.* 39 (1): 78, fig. 4 (1984). Type: Malaysia, Sarawak, Marudi District, Gunung Mulu National Park, Mount Mulu, 2250 m, 20 Oct. 1977, *Lewis* 366 (holotype K!). **Fig. 60.**

DESCRIPTION. Epiphyte. *Rhizome* up to 2.5 cm long. *Roots* branching, flexuous, smooth, 0.8–1 mm in diameter. *Cataphylls* 2, ovate-elliptic, acute, 1.5–2 cm long. *Pseudobulbs* fusiform, smooth, 1–2.2 × 0.5 cm. *Leaf*: petiole sulcate, 0.5–1 cm long; blade oblong-elliptic, obtuse and shortly mucronate, slightly constricted 0.7–1.5 cm below apex, rigid, coriaceous, 3.2–5 × 0.9–1.3 cm, main nerves 5, pale green. *Inflorescences* curving, subdensely many-flowered, flowers borne 2–3 mm apart, the lowermost remote, to 4 or 5 mm apart; peduncle 3.1–4.5 cm long; non-floriferous bract solitary, ovate, apiculate, 3 mm long; rachis quadrangular, canaliculate on two sides, 16–19 cm long; floral bracts ovate, obtuse, apiculate, 2–3 × 2.5–2.6 mm. *Flowers* cream. *Pedicel with ovary* slender, slightly curved above, 4–4.1 mm long. *Sepals* and *petals* 3-nerved. *Dorsal sepal* narrowly ovate-elliptic, acute, curving forward, 5.5 × 1.9 mm. *Lateral sepals* narrowly ovate-elliptic, slightly oblique, acute, spreading or curving forward, 5.8–6 × 1.9–2 mm. *Petals* ovate or narrowly elliptic, acute, 5–5.5 × 2–2.1 mm. *Labellum* firmly attached to column, sharply deflexed near base, oblong-lanceolate, acute to acuminate, margin minutely erose distally, 3-nerved, obscurely auriculate and somewhat erose at base, 3.8–4 × 1.2–1.5 mm; disc with an M-shaped basal callus consisting of a thickened, fleshy V-shaped transverse keel to which are joined 2 thickened keels confluent with the outer nerves. *Column* gently curved, 2–2.1 mm long; apical hood elongate, obscurely 3-lobed; stelidia basal, linear-ligulate, obtuse or subacute, 1.1–1.5 mm long; rostellum broadly ovate, obtuse; anther-cap ovoid, cucullate.

DISTRIBUTION. Borneo

SARAWAK. Marudi District, Gunung Mulu National Park, Mount Mulu, 20 Oct. 1977, *Lewis* 366 (holotype K!).

HABITAT. Ridge top upper montane forest. 2250 m.

D. lewisii is known only from the type collection found near the summit of Mount Mulu where it is recorded as being a locally common epiphyte.

The specific epithet honours Gwilym Lewis, a legume specialist at Kew, who collected the type.

28. DENDROCHILUM STACHYODES

Dendrochilum stachyodes (Ridl.) J.J. Sm. in *Recl. Trav. bot. néerl.* 1: 77 (1904). Type: Malaysia, Sabah, Mount Kinabalu, 3400 m, *Haviland* 1097 (holotype BM!, isotypes K!, SAR!, SING!). **Fig. 61, plates 15E & F; 16A & B.**

Platyclinis stachyodes Ridl. in *Trans. Linn. Soc. Lond.* 2, 4: 234 (1894).
Acoridium stachyodes (Ridl.) Rolfe in *Orchid Rev.* 12: 220 (1904).

DESCRIPTION. Lithophyte or terrestrial. *Rhizome* abbreviated or up to 6 cm long, branching. *Roots* forming dense clumps, branching, flexuous, smooth, 1–2 mm in diameter.

135

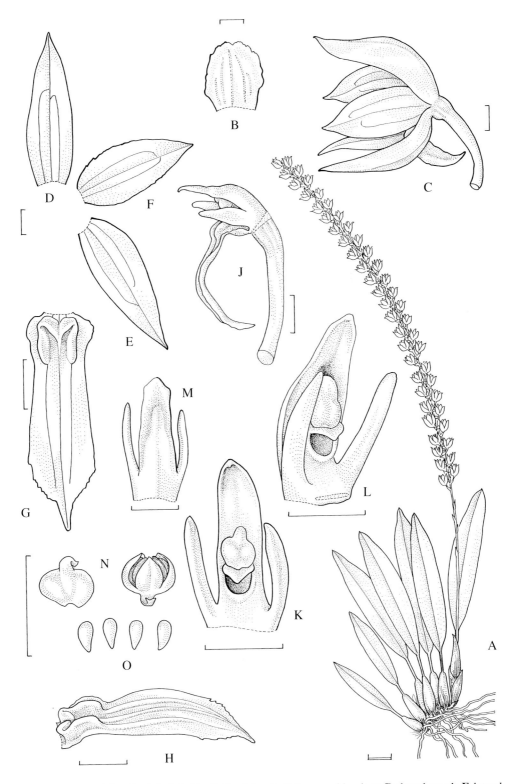

Fig. 60. *Dendrochilum lewisii*. **A** habit; **B** floral bract; **C** flower, side view; **D** dorsal sepal; **E** lateral sepal; **F** petal; **G** labellum, front view; **H** labellum, side view; **J** pedicel with ovary, labellum and column, side view; **K** column, front view; **L** column, oblique view; **M** column, back view; **N** anther-cap, front and back views; **O** pollinia. **A–O** from *Lewis* 366 (holotype). Scale: single bar = 1 mm; double bar = 1 cm. Drawn by Susanna Stuart-Smith.

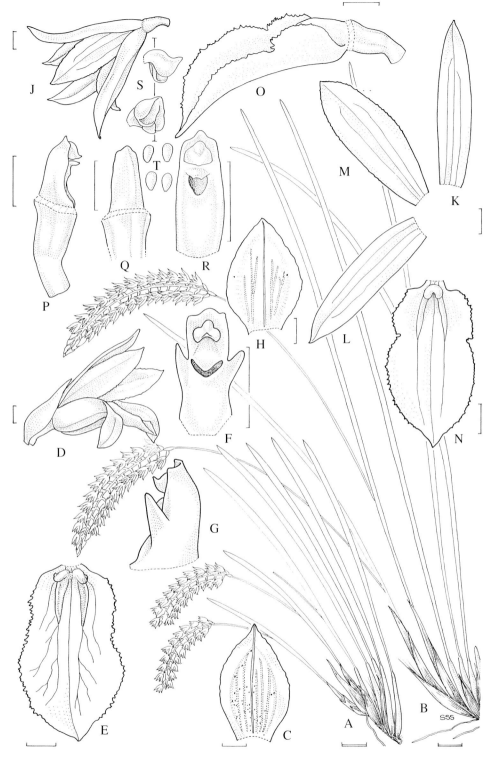

Fig. 61. *Dendrochilum stachyodes*. **A & B** habits; **C** floral bract; **D** flower with floral bract, side view; **E** labellum, front view; **F** column, front view; **G** column, side view; **H** floral bract; **J** flower, side view; **K** dorsal sepal; **L** lateral sepal; **M** petal; **N** labellum, front view; **O** pedicel with ovary, labellum and column, side view; **P** pedicel with ovary and column, side view; **Q** upper portion of ovary and column, back view; **R** column, front view; **S** anther-cap, interior and side views; **T** pollinia. **A** from *Gardner* 48, **B–G** from *J. & M.S. Clemens* 33177, **H–T** from *Wood* 605. Scale: single bar = 1 mm; double bar = 1 cm. Drawn by Susanna Stuart-Smith.

Cataphylls 3, ovate-elliptic, acute, slowly becoming fibrous, 1.4–5 cm long, speckled. *Pseudobulbs* densely crowded, cylindrical to conical or fusiform, sometimes broadly elliptic, often curved, smooth, (0.8–)1.2–5 × 0.3–0.8 cm, pale pinkish or brownish to orange salmon. *Leaf*: petiole sulcate, 0.8–2.5 cm long; blade linear-ligulate to narrowly elliptic, shortly apiculate or acute, falcate, often shallowly conduplicate, coriaceous, 6–12(–12) × 0.4–0.9 cm, main nerves 4–5, striolate. *Inflorescences* curving, densely many-flowered, superficially resembling an ear of wheat, flowers borne (1–)2–3 mm apart; peduncle slender, wiry, (4.5–)5–10(–16) cm long; non-floriferous bract solitary, ovate-elliptic, acute, 3–4 mm long; rachis quadrangular, 2.5–6 cm long, salmon-pink; floral bracts ovate to ovate-elliptic, acute, 3–4.6 × 3 mm, salmon-pink. *Flowers* reported as having a faint, sweet, spicy scent, not opening widely, pedicel with ovary pink, sepals, petals and lip creamy white, fading to lemon, column salmon-pink. *Pedicel with ovary* narrowly clavate, straight to slightly geniculate, 1.7–1.8 mm long. *Sepals* and *petals* 3-nerved, sepals often somewhat reflexed distally. *Dorsal sepal* linear-ligulate, acute, 6 × 1.5–1.6 mm. *Lateral sepals* linear-ligulate or narrowly oblong, acute, 6–6.1 × 1.5–1.6 mm. *Petals* narrowly elliptic, acute, margin slightly erose, 5.9–6 × 1.9–2 mm. *Labellum* obscurely 3-lobed or entire, elliptic, acute, margin erose-denticulate, 5 × 2.9–3 mm, side lobes (when present) rounded, subovate, erect; disc 3-nerved, with at the base a fleshy bilobed, often semi-circular callus, the outer nerves elevated and sometimes fleshy and extending as such to junction of side and mid-lobes. *Column* oblong, 1–1.1 mm long; apical hood truncate to obscurely 3-lobed; stelidia normally absent, rarely present, triangular, obtuse, *c.* 0.8 mm long; rostellum semi-ovate, fleshy, truncate or subacute; anther-cap cucullate. *Capsule* ovoid, 5 × 4–5 mm.

DISTRIBUTION. Borneo.

SABAH. Mount Kinabalu: near helipad, *Barkman* 8 & 9 (K!, Sabah Parks Herbarium, Kinabalu Park); Summit Trail, 16 Feb. 1962, *Bogle & Bogle* 532 (AMES!); Sayat Sayat to Panar Laban, 1976, *Buxton* s.n., cult. RBG Edinburgh (E!); Paka-paka Cave, June 1933, *Carr* 3521, SFN 27531 (K!, SING!); Summit Trail, Jan. 1987, *Chan & Gunsalam* 53/87 (K!, SING!); Paka-paka Cave, 15 Nov. 1915, *J. Clemens* 115 (AMES!, BM!, BO!, E!, K!, SING!); Lubang, *J. Clemens* 224 (AMES!); Paka-paka Cave, 14 Nov. 1931, *J. & M.S. Clemens* 27141 (BM!, E!, HBG!, K!, SING!); Kemburongoh, 8 Jan. 1932, *J. & M.S. Clemens* 27870 (BM!, BO!, E!, HBG!, K!, SING!) & 24 Mar. 1932, *J. & M.S. Clemens* 29120 (BM!, E!); upper Kinabalu, 14 Nov. 1931, *J. & M.S. Clemens* 30142 (AMES!, E!, HBG!, K!); Kilembun Basin, 1933, *J. & M.S. Clemens* 33177 (AMES!, BM!, E!, HBG!, K!); Gurulau Spur, 1 Dec. 1933, *J. & M.S. Clemens* 50653 (BM!, E!, K!) & 10 Dec. 1933, *J. & M.S. Clemens* 51012 (BM!, K!); Above Panar Laban, Sept. 1977, *Gardner* 104, cult. R.B.G. Edinburgh (E!, L!); summit area, Feb. 1910, *Gibbs* 4181 (BM!, K!); near Panar Laban, 26 Oct. 1983, *SNP* 1603 & 15 Feb. 1989, *SNP* 2705 (Sabah Parks Herbarium, Kinabalu Park); without precise locality, July-Aug. 1916, *Haslam* s.n. (BM!, E!, K!, SING!); without precise locality, *Haviland* 1097 (holotype BM!, isotypes K!, SAR!, SING!) Paka-paka Cave, 15 Nov. 1931, *Holttum* SFN 36570 (AMES!, SING!); locality unknown, July 1960, *Meijer* SAN 22067 (SAN!); above Paka-paka Cave, 14 Feb. 1962, *Meijer* SAN 28560 (K!, SAN!); Layang-Layang, 25 Sept. 1966, *Sidek bin Kiah* S. 38 (L!, SING!); summit area, 24 Sept. 1966, *Togashi* s.n. (TI!) & 25 Sept. 1966, *Togashi* s.n. (TI!); without precise locality, Mar. 1888, *Whitehead* s.n. (BM!); below Gunting Lagadan Hut, 17 Oct. 1985, *Wood* 605 (K!).

HABITAT. Growing almost exclusively in damp fissures of granite rocks, often in exposed places, sometimes among upper montane scrub, often associated with *Coelogyne papillosa* Ridl., *Eria grandis* Ridl., *Gahnia* spp., *Juncus* spp., *Trachymene saniculifolia* Stapf and mosses. (2400–)3200–3700 m.

This elegant species forms extensive colonies along rock crevices, usually in the open but sometimes under *Leptospermum recurvum* Hook.f., and is obviously able to withstand adverse climatic conditions. The mass flowering of *D. stachyodes*, which peaks in December and January, is one of the delights of the bleak granite slopes of the summit area of Mount Kinabalu where it is endemic. Such displays have, however, suffered in recent years from periodic droughts resulting from the El Niño/Southern Oscillation phenomenon, that of 1983 being particularly severe.

Clemens 33177 differs from other collections in having longer leaves and small stelidia on the column. Ridley (1894) described the inflorescences as "reminding one of an ear of wheat, whence the specific name".

29. DENDROCHILUM ACUIFERUM

Dendrochilum acuiferum Carr in *Gdns. Bull. Straits Settl.* 8: 227–228 (1935). Type: Malaysia, Sabah, Mount Kinabalu, near Paka-paka, 3000 m, 17 June 1933, *Carr* C. 3550 (SFN 27645) (holotype SING!, isotypes AMES, K!, L!). **Fig. 62, plate 1A & B.**

DESCRIPTION. Terrestrial. *Rhizome* creeping, branching, to 8 cm long, 3–4 mm in diameter. *Roots* much branched, flexuous, smooth, 0.8–1 mm in diameter. *Cataphylls* 3, ovate-elliptic, acute, becoming fibrous, 1–4.5 cm long, with a few speckles. *Pseudobulbs* densely crowded, up to 2 mm apart, cylindrical to narrowly ovoid, minutely rugulose, subapproximate, borne at an acute angle to the rhizome, 1.6–4.2 × 0.3–0.5 cm, orange. *Leaf*: petiole sulcate, 1.2–3 cm long; blade narrowly oblong-elliptic or oblong-oblanceolate, constricted below the acute apex, (6.5–)10.5–11 × 0.8–1.3 cm, main nerves 5, dark green. *Inflorescences* curving and nodding, laxly 10- to 18-flowered, flowers borne 3–5 mm apart, uppermost opening first; peduncle slender, 8.5–14 cm long; non-floriforous bract solitary, ovate-elliptic, acuminate, 6–6.5 mm long; rachis quadrangular, 5–8 cm long; floral bracts broadly elliptic or broadly ovate, subacute to acute, carinate, 3.7–5 × 3–3.4 mm. *Flowers* 6–8 mm across, faintly sweet scented, sepals and petals cream, pale yellow or pale salmon, labellum cream or pale yellow suffused very pale salmon-pink towards apex, keels white, and with a deep salmon-pink or peach-coloured basal spot, nerves peach-coloured, column deep salmon-pink, anther-cap pale salmon-pink. *Pedicel with ovary* narrowly clavate, 3.8 mm long. *Sepals* and *petals* 3-nerved. *Dorsal sepal* narrowly elliptic, acute, 7–8 × 2 mm. *Lateral sepals* obliquely narrowly-elliptic, subfalcate, acute, 6.5–7 × 2–2.3 mm. *Petals* narrowly elliptic, acute, margins minutely erose in upper half, 5.7–6.5 × 1.9–2 mm. *Labellum* shortly clawed, 3-lobed, cuneate below, (4.5–)4.7–5.1 mm long, 3.5–3.8 mm wide across mid-lobe, 2.5–3 mm wide across side lobes; side lobes very short, rounded or subtruncate, margins minutely erose; mid-lobe ovate or transversely elliptic, shortly acuminate or acute, margins minutely erose; disc with 2 thin, somewhat semi-ovate keels joined at the base by a fleshy, obscurely bilobed swelling, keels terminating at base of mid-lobe, 3-nerved, median nerve prominent and elevate at base, side nerves branching on mid-lobe. *Column* gently curving, 1.7–2.3 mm long, 0.5 mm wide above; apical hood broadly rounded or subtruncate; stelidia basal, slender, linear, acute, reaching to rostellum, 1–1.1 long; rostellum prominent, acute; stigmatic cavity ovate, anterior margin elevate; anther-cap cucullate.

DISTRIBUTION. Borneo.

139

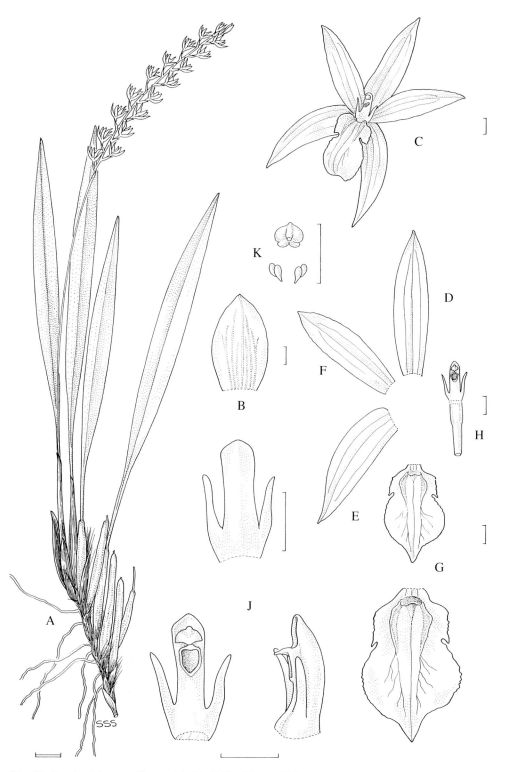

Fig. 62. *Dendrochilum acuiferum.* **A** habit; **B** floral bract; **C** flower, front view; **D** dorsal sepal; **E** lateral sepal; **F** petal; **G** labellum, front views; **H** pedicel with ovary and column, front view; **J** column, front, back and side views; **K** anther-cap and pollinia. **A–K** from *Carr* C. 3550, SFN 27645 (holotype). Scale: single bar = 1 mm; double bar = 1 cm. Drawn by Susanna Stuart-Smith.

SABAH. Mount Kinabalu, *Barkman* 193, 331, 332, 333, 334 & 335 (Sabah Parks Herbarium, Kinabalu Park). Mount Kinabalu, near Paka-paka, 17 June 1933, *Carr* C. 3550 (SFN 27645) (holotype SING!, isotypes AMES, K!, L).

HABITAT. Dwarf scrub composed of *Leptospermum recurvum* Hook.f. and open rocky places, associated with *D. stachyodes* or *D. grandiflorum*, but much less common than either. Mostly restricted to ultramafic substrate. 2950–3000 m.

Possible hybrid origin of *D. acuiferum*

Barkman (PhD dissertation, 1998) has undertaken tests based on nuclear ribosomal ITS 1 & 2 and *acc* D – *psa* I chloroplast DNA spacer sequence variation which seems to provide evidence of a hybrid origin for *D. acuiferum*. Short hybrid lineage branches were found in the molecular phylogeny suggesting that *D. acuiferum* was recently formed and/or has undergone little change since its origin. The putative maternal parent is thought to be *D. grandiflorum* (section *Cruciformia*) and the pollen source *D. stachyodes* (section *Eurybrachium*).

Populations of *D. acuiferum* are found only at sites where both putative parents also occur. *D. grandiflorum* and *D. stachyodes* usually grow on different substrates, the former most commonly on ultramafics and the latter almost exclusively on granite and, consequently, are rarely sympatric. However, several small granite outcrops occur among the ultramafics at around 3000 m along the summit trail on Mount Kinabalu where dense populations of *D. stachyodes* can be found. *D. acuiferum* grows between 2950 and 3000 m near these granite outcrops and the surrounding ultramafic substrates where *D. grandiflorum* is common.

It is unclear whether *D. acuiferum* is fertile, since all of the species of *Dendrochilum* propagate vegetatively and hybrid individuals could survive for several generations in the sterile state. However, capsules have been observed once and more than one hundred individuals in various stages of growth occur near the areas where both putative parents are sympatric.

Barkman concedes that hybridisation between *D. grandiflorum* and *D. stachyodes* may be limited by two factors. First, the flowering periods are not completely concurrent, *D. grandiflorum* occurring abundantly throughout the year, while *D. stachyodes* is more seasonal, peaking around December and January. Secondly, these species show a quite different floral morphology and fragrance, suggesting attraction to different pollinators. Earwigs have been observed in the flowers of all three species, but it is uncertain whether these effect pollination.

Despite the obvious breeding barriers mentioned above, molecular evidence for the hybrid origin of *D. acuiferum* appears strong. Barkman found that sequenced clones of the ITS 1 and ITS 2 from *D. acuiferum* yielded repeats that were identical to *D. stachyodes*, while repeats nearly identical to both *D. grandiflorum* and *D. stachyodes*, however, had several 'novel' substitutions, and repeats that were recombinant between both putative parents. It is uncertain whether this heterogenous assemblage of nr DNA ITS repeat types reflects an F1 hybrid or later generation status in *D. acuiferum*. Barkman favours the latter scenario. The biphyletic origin of ITS repeats and an identical chloroplast sequence shared with *D. grandiflorum* strongly suggests a hybrid origin for *D. acuiferum*. However, as Barkman points out, only after further study of additional nuclear markers can we be confident about the hybrid status of *D. acuiferum*.

In spite of its putative hybrid origin, Barkman believes "that *D. acuiferum* should be recognised as a species because it appears to reproduce sexually and is somewhat divergent from its parents at the molecular level."

The Latin specific epithet means to bear an *acumen*, i.e. a tapering point, and refers to the shortly acuminate labellum mid-lobe.

30. DENDROCHILUM TRUSMADIENSE

Dendrochilum trusmadiense J.J. Wood in *Lindleyana* 5(2): 93, fig. 8 (1990). Type: Malaysia, Sabah, Tambunan District, Mount Trus Madi, 2000 m, 15 June 1988, *Wood* 886 (holotype K!, isotypes L!, UKMS!). **Fig. 63, plate 17F.**

DESCRIPTION. Epiphyte. *Rhizome* creeping, up to 10 cm long, 2–3 mm in diameter. *Roots* numerous, much branched, smooth to very minutely papillose, 0.5–0.8 mm in diameter. *Cataphylls* 3–4, ovate-elliptic, obtuse to acute, 2–6 cm long, pale brown, speckled darker brown, becoming fibrous. *Pseudobulbs* crowded, fusiform or cylindrical, rugose and yellowish when dry, 1.5–3 × 0.6–0.8 cm. *Leaf*: petiole sulcate, 1–3(–4) cm long; blade oblong-elliptic to narrowly elliptic, obtuse, attenuate below, (6–)9–20 × 0.8–2 cm, thin-textured, main nerves 5–7. *Inflorescences* erect to curving, subdensely many flowered, flowers borne 1.5–2 mm apart, opening from top of inflorescence; peduncle slender, wiry, 10–20 cm long; non-floriferous bracts solitary, or rarely 2, ovate-elliptic, acute, 3–4 mm long; rachis quadrangular, 7–13 cm long; floral bracts ovate, acute, 3 × 2–2.5 mm, pale salmon-brown. *Flowers* sweetly scented, pedicel with ovary tan, sepals and petals translucent pale lemon-yellow, labellum chocolate-brown, column whitish. *Pedicel with ovary* slender to narrowly clavate, curved, 2.5–3 mm long. *Sepals* and *petals* spreading, 3-nerved. *Dorsal sepal* oblong-elliptic, acute, 5.5 × 2 mm. *Lateral sepals* ovate-elliptic, acute, slightly carinate at apex, 6 × 2.5 mm. *Petals* elliptic, mucronate, 5 × 2.8–3 mm. *Labellum* obscurely and shallowly 3-lobed, ovate, obtuse to subacute, somewhat concave, 2 × 2 mm, 3-nerved; side lobes rounded, obscurely crenulate; margin of mid-lobe incurved; disc bicallose, calli thick and fleshy at base, extending as raised keels confluent with outer nerves to halfway along labellum. *Column* oblong, 0.8 mm long; apical hood rounded, entire; stelidia basal, oblong, obtuse, somewhat falcate, 0.8 mm long; rostellum broadly triangular, acute; anther-cap ovate, cucullate.

DISTRIBUTION. Borneo.

SABAH. Tambunan District, Mount Trus Madi: 3 Oct. 1995, *Barkman et al.* 154 (K!, Sabah Parks Herbarium, Kinabalu Park!); 28 June 1986, *Suhaili, Kamarudin & Jumaat* TM17 (UKMS!) & 28 June 1986, *Suhaili, Kamarudin & Jumaat* TM18 (UKMS!); 15 June 1988, *Wood* 886 (holotype K!, isotypes L!, UKMS!).

HABITAT. Twig epiphyte in upper montane mossy ericaceous forest, often on ridges. 1900–2300 m.

D. trusmadiense is one of the most attractive Bornean species to be described in recent years. Superficially recalling *D. kamborangense* (section *Platyclinis*) from Mount Kinabalu, it is easily distinguished by the small, subentire, ovate labellum and column lacking a foot. It is so far only recorded from the upper slopes of Mount Trus Madi (2642 m), the third highest mountain in Malaysia.

Section **Mammosa**

Section **Mammosa** *J.J. Wood & H. Ae. Peders.* in *Opera Bot.* 130: 41 (1997). Type species: *Dendrochilum saccatum* J.J. Wood.

Labellum firmly attached to the column, making a right angle to the latter; distinctly 3-lobed (not cruciform), neither replicate nor coiled-up, but divided into a distinctly saccate or somewhat saccate hypochile and a flat epichile by two free, prominent calli. *Column* shorter than the dorsal sepal, relatively slender or somewhat stout, usually strongly incurved, not tapering from the base upwards; foot short or rudimentary; stelidia present; rostellum prominent.

DISTRIBUTION. Peninsular Malaysia, Borneo and the Philippines.

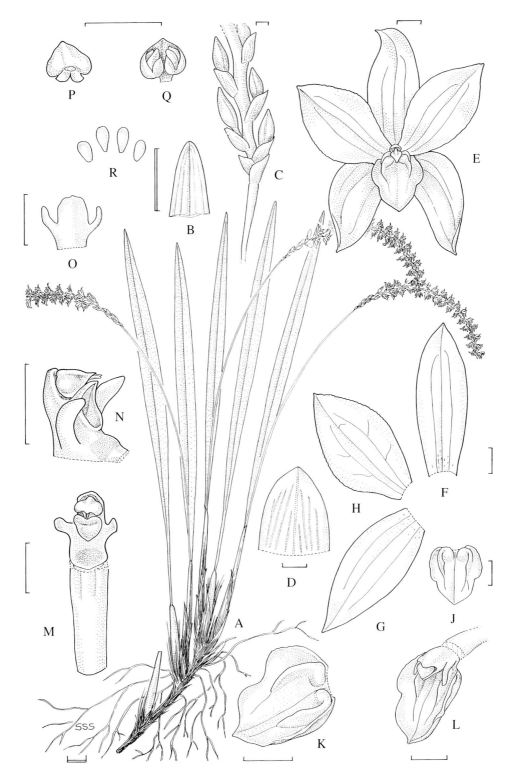

Fig. 63. *Dendrochilum trusmadiense.* **A** habit; **B** leaf apex, adaxial surface; **C** junction of rachis with peduncle showing solitary non-floriferous bract, six floral bracts and five buds; **D** floral bract; **E** flower, front view; **F** dorsal sepal; **G** lateral sepal; **H** petal; **J** labellum, front view; **K** labellum, oblique view; **L** ovary, labellum and column, oblique view; **M** pedicel with ovary and column, front view; **N** column, side view; **O** column, back view; **P** anther-cap, back view; **Q** anther-cap, interior view showing pollinia; **R** pollinia. **A–R** from *Wood* 886 (holotype). Scale: single bar = 1 mm; double bar = 1 cm. Drawn by Susanna Stuart-Smith.

KEY TO THE SPECIES OF SUBGENUS *PLATYCLINIS* SECTION *MAMMOSA* IN BORNEO

1. Hypochile of labellum strongly saccate. Pedicel with ovary elongate, 5–6 mm long. Epichile of labellum strongly recurved, broadly lanceolate, narrowly acute ....
    .... 31. *saccatum* J.J. Wood
    Hypochile of labellum not saccate, or only subsaccate. Pedicel with ovary 2–4.5 mm long. Epichile of labellum not strongly recurved, although often somewhat deflexed, either narrowly oblong, oblong-ovate, broadly ovate to sub-elliptic, or triangular-ovate, apex obtuse, subtruncate and shortly apiculate-cuspidate, or acuminate 2

2. Flowers *c.* 9–10.1 mm across, variable in colour, sepals and petals pale apple-green, lemon-yellow or greenish orange to pale brown, pinkish brown or ochre. Sepals (4.8–)5.5–8 mm long. Labellum 3.9–5.5 mm long, epichile obtuse to acuminate ....
    .... 32. *kingii* (Hook.f.) J.J. Sm.
    Flowers 5–6 mm across, sepals and petals pale dull red or dull reddish brown. Sepals 3.1–4.8 mm long. Labellum 2.1–3.5 mm long, epichile subobtuse to acute ....
    .... 33. *rufum* (Rolfe) J.J. Sm.

31. DENDROCHILUM SACCATUM

Dendrochilum saccatum J.J. Wood in *Opera Bot.* 130: 41 (1997). Type: Malaysia, Sarawak, Marudi District, Tama Abu Range (Mount Temabok), Upper Baram (Barami, sphalm.) Valley, 900 m, 8 Nov. 1920, *Moulton* SFN 6762 (holotype AMES!, isotype SING!). **Fig. 64.**

Pholidota gracilis L.O. Williams in *Bot. Mus. Leafl. Harv. Univ.* 6: 59–60 (1938).

DESCRIPTION. Epiphyte. *Rhizome* not collected. *Roots* branching, flexuous, smooth, *c.* 0.8 mm in diameter. *Cataphylls* 3, ovate-elliptic, obtuse to acute, 2–7 cm long; speckled. *Pseudobulbs* crowded, cylindrical to oblong-ovoid, 1–2 × 0.4–0.5 cm. *Leaf*: petiole sulcate, 1.7–4.5 cm long; blade linear, ligulate, acute or acuminate, somewhat constricted just below apex, thin-textured, 17–25(–30) × 0.8–1 cm, main nerves 4–5. *Inflorescences* laxly 10- to 18-flowered, flowers borne 2.5–3.5 mm apart, distichous; peduncle slender, 18–25 cm long; non-floriferous bracts 2, ovate-elliptic, acute, 4–5 mm long; rachis quadrangular, 4–6 cm long; floral bracts oblong-ovate or oblong-elliptic, obtuse and mucronate, 3.5–5 × 3.5 mm. *Flower* colour not noted. *Pedicel with ovary* slender, 5–6 mm long. *Sepals* and *petals* 3-nerved. *Dorsal sepal* broadly elliptic, acute to shortly acuminate, concave, 4 × 2 mm. *Lateral sepals* ovate-elliptic, slightly oblique, acuminate, 5 × 2–2.5 mm. *Petals* oblong-elliptic or lanceolate-rhomboid, acute to shortly acuminate, concave, 4 × 1.5 mm. *Labellum* 3-nerved; hypochile saccate, side lobes erect, oblong, truncate, 2 × 1.8–2 mm; epichile strongly recurved, broadly lanceolate, narrowly acute, 2.5 × 1–1.3 mm; disc with 2 prominent, free, mammose, broadly transverse, obtuse calli between hypochile and epichile. *Column* gently incurving, 2 mm long; foot rudimentary; apical hood ovate, acute, entire, margin slightly recurved; stelidia median, oblong-triangular, irregularly bifid; rostellum prominent, linear-lanceolate, *c.* 0.7 mm long; stigmatic cavity narrowly oblong; anther-cap cucullate, cordate, acute.

DISTRIBUTION. Borneo.

SARAWAK. Marudi District, Tama Abu Range (Mount Temabok), Upper Baram Valley, 8 Nov. 1920, *Moulton* SFN 6762 (holotype of *Pholidota gracilis* AMES!, isotype SING!).

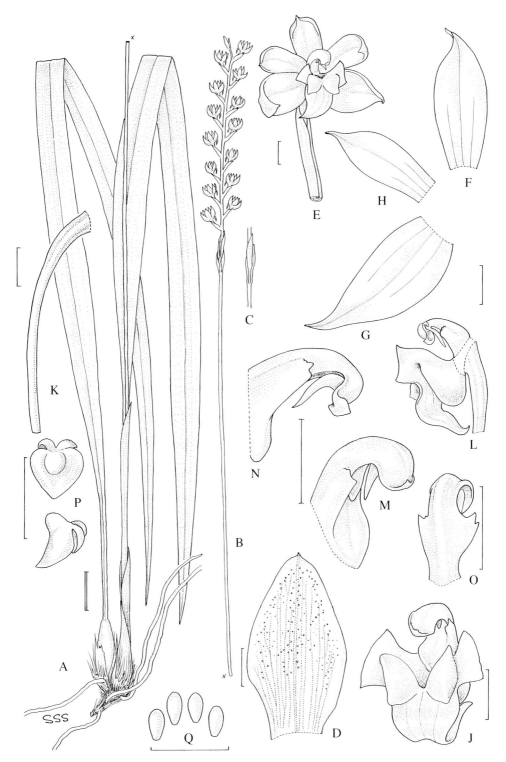

Fig. 64. *Dendrochilum saccatum.* **A** habit; **B** inflorescence; **C** junction of rachis with peduncle showing two non-floriferous bracts; **D** floral bract; **E** flower with floral bract, front view; **F** dorsal sepal; **G** lateral sepal; **H** petal; **J** labellum and column, oblique view; **K** pedicel with ovary, side view; **L** ovary, labellum and column, side view; **M** column, anther-cap removed, oblique view; **N** column, anther-cap removed, side view; **O** column, anther-cap removed, back view; **P** anther-cap, back and side views; **Q** pollinia. **A–Q** from *Moulton* 6762 (holotype of *Pholidota gracilis)*. Scale: single bar = 1 mm; double bar = 1 cm. Drawn by Susanna Stuart-Smith.

HABITAT. Not recorded. 900 m.

De Vogel (1988), in his monograph of *Pholidota,* listed *P. gracilis* among several dubious taxa with the comment that it may belong to *Dendrochilum* subgenus *Aphanostelidion* (now included in subgenus *Platyclinis* section *Platyclinis*). *D. saccatum* (and to a lesser extent *D. rufum*) occupies a somewhat isolated position in the genus on account of the saccate hypochile similar to that of *Pholidota*. However, the presence of stelidia similar to those of *D. kingii* and *D. rufum* on the column confirms its inclusion within *Dendrochilum*. The column appears swollen at the base, resembling a rudimentary foot. Fresh material of this species is required before an accurate description of the labellum and column morphology is possible.

The Latin specific epithet refers to the pouched, bag-shaped labellum hypochile.

32. DENDROCHILUM KINGII

Dendrochilum kingii (Hook.f.) J.J. Sm. in *Recl. Trav. bot. néerl.* 1: 76 (1904). Types: Peninsular Malaysia, Perak, received Aug. 1888, *Scortechini* 608a (lectotype K, herbarium material and sketch!); Perak, April 1885, *King's collector* 7455(syntype K!).

DESCRIPTION. Clump forming epiphyte or terrestrial. *Rhizome* branching, 1–10 cm long, covered in fibrous remains of sheaths. *Roots* wiry, smooth, 0.5–1 mm in diameter. *Cataphylls* 3–4, acute, minutely furfuraceous-punctate, 2–7.5 cm long. *Pseudobulbs* oblong to ovoid, often ampulliform, smooth to wrinkled, aggregated or up to 1.5 cm apart on rhizome, 1.8–4 × 0.5–1.2 cm. *Leaf:* petiole sulcate, (0.3–)1.5–8 cm long; blade linear-lanceolate to narrowly elliptic, ligulate, sometimes distally constricted *c.* 3 cm below apex, obtuse to acute, thin to rather thick and leathery-textured, shiny, (8–)12–40 × 1–2.8(–3.5) cm, main nerves 5–6. *Inflorescences* laxly *c.* 6- to 3-flowered, flowers borne 3–7 mm apart; peduncle terete, erect to curving, glabrous or sparsely minutely furfuraceous-punctate, 12–33 cm long; non-floriferous bracts 1–3, ovate-elliptic, acute, 5–7 mm long; rachis somewhat fractiflex, quadrangular, sulcate, decurved, minutely furfuraceous-punctate to almost glabrous, (2.5–)9–11(–13) cm long; floral bracts linear-lanceolate, ovate-elliptic or ovate, acute to shortly acuminate, convolute, 5–8 × 4–5 mm. *Flowers* 9–10.1 mm across, sweetly scented or unscented, sepals green, greenish orange, pale brown, pinkish brown, ochre-brown, translucent lemon-yellow, flushed green or pale apple-green and pale pink at base, petals straw-coloured or any combination of the above, labellum orange-brown, yellow, salmon-pink, dark pinkish brown or greenish ochre, the calli olive-green, sometimes with a green median line on the mid-lobe, column yellowish pink or orange-brown, stelidia often cream or white. *Pedicel with ovary* clavate, 2–4.5 mm long. *Sepals* and *petals* spreading, 3-nerved. *Dorsal sepal* oblong-elliptic, ovate-elliptic to narrowly elliptic, acute to acuminate, (4.8–)5.5–8 × 2–3 mm. *Lateral sepals* oblong-elliptic or ovate-elliptic, slightly oblique, acuminate to narrowly acuminate, (5–)6–8 × 2–3.1 mm. *Petals* obliquely ovate-elliptic, oblong-elliptic or narrowly elliptic, shortly narrowed at base, subacute or acute, less often acuminate, margin often minutely erose, 4–6.6 × (1.7–)2–3 mm. *Labellum* (3.9–)4–5.5 mm long, 2.6–5 mm wide across side lobes when flattened; hypochile lobes erect, auriculate, subquadrate or obliquely rounded, obtuse, crenulate to erose, 0.8–1.8 × 1–2 mm wide at base; epichile somewhat deflexed, oblong-ovate, broadly ovate to sub-elliptic, obtuse and shortly cuspidate or acute to acuminate, sometimes erose at base, (1–)2.6–4 × 2–3 mm; disc with 2 fleshy, rounded, depressed flap-like calli between side lobes, their forward ends covering and often touching, with a sulcate area between and often extending a short way on to the swollen base of the mid-lobe. *Column* decurved, 2–3.5 mm long; foot short, but distinct; apical hood large, ovate, suborbicular, entire or minutely unevenly denticulate *c.* 0.9–1.5 mm wide; stelidia broad, quadrate, wing-like, extending from base to below apical hood, regularly or irregularly and acutely bidentate or tridentate, the teeth shorter than or just reaching base of apical hood, sometimes less wing-like and reduced to 2 subulate narrowly triangular teeth; stigmatic cavity margin prominent and fleshy; rostellum large, narrowly to broadly rectangular to quadrate, sometimes triangular, incurved, 0.6–0.9 × 0.6–1.4 mm; anther-cap galeate, umbonate, 1.4–1.8 × 1.1–1.5 mm.

KEY TO THE VARIETIES OF *D. KINGII*

Epichile of labellum 2–3 mm wide .... var. *kingii*
Epichile of labellum 1–1.6 mm wide .... var. *tenuichilum* J.J. Wood

a. var. **kingii** . Figs. 65 & 66, plate 10B.

Platyclinis kingii Hook.f., *Hooker's Icon. Pl.* 21: pl. 2015 (1880); Fl. Brit. Ind. 5: 708 (1890).
P. sarawakensis Ridl. in *J. Linn. Soc., Botany* 31: 267 (1896). Type: Malaysia, Sarawak, *Biggs* s.n., cult. Hort. Bot. Penang (not located).
Dendrochilum sarawakense (Ridl.) J.J. Sm. in *Recl. Trav. bot. néerl.* 1: 66 (1904).
Acoridium kingii (Hook.f.) Rolfe in *Orchid Rev.* 12: 220 (1904).
A. sarawakense (Ridl.) Rolfe in *Orchid Rev.* 12: 220 (1904).
Dendrochilum palawanense Ames, *Orchidaceae* 2: 103–104, fig. s.n. (1908). Type: Philippines, Palawan, Mount Pulgar, 1250 m, *Foxworthy* in Bur. Sci. 553 (holotype AMES, isotype K!).
D. bicallosum J.J. Sm. in *Bull. Dép. Agric. Indes Néerl.* 22: 17–18 (1909), *nom. illeg.* (non Ames).
D. bigibbosum J.J. Sm. in *Bull. Dép. Agric. Indes Néerl.* 45: 13 (1911). Types: Indonesia, Kalimantan, Keribung (Keriboeng) River, *Hallier* 1312 (syntype BO!, isosyntypes K!, L!). Malaysia, Sarawak, Kuching District, Mount Bengoh (Mount Bengkaum), *Brooks* s.n. (syntype BO, isosyntypes K!, L!, SING!).

Distribution. Peninsular Malaysia, Borneo and the Philippines (Palawan).
Sarawak. Locality unknown, *Beccari* 2095 (FI!, K!). Locality unknown, *Biggs* s.n., cult. Hort. Bot. Penang (holotype of *Platyclinis sarawakensis*, not located). Kapit District, Ng. Meranu, upper Mengiong River, Balleh, 31 Jan. 1991, *Blicher et al.* S.59856 (L!, SAR). Kuching District, Mount Bengoh (Bengkaum), Oct. 1908, Brooks s.n. (syntype of D. bigibbosum BO, isosyntypes K!, L!, SING!). Hose Mountains, Semako Hill, 16 Aug. 1967, *Burtt* B. 4943 (E!). Belaga District, Iban River, Linau, 29 Oct. 1982, *Lee* S. 45322 (K!, KEP!, L!, MEL, SAR!, SING!) & 11 Nov. 1982, *Lee* S. 45514 (AAU, MEL, SAR!). Kapit District, Entulah/Mengiong Rivers, Batang Balleh, 15 July 1987, *Lee* S. 54569 (K!, L, SAR) & 20 July 1987, *Lee* S.54688 (L!, SAR). Bintulu District, Merurong Plateau, Sekiwa Hill, Tubau, 25 Aug. 1986, *Mohtar et al.* S. 53913 (K!, SAR, SING). Marudi District, Tama Abu Range (Mount Temabok), Upper Baram Valley, 8 Nov. 1920, *Moulton* SFN 6761 (K!, SING!). Belaga District, Kenaban River, Upper Pelieran, 21 Sept. 1955, *Pickles & Ahmad bin Topin* S. 2911 (L!, SING!). Kuching district, Mount Matang, July 1893, *Ridley* s.n. (SING!). Marudi

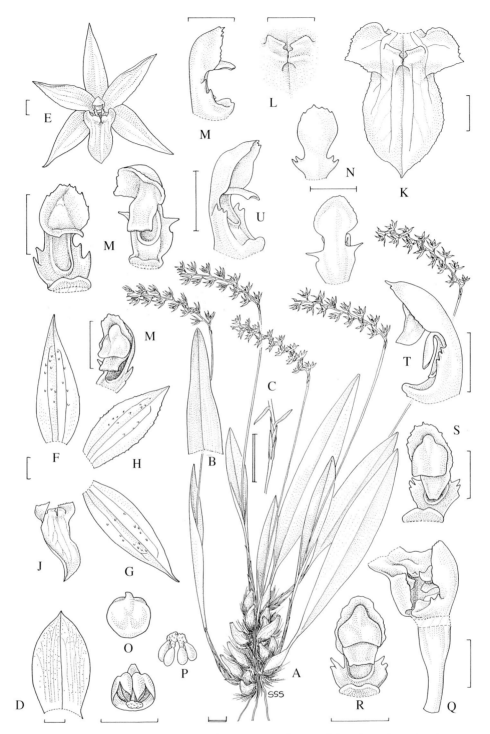

Fig. 65. *Dendrochilum kingii* var. *kingii*. **A** habit; **B** leaf apex; **C** junction of rachis with peduncle showing three non-floriferous bracts and two floral bracts; **D** floral bract; **E** flower, front view; **F** dorsal sepal; **G** lateral sepal; **H** petal; **J** labellum, oblique view; **K** labellum front view; **L** disc of labellum showing basal calli; **M** columns, oblique and side views; **N** columns, back view; **O** anther-cap, back and interior views; **P** pollinia; **Q** pedicel with ovary, basal portion of labellum, and column, side view; **R** column with anther-cap, front view; **S** column with anther-cap, oblique view; **T** column with anther-cap, side view; **U** column without anther-cap, oblique view. **A–C** from *Ridley* s.n., **D–P** from *Cribb* 89/65, **Q & R** from *Lamb* s.n., **S & T** from *Lamb* AL 359/85, **U** from *de Vogel* 919. Scale: single bar = 1 mm; double bar = 1 cm. Drawn by Susanna Stuart-Smith.

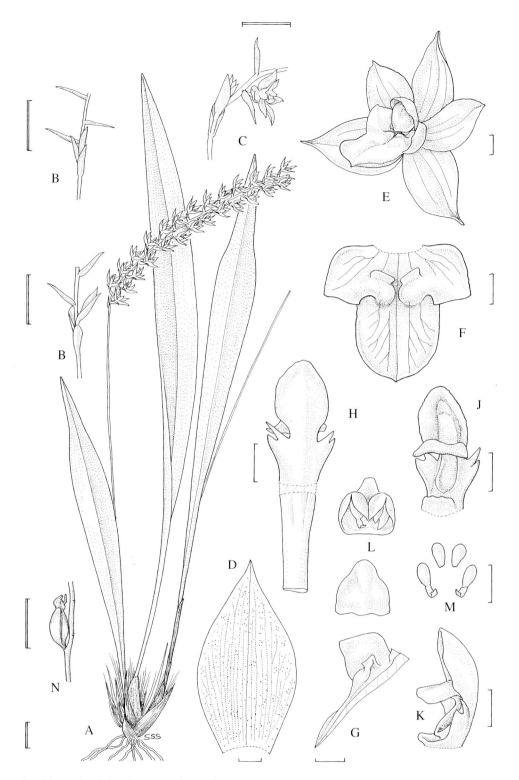

Fig. 66. *Dendrochilum kingii* var. *kingii*. **A** habit; **B** junctions of rachis with peduncle showing two and three non-floriferous bracts, and three floral bracts; **C** lower portion of inflorescence; **D** floral bract; **E** flower, oblique view; **F** labellum, front view; **G** labellum, longitudinal section; **H** pedicel with ovary and column, back view; **J** column, anther-cap removed, oblique view; **K** column, anther-cap removed, side view; **L** anther-cap, interior and back views; **M** pollinia; **N** fruit capsule. **A & B** from *Hallier* 1312 (syntype of *D. bigibbosum*), **C–M** from *de Vogel* 1584, **N** from *Ahmad Talip* SAN 70988. Scale: single bar = 1 mm; double bar = 1 cm. Drawn by Susanna Stuart-Smith.

District, Mount Dulit, 1932, *Synge* S. 397 (K!). Marudi District, Mount Dulit, near Long Kapa, Upper Tinjar, 10 Aug. 1932, *McLeod* in *Synge* S. 156 (K!). Kapit District, northern Hose Mountains, base of ridge leading to Batu Hill, 10 Dec. 1991, *de Vogel* 1584 (K!, L!) & *de Vogel* 1585 (L!). Kuching District, Woen Hill, Padawan, 2 Oct. 1987, *Yii* S. 61421 (AAU, K!, L, SAR!, SING).

SABAH. Sipitang District, Karamuak River, 11 Dec. 1985, *Amin et al.* SAN 111793 (SAN!). Sipitang District, Long Pa Sia, *Bacon* 80, cult. RBG Edinburgh (E!). Nabawan, 21 Nov. 1989, *Cribb* 89/65 (K!). Nabawan, *Cribb & Lamb* s.n., cult. RBG Kew (K!). Kalabakan District, Seranum, Mount Rara, 25 June 1983, *Fidilis & Sumbing* SAN 96124 (SAN!). Nabawan, 1977, *Lamb* SAN 88950 (K, sketch only!). Nabawan, 15 May 1985, *Lamb* AL 359/85 (K!), Dec. 1985, *Lamb* AL 502/85 (K!) & Nov. 1988, *Lamb* AL 1119/89 (K!). Lahad Datu District, Ulu Segama Forest Reserve, 25 May 1985, *Madani & Ismail* SAN 108904 (K!, SAN). Sandakan District, Kun-Kun River, 8 Dec. 1984, *Madani & Sigin* SAN 107709 (K!, L!, SAN!, SING!). Millian River, Nabawan, 26 Nov. 1986, *Mantor* SAN 118767 (K!, SAN!). Lahad Datu District, Ulu Segama, 19 July 1970, *Talip* SAN 70988 (L!, SAN). Nabawan, near old airstrip, Oct. 1986, cult. *Tenom Orchid Centre* TOC 863 (L!). Pun Batu, Sept. 1986, *Vermeulen* 530 (L!). Keningau to Sapulot road, 6 km from Nabawan, June 1986, *Vermeulen & Lamb* 337 (K!, L!).

KALIMANTAN BARAT. Locality unknown, comm. Jan. 1999, cult. *Bogor* 997.II.457 (BO, K!) & 3897 (BO, K!).

KALIMANTAN TIMUR. Sangkulirang, P.T. Sangkulirang, 42 km inland from Pengadan to the west, Nov. 1992, *de Vogel* 919 (L!).

KALIMANTAN (province unknown). Keribung (Keriboeng) River, *Hallier* 1312 (syntype of *D. bigibbosum* BO!, isosyntypes K!, L!).

HABITAT. Lowland and hill-dipterocarp forest; podsol forest on very wet sandy soil, with dipterocarps and *Dacrydium* spp., etc.; hill forest on limestone; recorded as growing on *Syzygium rejangense* Merrill & Perry on a stream bank; epiphytic on the trunks and boles of trees, more rarely in the crown; also observed as a terrestrial in podsol forest. 200–900 m.

A lowland species superficially resembling a large flowered *D. rufum* (Rolfe) J.J. Sm. Although the column structure is similar to *D. rufum*, the labellum is quite different, lacking the saccate hypochile and more prominent calli of that species. Populations from Borneo show much variability, particularly in leaf length and texture, inflorescence density, number of non-floriferous bracts, flower size, labellum mid-lobe shape and the degree of toothing of the stelidia.

Vernacular names include *Bunga Tupan* (Kayan dialect) and *Darchang* (Murut dialect).

The specific epithet honours Sir George King (1840–1909) who, with Robert Pantling, produced the authoritative two volume work entitled The Orchids of the Sikkim Himalaya.

b. var. **tenuichilum** J.J. Wood **var. nov.** a varietate typica lobo medio labelli angusto tantum 1–1.6 mm lato metienti. Typus: Malaysia, Sarawak, Sarikei Division, Julau District, Lanjak Entimau Protected Forest, along Upper Ensirieng River, 1993, *Leiden* cult. (*Schuiteman, Mulder & Vogel*) 932946 (holotypus L, isotypus K). **Fig. 67, plate 10A.**

Peduncle slender, filiform, up to 17 cm long; *rachis* 2.5–11.5 cm long. *Flowers* brownish orange, column lighter salmon-pink, pollinia pale yellow. *Pedicel with ovary* 2–2.9 mm long. *Sepals* ovate-elliptic, subulate-acuminate. *Dorsal sepal* 4.8–6 × 2–2.6 mm. *Lateral sepals*

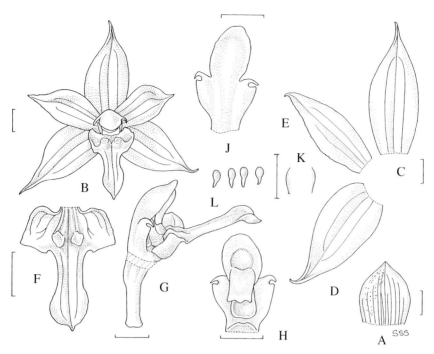

Fig. 67. *Dendrochilum kingii* var. *tenuichilum*. **A** floral bract; **B** flower, front view; **C** dorsal sepal; **D** lateral sepal; **E** petal; **F** labellum, front view; **G** pedicel with ovary, labellum and column, side view; **H** column, anther-cap removed, front view; **J** column, back view; **K** anther-cap, back view; **L** pollinia. **A–L** from *Leiden* cult. (*Schuiteman et al.*) 932946 (holotype). Scale: single bar = 1 mm. Drawn by Susanna Stuart-Smith.

5–6.1 × 2–2.6 mm. *Petals* oblong-elliptic, acute, 4–5.2 × 1.7–2.6 mm. *Labellum* 3.9–4 mm long; mid-lobe narrowly oblong-subspathulate, rounded to subtruncate, apiculate, 2–3 × 1–1.6 mm. *Column* 2.5–2.6 mm long; stelidia bifid, with one tooth elongate and falcate.

DISTRIBUTION. Borneo.

SARAWAK. Locality unknown, 1904, *Hose* s.n. (SING!). Lawas District, Puteh, Upper Lawas, near Kenaya River, 31 Oct. 1971, *Paie* S.31566 (A, BO!, K!, KEP!, L!, SAN, SAR, SING). Sarikei Division, Julau District, Lanjak Entimau Protected Forest, along Upper Ensirieng River, 1993, *Leiden* cult. (*Schuiteman, Mulder & Vogel*) 932777 (K!, L!) & 1993, *Leiden* cult. (*Schuiteman, Mulder & Vogel*) 932946 (holotype L!, isotype K!).

HABITAT. Primary mixed dipterocarp forest. 200 m.

The varietal epithet is derived from the Latin *tenuis*, slender, and the Greek *chilus*, lipped, in reference to the narrow labellum mid-lobe.

33. DENDROCHILUM RUFUM

Dendrochilum rufum (Rolfe) J.J. Sm. in *Rec. Trav. bot. néerl.* 1: 76–77 (1904). Type: 'Tropical Asia', sine coll., cult. Hort. Bot. *Glasnevin*, flowered February 1894 (holotype K!, isotype BM!). **Fig. 68, plate 15A.**

Platyclinis rufa Rolfe in *Kew Bull.* (1898): 192 (1898).
Acoridium rufum (Rolfe) Rolfe in *Orchid Rev.* 12: 220 (1904).

Epiphyte. *Rhizome* not collected. *Roots* branching, flexuous, smooth, *c.* 0.6–1 mm in diameter. *Cataphylls* not collected. *Pseudobulbs* crowded, ovoid-oblong, 1–1.2 × 0.3–0.4 cm. *Leaf*: petiole sulcate, 2–2.5 cm long; blade linear, ligulate, acute, slightly constricted below apex, 14.5–22 × 0.8–0.9 cm, thin-textured, main nerves 5. *Inflorescences* laxly 8- to 22-flowered, flowers borne 2–2.5 mm apart; peduncle curving to pendulous, filiform, 15.5–16 cm long; non-floriferous bracts 1–3, ovate to ovate-elliptic, obtuse to subacute, 2–4 mm long; rachis arcuate to decurved, quadrangular, (2–)3–5 cm long; floral bracts oblong-elliptic or oblong-ovate, apiculate, 3.5–5 mm long. *Flowers* 5–6 mm across, sepals and petals pale dull red or dull reddish brown, labellum dull brownish red with paler margins, disc green, keels dark reddish brown, column orange-red to brick-red, anther-cap pale yellow. *Pedicel with ovary* clavate, gently curved, 2–3 mm long. *Sepals* and *petals* 3-nerved. *Dorsal sepal* porrect to spreading, ovate-elliptic, acute to shortly acuminate, concave, 3.1–4.8 × 1.8–1.9 mm. *Lateral sepals* spreading, slightly obliquely ovate-elliptic, shortly acuminate, concave, 3.1–4.8 × 2 mm. *Petals* oblong-elliptic, acute to shortly acuminate, slightly minutely erose, slightly concave, 3.4–3.5 × 1.5–1.6 mm. *Labellum* 3-lobed, 3-nerved, 2.1–3.5 mm long; hypochile subsaccate, 1.2–1.3 mm long, lobes erect to spreading, semi-circular, auriculate, minutely erose to serrulate, *c.* 0.4 mm wide; epichile strongly recurved, broadly oblong-ovate, rounded, sometimes shortly apiculate, minutely erose to serrulate, 1–1.1 × 1.5–1.6 mm; disc with 2 free, fleshy boss-like calli at base of hypochile lobes, terminating at base of epichile, their surface uneven and slightly warty in places particularly along junction with hypochile lobes. *Column* straight to gently curving, 2–2.1 mm long; foot short; apical hood ovate, rounded, entire or with an irregular margin; stelidia median, borne opposite upper part of stigmatic cavity, small, triangular-acute, tooth-like, or bifid; stigmatic cavity narrowly oblong, lower margin elevate; rostellum prominent, narrowly triangular-ovate, subacute; anther-cap ovate, cucullate.

DISTRIBUTION. Borneo.

SARAWAK. Sarikei Division, Julau District, Lanjak Entimau Protected Forest, along Upper Ensirieng River, 1993, *Leiden* cult. (*Schuiteman, Mulder & Vogel*) 932959 (L!) & 933008 (L!). Sarikei Division, Julau District, Lanjak Entimau Protected Forest, between Mujok and Upper Ensirieng rivers, 1993, *Leiden* cult. (*Schuiteman, Mulder & Vogel*) 933028 (L!).

'TROPICAL ASIA'. Flowered Feb. 1894, sine coll., cult. Hort. Bot. *Glasnevin* (holotype K!, isotype BM!). Cult. Hort. Bot. *Glasnevin*, 16 Jan. 1895 (K!), Dec. 1895 (K!), 19 Feb. 1896 (K!) & Feb. 1896 (K!).

HABITAT. Primary mixed dipterocarp forest. 200– 300 m.

Until recently known only from material of unknown origin cultivated in Victorian times at Glasnevin Botanic Garden, Dublin, Ireland, *D. rufum* in many respects resembles a delicate, miniature dark-flowered version of *D. kingii*. It clearly has a closer affinity with the little known *D. saccatum* from which it can be distinguished primarily by the very short pedicel with ovary, less saccate hypochile and shorter epichile.

The specific epithet is derived from the Latin *rufus*, reddish, referring to the flower colour.

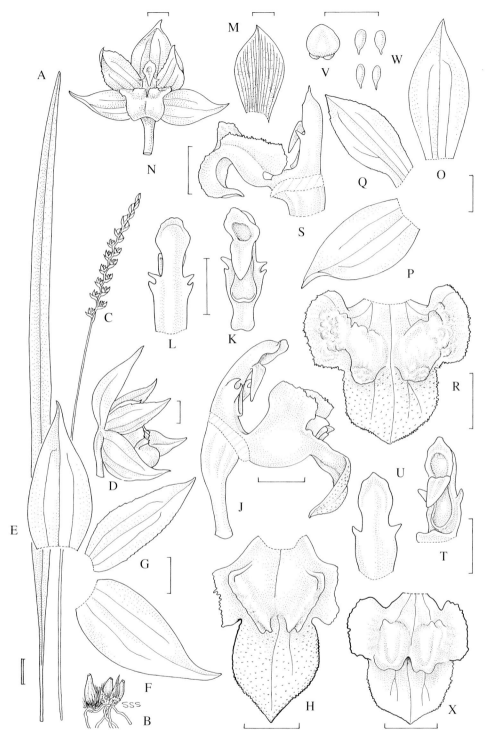

Fig. 68. *Dendrochilum rufum.* **A** leaf; **B** pseudobulbs; **C** inflorescence; **D** flower, side view; **E** dorsal sepal; **F** lateral sepal; **G** petal; **H** labellum, front view; **J** pedicel with ovary, labellum and column, anther-cap lost, side view; **K** column, anther-cap lost, front view; **L** column, back view; **M** floral bract; **N** flower, front view; **O** dorsal sepal; **P** lateral sepal; **Q** petal; **R** labellum, front view; **S** portion of ovary, labellum and column, side view; **T** column, anther-cap removed, oblique view; **U** column, back view; **V** anther-cap, back view; **W** pollinia; **X** labellum, front view. A–L from sine coll., cult. *Glasnevin*, flowered February 1894 (holotype), M–W from *Leiden* cult. (*Schuiteman et al.*) 933008, X from *Leiden* cult. (*Schuiteman et al.*) 933028. Scale: single bar = 1 mm; double bar = 1 cm. Drawn by Susanna Stuart-Smith.

Section **Platyclinis**

Dendrochilum subgen. *Aphanostelidion* Pfitzer in Engler, *Pflanzenreich* 4, 50, 2 B 7: 87, 110,
    1907. - Rosinski 1992: 105. - Type species not designated.
Dendrochilum subgen. *Platyclinis* ser. *Arachnites* Pfitzer in Engler, *Pflanzenreich* 4, 50, 2 B
    7: 91, 92, 1907. - Rosinski 1992: 93. - Type species: *Dendrochilum arachnites* Rchb. f.
Dendrochilum subgen. *Platyclinis* ser. *Similia* Pfitzer in Engler, *Pflanzenreich* 4, 50, 2 B 7: 91,
    92, 1907. - Rosinski 1992: 105. - Type species not designated.
Dendrochilum subgen. *Platyclinis* ser. *Edentula* Pfitzer in Engler, *Pflanzenreich* 4, 50, 2 B 7:
    91, 95, 1907, p.p. - Rosinski 1992: 93, 104 (p.p.). - Type species not designated.
Dendrochilum subgen. *Platyclinis* ser. *Dilatata* Pfitzer in Engler, *Pflanzenreich* 4, 50, 2 B 7:
    92, 100, 1907. - Rosinski 1992: 93, 104. - Type species not designated.
Dendrochilum subgen. *Platyclinis* ser. *Angustata* Pfitzer in Engler, *Pflanzenreich* 4, 50, 2 B 7:
    92, 105, 1907, p.p. - Rosinski 1992: 93, 104 (p.p.). - Type species not designated.

Labellum elastically attached to the column, making an obtuse to acute angle to the latter;
entire to distinctly 3-lobed (never cruciform), never replicate but rarely coiled-up, not divided
into a hypochile and epichile by two prominent calli. *Column* shorter than (very rarely
subequal to) the dorsal sepal, slender (stout in *D. glumaceum* and in the Philippine *D. niveum*),
more or less incurved (straight in the Philippine *D. erectilabium*), not tapering from the base
upwards; foot present, usually short, flat but sometimes apically incurved; stelidia usually
present.

DISTRIBUTION. Taiwan, Myanmar (Burma), Peninsular Malaysia, Singapore, Sumatra,
Java, Borneo, Lesser Sunda Islands, Sulawesi, Maluku, Philippines and New Guinea (east to
New Britain).

The primary centre of diversity of *Platyclinis*, the most widely distributed section in the
genus, is Borneo (50 species), although it is also well represented in Sumatra and the
Philippines. The section contains virtually all of the relatively few species seen in cultivation,
viz. D. cobbianum Rchb. f., *D. filiforme* Lindl., *D. glumaceum* Lindl., *D. gracile* (Hook. f) J.J.
Sm. and *D. longifolium* Rchb. f.

KEY TO THE SPECIES OF SUBGENUS *PLATYCLINIS* SECTION *PLATYCLINIS* IN BORNEO

1. Leaves, cataphylls, inflorescences and sometimes sepals covered with a black or
    brownish, finely setose indumentum. Labellum never 3-lobed, margin lacerate,
    serrulate to obscurely erose 47
    Leaves, cataphylls and inflorescences lacking a black or brownish, finely setose
    indumentum (vegetative parts and sometimes the sepals, petals and ovary provided with
    very minute, scattered trichomes). Labellum entire to distinctly 3-lobed. 2

2. Labellum entire, or obscurely shallowly 3-lobed (margins sometimes erose or toothed),
    free apical portion of side lobes (when present) usually less than 2 mm long (rarely up
    to 2 mm long in forms of *D. dewindtianum*) 3
    Labellum more distinctly 3-lobed, free apical portion of side lobes usually longer than 2
    mm (shorter in some species) 28

3. Labellum with a papillose-hairy disc and ridges. Column with a bilobed flange at base of stigmatic cavity ...... 69. *D. tenompokense* Carr var. *papillilabium* (J.J. Wood) J.J. Wood
    Labellum glabrous or minutely papillose, never papillose-hairy. Column lacking a bilobed flange on the stigma 4

4. Labellum entire, the side lobes absent, margins not erose or toothed 5
    Labellum obscurely 3-lobed, the side lobes often rudimentary; or side lobes absent, but lower margins of labellum erose or toothed 11

5. Pseudobulbs borne 1–4.5 cm apart on the rhizome. Rhizome up to 40 cm long. Plants often terrestrial. Flowers white, sometimes flushed pale yellowish green, or yellowish cream, fragrant .... 78. *D. simplex* J.J. Sm. (see also couplet 16)
    Pseudobulbs crowded on the rhizome, at most 0.8 cm apart, usually much less. Rhizome shortly creeping, usually less than 3 cm long, rarely up to 15 cm long. Plants usually epiphytic. Flowers variously coloured, unscented or fragrant 6

6. Peduncle strongly flattened. Flowers bright yellowish green or citron-yellow, with a dark brown or maroon-purple labellum and pink stelidia. Sepals and petals with revolute margins. Labellum strongly coiled-up distally .... 40. *D. planiscapum* Carr
    Peduncle terete. Flowers coloured otherwise. Sepals and petals with flat margins. Labellum not strongly coiled-up distally. 7

7. Leaf-blades 10.5–21.5 cm long, linear-ligulate. Sepals and petals 6–7.1 mm long, narrowly lanceolate, acuminate. Labellum surface minutely papillose. Stelidia longer than the column apex .... 68. *D. tenuitepalum* J.J. Wood
    Leaf-blades 3–8 cm long, narrowly linear to elliptic. Sepals usually shorter (if as long as *D. tenuitepalum*, then sepals acute, petals spathulate and labellum obtuse). Labellum surface glabrous. Stelidia equal to or shorter than the column apex 8

8. Flowers yellow, flushed plum-purple. Stelidia basal, shorter than the column, not reaching the stigma, obtuse, truncate. Rhizome up to 15 cm or more long. Pseudobulbs 6–8 mm apart .... 42. *D. suratii* J.J. Wood
    Flowers cream, white or green. Stelidia subapical, either borne opposite the stigma, between it and the apical hood, or basal and almost equal to the column apex. Rhizome abbreviated, rarely up to *c.* 6 cm long. Pseudobulbs crowded, at most 3–4 mm apart 9

9. Leaves narrowly linear, 2–3(–4) mm wide. Stelidia basal, almost equal to the column apex, obtuse to acute. Sepals 2–2.3 mm long. Labellum 1.3 mm long. Rhizome up to 6 cm long. Pseudobulbs 3–4 mm apart .... 37. *D. integrilabium* Carr
    Leaves elliptic to ligulate, (5–)7–12(–15) mm wide. Stelidia subapical to median, almost equal to the column apex, acuminate. Sepals 5–8 mm long. Labellum around 3 mm long. Rhizome abbreviated. Pseudobulbs crowded 10

10. Flowers white with a brown labellum and column, sweetly scented. Rachis 3–4.5 cm long. Petals 2.5–3 mm long. Labellum oblanceolate, narrowest at base, broadest towards apex .... 77. *D. globigerum* (Ridl.) J.J. Sm.

Flowers green, unscented. Rachis 8–18 cm long. Petals 6–6.8 mm long. Labellum oblong-pandurate, constricted at middle, not broadest towards the apex ....
.... 70. *D. lumakuense* J.J. Wood

11. Plants loose in habit, the rhizomes elongated, pseudobulbs spaced 1–6.5 cm apart, sometimes contiguous 12
Plants caespitose, forming tight clumps, the rhizomes abbreviated, pseudobulbs usually crowded together, never more than 1 cm apart 17

12. Pseudobulbs (1.5–) 5–16 (–22) cm long 13
Pseudobulbs 0.8–1.5 cm long 16

13. Labellum lanceolate, acuminate, simple, margins minutely erose, fimbriate to denticulate, not strongly decurved above the base or about the middle, weakly attached to the column-foot, versatile 14
Labellum spathulate-elliptic with very small side lobes, or subentire and broadly ovate-triangular, with obscure, erect, rounded side lobes, strongly decurved above the base or about the middle, less mobile or immobile 15

14. Leaves 0.6–1 (–1.5) cm wide, linear-ligulate or narrowly elliptic. Rhizome cataphylls without dark green to brown apical margins. Labellum simple, widest across the middle
.... 44. *D. lancilabium* Ames
Leaves much wider, usually 2.5–4.5 cm wide, broadly elliptic. Rhizome cataphylls with dark green to brown apical margins. Labellum subentire, with obscure side lobes, widest towards base .... 46. *D. subintegrum* Ames

15. Stelidia arising from just below the middle of the column, almost equalling the apical hood, subacute or obtuse, never dilated, obliquely obtuse and spathulate-truncate at the apex. Labellum versatile, spathulate-elliptic, with irregularly serrulate decurved margins, the side lobes very small, triangular and tooth like, the disc with a fleshy M-shaped basal callus .... 43. *D. gracilipes* Carr
Stelidia basal, longer than the apical hood, apex dilated, obliquely obtuse, spathulate-truncate. Labellum immobile, subentire, mid-lobe broadly ovate-triangular, margins undulate and not decurved, side lobes obscure, erect, rounded, the disc with 2 prominent fleshy keels adnate above the base to the margin of the side lobes ....
.... 45. *D. longipes* J.J. Sm.

16. Labellum oblong-ligulate to subpandurate, narrowest at middle, the proximal margins irregularly toothed, 3–3.5 mm long. Stelidia subulate, hamate, decurved. Rhizome much less than 40 cm long. Leaves linear-ligulate, 0.3–0.4 cm wide ....
.... 79. *D. hamatum* Schltr.
Labellum semiorbicular-ovate, with obscure auriculate side lobes, versatile. Stelidia porrect, neither hamate nor decurved. Rhizome often up to 40 cm long. Leaves narrowly elliptic to ovate-elliptic, (0.5–)0.8–1.7 cm wide .... 78. *D. simplex* J.J. Sm.

17. Leaves narrowly linear to linear, grass-like, the blade (6–)10–18 × 0.2–0.5 cm 18
Leaves broader; if linear, then shorter or wider 20

18. Pseudobulbs borne 5–7 mm apart on rhizome. Leaves rigid and coriaceous, generally 3–4 mm wide. Rachis 12–16 cm long. Flowers cream. Stelidia lanceolate, slightly exceeding column apex .... 67. *D. mucronatum* J.J. Sm.
   Pseudobulbs closely aggregated on rhizome. Leaves, inflorescence and flowers lacking this combination of characters 19

19. Sepals and petals 4–4.3 mm long. Labellum broadest across the mid-lobe, 3 mm long, the side lobes minutely erose, the mid-lobe elliptic. Stelidia arising just below stigma at the middle of the column, subulate, reaching or almost reaching the rostellum. Flowers lemon-yellow .... 56. *D. graminoides* Carr
   Sepals and petals 2–2.5 mm long. Labellum broadest at the base, 1.5 mm long, the side lobes obscure and rounded, the mid-lobe triangular, acute to acuminate. Stelidia basal, linear, obtuse, slightly shorter than the apical hood. Flowers greenish yellow .... 41. *D. sublobatum* Carr

20. Leaves linear-subspathulate, usually distinctly broadest towards the apex, obtuse and retuse. Labellum simple, central margins slightly irregularly toothed .... 38. *D. johannis-winkleri* J.J. Sm.
   Leaves otherwise, either of equal width or broadest at the middle. Labellum simple or obscurely lobed 21

21. Plants very small, 5 cm or less high 22
   Plants larger, 8–45 cm high 23

22. Leaves linear (margins often revolute in dried material), *c.* 2 mm wide. Flowers pale salmon-pink with a yellow labellum. Sepals and petals minutely papillose at base. Labellum *c.* 4.3 × 3.7 mm. Stelidia cuneate, rounded and subtruncate .... 53. *D. dulitense* Carr
   Leaves narrowly elliptic, with very minute black, spot-like trichomes, particularly on abaxial surface, 3–4 mm wide. Flowers pale green, labellum brownish ochre. Sepals and petals glabrous. Labellum 2.8 × 1.2–1.3 mm. Stelidia ligulate, acute .... 54. *D. ochrolabium* J.J. Wood

23. Leaves linear to linear-ligulate, 3–6 mm wide. Labellum elliptic, acute, simple, erose towards the base. Flowers greenish yellow or bright yellow, the labellum yellow, brown centrally, the callus red .... 55. *D. galbanum* J.J. Wood
   Leaves narrowly elliptic to oblong-elliptic, 0.8–3.5 cm wide. Not this combination of characters 24

24. Leaf-blade 8–30 × 2–3.8 cm, obtuse or acute. Flowers cream or white, rarely greenish white. Floral bracts conspicuous, glumaceous, spreading, (5–)7–12 mm long. Labellum obscurely lobed, not sharply deflexed above the base .... 48. *D. havilandii* Pfitzer
   Not this combination of characters 25

25. Leaves thick and fleshy, narrowly linear-elliptic, acute, sulcate above, curved, the blade 2–7 × 0.3–0.6 cm. Flowers salmon-pink to pale brownish, the labellum white or dull yellow, reddish and rather fleshy at the centre .... 36. *D. pachyphyllum* J.J. Wood & A. Lamb

Leaves thin-textured, often coriaceous, never thick and fleshy, variously shaped. Flowers variously coloured 26

26. Pseudobulbs 6–9 cm long. Stelidia slightly longer than the apical hood. Sepals, petals and labellum pale green. Floral bracts 5–7 mm long .... 62. *D. geesinkii* J.J. Wood
Pseudobulbs *c*. 0.6–4.5 cm long. Stelidia slightly shorter than or equalling the apical hood. Sepals, petals and labellum otherwise. Floral bracts 2–4 mm long 27

27. Stelidia usually arising opposite or just below lower portion of stigmatic cavity, slightly shorter than the apical hood. Sepals and petals lemon-yellow or greenish yellow, rarely whitish, centre of the labellum often suffused bright green, the keels usually bright green. Mid-lobe of labellum (2–)3–4.5 mm wide, rounded, obtuse ....
.... 60. *D. dewindtianum* W.W. Sm.
Stelidia basal, equalling or shorter than the apical hood. Details of flower colour unknown. Mid-lobe of labellum (1–)1.2–1.3 mm wide, acute to acuminate ....
.... 65. *D. jiewhoei* J.J. Wood

28. Flowers white, labellum with a yellowish orange or brownish blotch, non-resupinate. Rachis stiffly erect to porrect. Leaves 0.4–1.1 cm wide. Side lobes of labellum entire, never erose or toothed .... 39. *D. muluense* J.J. Wood
Not this combination of characters 29

29. Flowering plants small, 15(–20) cm or less high (usually much less). Leaves narrowly linear-ligulate to linear-lanceolate 30
Flowering plants much larger. Leaves lanceolate, oblong-elliptic to elliptic 33

30. Flowers brownish salmon with brownish olive-green column. Labellum mid-lobe thick and fleshy, oblong-spathulate and shallowly retuse. Sepals obtuse ....
.... 35. .*D. crassilabium* J.J. Wood
Flowers greenish, greenish cream, pale yellow to pale ochre, or salmon-pink to buff. Labellum mid-lobe thin, not fleshy, broadly elliptic or broadly ovate from a cuneate base, obtuse, truncate or acute. Sepals acute to acuminate 31

31. Stelidia borne at each side of or just below the stigmatic cavity, lanceolate, falcate, usually rather obtuse. Mid-lobe of labellum acute ....
.... 69. *D. tenompokense* Carr var. *tenompokense*
Stelidia borne at or just above the base of the column, subulate, acicular. Mid-lobe of labellum obtuse to truncate 32

32. Flowers salmon-pink to buff, or cream with an orange to pale ochre central area on sepals, petals and labellum. Petals elliptic, obtuse or mucronate. Mid-lobe of labellum broadly elliptic, obtuse .... 57. *D. magaense* J.J. Wood
Flowers yellow, yellowish green or pale green, labellum with 2 pale brown longitudinal central streaks. Petals acuminate. Mid-lobe of labellum broadly ovate to obscurely 6-angled, truncate .... 58. *D. subulibrachium* J.J. Sm.

33. Floral bracts large, ovate-elliptic, cymbiform, 8–13 mm long, almost concealing the flowers. Flowers translucent greenish yellow with brown areas on labellum, the column cream, orange at base. Stelidia subulate, borne just below the stigmatic cavity, barely reaching the rostellum .... 73. *D. imbricatum* Ames
Floral bracts much smaller, never almost concealing the flowers. Flowers lacking this combination of characters 34

34. Column distinctly papillose, stelidia arising from just below the stigmatic cavity, short, falcate, papillose. Sepals and petals pale green, tipped pink to dull orange-yellow, the labellum and the column orange to salmon-pink or reddish. Labellum *Coelogyne*-like, posterior margin of side lobes erect, roundly dilate below apex, recurved towards apex and with a short transverse fold; mid-lobe abruptly and strongly recurved, ovate, subacute, fleshy, erose, provided with 3 inconspicuous broad rounded keels; disc with a V-shaped keel between side lobes, with a longer, thinner keel either side. Leaves oblong-elliptic, margins slightly recurved, 9–15 × 3.3–4 cm .... 34. *D. anomalum* Carr
Plants lacking this combination of characters 35

35. Stelidia bifid. Rhizome pendulous, slender, up to 18 cm long, bearing distinctly spotted cataphylls. Pseudobulbs narrowly cylindrical or narrowly fusiform, 2–3 mm wide. Lateral sepals reflexed. Side lobes of labellum triangular, acute; mid-lobe obovate. Disc of labellum with 3 fleshy papillose keels. Column 3.5 mm long. ....
.... 51. *D. imitator* J.J. Wood
Stelidia normally entire. Not this combination of characters 36

36. Leaves ovate to oblong-oblanceolate, obtuse, 3.5–10 × 0.9–3 cm, petiole 2–5 mm long. Inflorescence from apex of nearly mature pseudobulb; peduncle 3–5.5 cm long, filiform, with several adpressed non-floriferous bracts at junction with rachis; rachis 20–35 cm long, pendulous, with up to 80 or more flowers. Flowers yellowish green; labellum pale greenish white, the mid-lobe yellowish green with 2 converging brown streaks. Labellum stipitate to column-foot by a narrow claw, mid-lobe oblong-oblanceolate or oblong above a cuneate base, shortly apiculate, *c.* 2.3 × 0.2–0.5 mm, disc with 2 papillose keels. Stelidia sometimes bifid at apex ... 49. *D. angustilobum* Carr
Not this combination of characters 37

37. Side lobes of labellum irregularly laciniate, mid-lobe strongly recurved, ovate-elliptic, rounded or subacute, minutely papillose. Disc of labellum with 2 flange-like basal keels, each *c.* 0.4 mm wide, which curve toward the middle and meet. Stelidia borne either side of stigmatic cavity, shorter than apical hood. Sepals and petals creamy white; labellum sometimes very pale green, with yellowish cinnamon or ochre keels ....
.... 66. *D. lacinilobum* J.J. Wood & A. Lamb
Plants lacking this combination of characters 38

38. Labellum entirely dark chocolate-brown or reddish brown 39
Labellum never entirely brown, but often with brown markings 40

39. Side lobes of labellum distinctly erose-fimbriate, rounded. Disc of labellum with 2 fleshy keels, connected at base by a transverse thickening with a prominent median nerve between. Petals minutely denticulate or erose. Sepals and petals yellowish green or lemon-yellow. .... 64. *D. kamborangense* Ames

159

    Side lobes of labellum entire or a little uneven, never erose-fimbriate, narrowly triangular-subulate, acute. Disc of labellum with 2 main fleshy keels, not united at base, and 2 shorter keels in between higher up. Petals entire. Sepals and petals pale green suffused brown, cinnamon to orange brown or reddish brown .... 75. *D. oxylobum* Schltr.

40. Leaves with a tough, coriaceous texture, rigid when dry .... 61. *D. crassifolium* Ames
    Leaves thin-textured, often chartaceous 41

41. Flowers white, labellum orange-yellow, fragrant. Side lobes of labellum falcate, acute, erect, mid-lobe ovate-orbicular, recurved. Disc of labellum 2-keeled. Column with a tridentate apical hood .... 47. *D. glumaceum* Lindl.
    Flowers lacking this combination of characters 42

42. Leaves ligulate, blade 12–22 × 0.8–1.2 cm, gradually attenuated into a slender petiole. Petals linear-oblong, acute, *c*. 0.7–0.8 mm wide. Sepals and petals pale yellow or greenish yellow, labellum pale yellow or orange with whitish margins; keels and elevated median nerve bright salmon-pink; column and anther-cap salmon-pink, the stelidia and apical hood whitish. Stelidia arising from the middle of the column just below stigmatic cavity, equalling apical hood .... 59. *D. angustipetalum* Ames
    Leaves broader, blade (1–)1.5–6.5 cm wide, not ligulate. Not this combination of characters 43

43. Flowers large, sepals 13–17 mm long, petals 12–15 mm long, labellum 11–13.8 mm long. Keels on disc of labellum separated by a deep groove which is narrow proximally, becoming much wider distally, resulting in a low raised keel on underside of labellum .... 72. *D. flos-susannae* J.J. Wood
    Flowers much smaller, sepals 4–8.7 mm long, petals 3.3–8.2 mm long, labellum 2.5–7.7 mm long. Deep groove between keels absent 44

44. Leaves (22–)25–60 cm long, variable in width, usually between 3.5 and 6.5 cm wide. Pseudobulbs narrowly conical, 3–10 × 2–3 cm wide at base. Disc of labellum with 2 small basal keels. Side lobes of labellum short, tooth-like, acute, falcate, margins entire; mid-lobe ovate-rhomboid. Sepals and petals yellow or yellowish green, tipped ochre; labellum green with 2 sepia-brown streaks or a sepia-brown central area ....
.... 74. *D. longifolium* Rchb. f.
    Leaves shorter, between 2.3 and 18 cm long. Not this combination of characters 45

45. Sepals and petals semi-translucent cream, suffused pale salmon; labellum cream, the base and keels pale salmon; column pale salmon. Plants often pendulous. Mid-lobe of labellum obovate or suborbicular, obtuse, often as long as wide. Disc of labellum with 2 prominent papillose keels, incurved and touching at base. Petals erose. Stelidia lanceolate, shortly acuminate .... 52. *D. lacteum* Carr
    Sepals and petals pale green, greenish cream to yellow; labellum similar, with 2 brown streaks. Plants lacking this combination of characters 46

46. Sepals and petals papillate on adaxial surface. Stelidia spathulate, rounded to truncate, only reaching to base of stigmatic cavity. Labellum 2.5–2.6 mm long ....
.... 76. *D. papillitepalum* J.J. Wood

Sepals and petals lacking papillae. Stelidia linear-subulate, acute, slightly exceeding anther-cap but shorter than apical hood. Labellum 3.5–7.7 mm long 47

47. Rachis (4–)8–12.5(–30) cm long. Free apical portion of side lobes of labellum 0.7–1 mm long, not reaching middle of mid-lobe. Keels on disc of labellum glabrous, occasionally each with a thorn-like, acute basal projection .... 71. *D. gracile* (Hook. f.) J.J. Sm.
Rachis 20–45 cm long. Free apical portion of side lobes of labellum 1.5–1.6 mm long, reaching middle of mid-lobe. Keels on disc of labellum partially papillose, never with basal projections .... 50. *D. longirachis* Ames

48. Leaf-blades (6.5–)9–17(–25) × 0.7–2.5(–3.5) cm. Inflorescence lax, 2– to 7(–12)-flowered. Sepals 7–10 × 3–3.5 mm. Petals 6–7.5 × 2.5–3.2 mm. Labellum 5–6 × 2.5–4.1 mm. Column apex shortly tridentate or unevenly toothed. Stelidia triangular, acute, falcate, equalling or slightly shorter than rostellum .... 80. *D. pubescens* L.O. Williams
Leaf-blades 3–5.4 × 0.8–1 cm. Inflorescences rather dense, 15– to 17–flowered. Sepals 5.3–5.5 × 1.3–1.7 mm. Petals 4.5 × 1.5 mm. Labellum 3.4 × 2.4 mm. Column apex deeply tridentate. Stelidia linear-subulate, the apex filiform, slightly longer than column apex. .... 81. *D. vestitum* J.J. Sm.

34. DENDROCHILUM ANOMALUM

Dendrochilum anomalum Carr in *Gdns. Bull. Straits Settl.* 8: 87–88 (1935). Type: Malaysia, Sarawak, Belaga District, Ulu Koyan, Mount Dulit, 7 Nov. 1932, *Oxford University Expedition native collector* 2497 (holotype K!, isotype SING!). **Fig. 69.**

DESCRIPTION. Epiphyte. *Rhizome* abbreviated, 2–3 mm in diameter. *Roots* 1–1.5 mm in diameter, smooth. *Cataphylls* 4, 1–7 cm long, obtuse to acute, finely nerved, rather fleshy, dark brown to purplish, finely speckled darker brown, becoming fibrous. *Pseudobulbs* crowded on rhizome, ovoid, smooth to finely wrinkled, 2.3–4 × 0.7–2 cm, yellowish olive. *Leaf*: petiole 1.3–2.5 cm long, to 2.9 mm wide, sulcate, tough; blade oblong-elliptic, abruptly narrowed below, obtuse, shortly apiculate, 9.3–15.5 × 3.3–4.5 cm, coriaceous, margins slightly recurved, main nerves 9. *Inflorescences* from within the unexpanded leaf, many-flowered, lax to subdense, flowers borne 3–4 mm apart, laterally arranged, but twisting to form a spiral; peduncle terete, rather stout, 16–25 cm long, 1–2 mm in diameter, yellowish green; non-floriferous bracts 3–4, ovate-elliptic, acute, 5–6 mm long; rachis quadrangular, 17.5–27 cm long, 1–1.5 mm in diameter, pale brownish green, sparsely ramentaceous; floral bracts broadly ovate, obtuse to subacute, concave, 5–5.5 × 4.8–6 mm, multi-nerved, ramentaceous, yellowish green or pink, speckled brown. *Flowers* described as dull orange-yellow, labellum orange, central parts red, or sepals greenish cream or pale green to yellowish green, tipped brownish pink, labellum dark orange buff, pale brownish orange or pale reddish brown, column pale reddish brown or orange, anther-cap orange to salmon-pink. *Pedicel with ovary* clavate, curved, 2.5–3.7(–4) mm long, minutely papillose, ovary slightly ramentaceous. *Sepals* and *petals* 3–nerved. *Sepals* sparsely ramentaceous on abaxial surface. *Dorsal sepal* narrowly oblong, acute, 5–6.3 × 1.5–1.9 mm. *Lateral sepals* obliquely oblong-elliptic, shortly acuminate, acute, inconspicuously sigmoidly curved, dorsally carinate near apex, 6–7 × 1.6–2.1 mm. *Petals* linear-ligulate or narrowly oblong-elliptic, acute, sometimes subfalcate, 4.8–6 × 1–1.2 mm, very sparsely ramentaceous on abaxial surface. *Labellum* 3–lobed from the

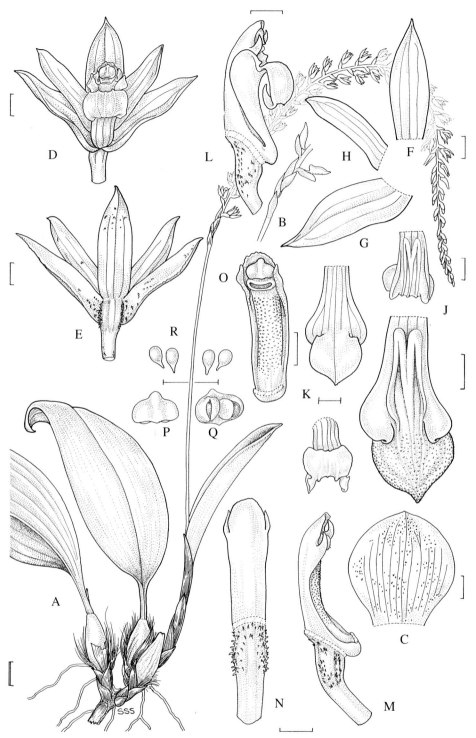

Fig. 69. *Dendrochilum anomalum.* **A** habit; **B** junction of rachis with peduncle showing four non-floriferous bracts, two floral bracts and two buds; **C** floral bract; **D** flower, front view; **E** flower, back view; **F** dorsal sepal; **G** lateral sepal; **H** petal; **J** labellum, front views, natural position (above), flattened (below); **K** labellum, back views, natural position (below), flattened (above); **L** pedicel with ovary, labellum and column, side view; **M** pedicel with ovary and column, side view; **N** pedicel with ovary and column, back view; **O** column, front view; **P** anther-cap, back view; **Q** anther-cap, interior view showing loculi; **R** pollinia. A–R from *Giles* 964A. Scale: single bar = 1 mm; double bar = 1 cm. Drawn by Susanna Stuart-Smith.

middle, 5–6 mm long when flattened, 2–3 mm wide across side lobes, 3 mm wide across mid-lobe, immobile, erect and parallel with column, very minutely papillose, especially beneath; side lobes clasping column, triangular, subacute, margins minutely papillose towards apex, posterior margin erect, roundly dilate below apex, recurved towards apex with a short transverse fold; mid-lobe abruptly and strongly reflexed, ovate, subacute, fleshy, *c.* 3 mm long, minutely transversely rugose, margins minutely erose, provided with 3 obscure broad, rounded keels extending from base to below apex; disc between side lobes provided with a fleshy V-shaped keel commencing from a little above the base to junction with mid-lobe and with a thinner keel, dilated towards apex, on either side. *Column* papillose, 4–4.5 × 1–1.1 mm, slightly curved; foot papillose, 0.6–1 mm long; apical hood transversely oblong above the triangular base, truncate, minutely erose; stelidia borne at or just below stigmatic cavity, narrowly triangular-falcate, upcurved, acute, papillose, *c.* 0.7 mm long; stigmatic cavity transversely oblong, anterior margin elevate; anther-cap ovate, cucullate, minutely papillose, 0.5–0.6 × 1 mm.

DISTRIBUTION. Borneo.

SARAWAK. Belaga District, Ulu Koyan, Mount Dulit, 7 Nov. 1932, *Oxford University Expedition native collector* 2497 (holotype SING!, isotype K!).

SABAH. Tenom District, near Sapong, 1965, cult. Kew, coll. 28 Jan. 1976, EN 178–65, *Giles* GPS 659 (K!) & 1965, cult. Kew, coll. 25 Jan. 1968, EN 543–65, *Giles* 964A (K!). Mount Kinabalu, Hempuen Hill, 18 Aug. 1991, *Lamb & Surat* in *Lamb* AL 1365/91 (K!).

KALIMANTAN TIMUR. Apo Kayan, Mount Sungai Pendan, near base of east ridge, east of Long Nawan, 14 Oct. 1991, *Leiden* cult. (*de Vogel*) 913452 (K!, L!). Apo Kayan, Kayan River between Long Ampung and Long Nawan, Oct 1991, *Leiden* cult. (*de Vogel*) 913346 (L!).

HABITAT. Heath forest; hill forest with *Gymnostoma sumatranum* (Jungh. ex de Vriese) L.A.S.Johnson on ultramafic substrate; kerangas-like forest 15–20 m high, thin boles, close together, light, with little undergrowth, many light green moss cushions on ground, soil sandy. 600–1000 m.

A quite distinct species which appears to have no close allies, hence the specific epithet. The short, upcurved, falcate stelidia borne at or just below the stigmatic cavity are similar to those of *D. muluense* J.J. Wood. It seems to favour forest types with a rather open canopy allowing higher light levels to penetrate.

35. DENDROCHILUM CRASSILABIUM

Dendrochilum crassilabium J.J. Wood in Wood & Cribb, *Checklist orch. Borneo*: 168–169, fig. 22 C & D (1994). Type: Indonesia, Kalimantan Timur, Apo Kayan, Mount Sungai Pendan, east ridge, east of Long Nawan, 1300 m, 13 Oct. 1991, *Leiden* cult. (*de Vogel & Cribb*) 913205A (holotype L!, isotype K!). **Fig. 70.**

DESCRIPTION. Epiphyte. *Rhizome* up to 4 cm long. *Roots* branching, flexuous, very minutely papillose, up to 1 mm in diameter. *Cataphylls* 3, ovate-elliptic, acute, becoming fibrous, 0.8–1.5 cm long. *Pseudobulbs* crowded on rhizome, ovoid-elliptic or narrowly fusiform, smooth, 1–1.6 × 0.3–0.4 cm. *Leaf*: petiole sulcate, 0.2–1.5 cm long; blade narrowly linear-ligulate, acute, coriaceous, 5–10.5 × 0.45–0.5 cm, main nerves 3. *Inflorescences* pendulous, usually longer than leaves, 2– to 6–flowered, lax, flowers borne 3–5 mm apart; peduncle filiform, 5.5–12(–14) cm long; non-floriferous bracts 1 or 2, ovate-elliptic, acute to

163

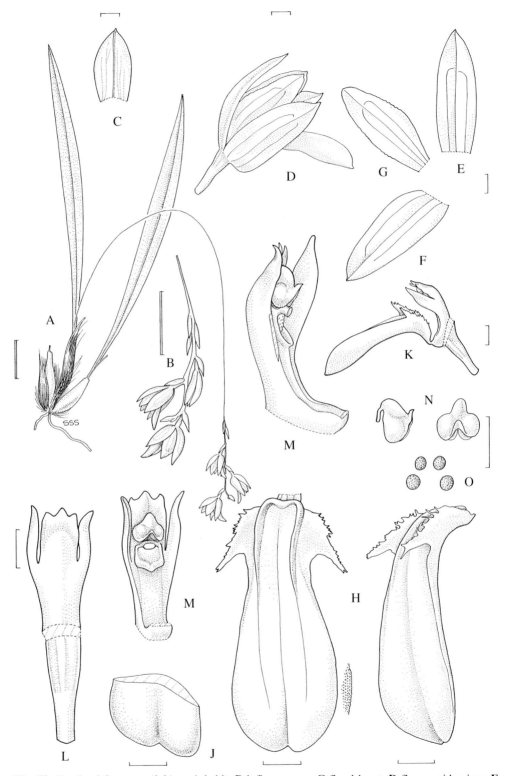

Fig. 70. *Dendrochilum crassilabium*. **A** habit; **B** inflorescence; **C** floral bract; **D** flower, side view; **E** dorsal sepal; **F** lateral sepal; **G** petal; **H** labellum, front and oblique views, with close-up of marginal zone of mid-lobe; **J** apical portion of mid-lobe of labellum, transverse section; **K** pedicel with ovary, labellum and column, side view; **L** pedicel with ovary and column, back view; **M** column, front view, and side view with stelidium removed; **N** anther-cap, front and side views; **O** pollinia. **A–O** from *Leiden* cult. (*de Vogel & Cribb*) 913205A (holotype). Scale: single bar = 1 mm; double bar = 1 cm. Drawn by Susanna Stuart-Smith.

acuminate, 4 mm long; rachis quadrangular, slightly fractiflex, 0.7–1.8 cm long; floral bracts ovate, acuminate, brown, 3–5 mm long. *Flowers* with dull brownish salmon, translucent sepals and petals, labellum paler brownish salmon, column brownish olive-green, anther-cap cream. *Pedicel with ovary* clavate, 2 mm long. *Sepals* and *petals* 3–nerved. *Dorsal sepal* oblong-elliptic, obtuse, curved, 7 × 2.1 mm. *Lateral sepals* obliquely ovate-elliptic, obtuse or subacute, 7.2–7.5 × 2.8 mm. *Petals* oblong to oblong-elliptic, obtuse, slightly erose, 7 × 2 mm. *Labellum* 3–lobed, minutely papillose, 7–7.2 mm long, 3–nerved; side lobes narrowly triangular, acuminate, lacerate, 1.5–2 mm long; mid-lobe spathulate, shallowly retuse or narrowly ovate-elliptic, often longitudinally concave, thick and fleshy, margin thinner, gently curved at base, 5.5–5.6 × 2–3.5 mm; disc with 2 low, fleshy basal ridges joined to form a U-shape. *Column* 3–3.5 mm long; foot prominent; apical hood truncate, often irregularly 3– or more-toothed, 1 mm wide; stelidia ligulate or oblong-ligulate, sometimes falcate, obtuse, borne opposite stigmatic cavity, slightly shorter than to slightly longer than apical hood, 2 mm long; stigmatic cavity quadrate, lower rim somewhat thickened; anther-cap cucullate, 0.8–0.9 × 1 mm.

DISTRIBUTION. Borneo.

SARAWAK. Limbang District, route from Ulu Sungai Limbang to Batu Buli (Bukit Buli), Oct. 1987, *Awa & Lee* S.50774 (K mixed collection with *D. pachyphyllum*!, SAR). Marudi District, trail to Mount Mulu, 21 March 1998, *Leiden* cult. (*Vogel et al.*) 980410 (K!, L!).

SABAH. Kinabalu Park, Mount Tembuyuken, 11 March 1993, *Nais et al.* SNP 5473 (Sabah Parks Herbarium, Kinabalu Park!).

KALIMANTAN TIMUR. Apo Kayan, Mount Sungai Pendan, east ridge, east of Long Nawan, 13 Oct. 1991, *Leiden* cult. (*de Vogel & Cribb*) 913205A (holotype L!, isotype K!).

HABITAT. Mixed hill dipterocarp forest; upper montane ridge forest to 30 m high, with little undergrowth; recorded from forest on ultramafic substrate on Mount Tembuyuken. 1300–2500 m.

D. crassilabium is similar in habit to *D. pachyphyllum* J.J. Wood and has similarly coloured flowers. The leaves of *D. crassilabium*, however, are coriaceous, never thick and fleshy, and normally proportionately longer. The inflorescence is usually longer than the leaves. The sepals and petals are obtuse and the thick, fleshy labellum is slightly longer and usually has a shallowly retuse, obovate-spathulate mid-lobe and lacerate side lobes.

The material of *Awa & Lee* S.50774 at K appears to be a mixed collection of both species. The plant mounted at the top of the sheet has longer leaves and matches *D. crassilabium*, although the inflorescence is shorter. The flowers are rather badly preserved.

Nais et al. SNP 5473, from Mount Tembuyuken, differs in having a narrower ovate-elliptic labellum mid-lobe.

The thin-textured, coriaceous leaves and thick, fleshy, obovate-spathulate, retuse labellum of *D. crassilabium* are distinctive. However, intermediates between it and *D. pachyphyllum* may well exist.

The specific epithet is derived from the Latin *crassus*, thick and *labiatus*, lipped, referring to the fleshy labellum.

36. DENDROCHILUM PACHYPHYLLUM

Dendrochilum pachyphyllum J.J. Wood & A. Lamb in Wood & Cribb, *Checklist orch. Borneo*: 189–190, fig. 21, plate 9E (1994). Type: Malaysia, Sabah, Crocker Range, Mount

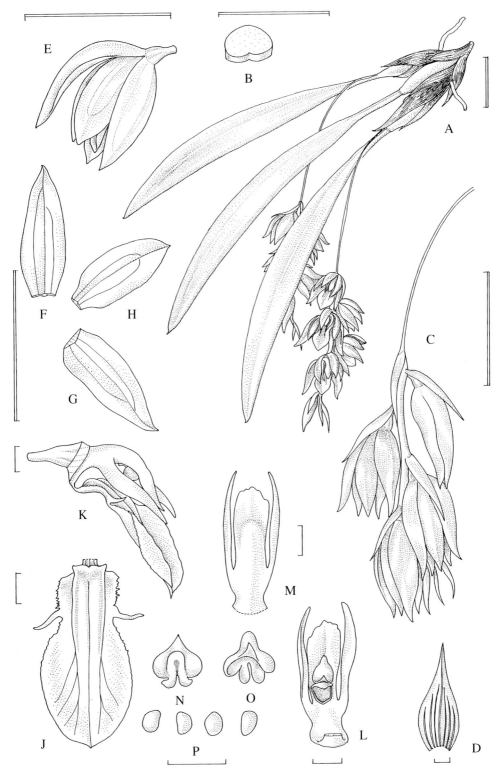

Fig. 71. *Dendrochilum pachyphyllum.* **A** habit; **B** portion of leaf, transverse section; **C** inflorescence; **D** floral bract; **E** flower, side view; **F** dorsal sepal; **G** lateral sepal; **H** petal; **J** labellum, front view; **K** pedicel with ovary, labellum and column, side view; **L** column, front view; **M** column, back view; **N** anther-cap, back view; **O** anther-cap, front view; **P** pollinia. **A–P** from *Lamb* AL 674/86 (holotype). Scale: single bar = 1 mm; double bar = 1 cm. Drawn by Susanna Stuart-Smith.

Alab, Sinsuron road, 1650 m, 7 Nov. 1986, *Lamb* AL 674/86 (holotype K!). **Fig. 71, plate 12C.**

DESCRIPTION. Epiphyte, occasionally terrestrial. *Rhizome* to 3 cm long. *Roots* branching, flexuous, minutely papillose, 0.3–1 mm in diameter. *Cataphylls* 3, ovate-elliptic, acute, becoming fibrous, 0.4–1.8 cm long. *Pseudobulbs* crowded and clump-forming, oblong-elliptic to narrowly fusiform, 0.7–1.2 × 0.3–0.5 to 1.8 × 0.2–0.3 cm. *Leaf*: petiole narrow, sulcate, 0.6–1.3 cm long; blade narrowly linear-elliptic, acute, thick and fleshy, sulcate above, curved, 2–7 × 0.3–0.6 cm, main nerves 3. *Inflorescences* pendulous, shorter than leaves, 3– to 5–flowered, lax, flowers borne 3–4 mm apart; peduncle filiform, yellowish green, 1–7 cm long; non-floriferous bracts 1 or 2, ovate-elliptic, acuminate, 3–6 mm long; rachis quadrangular, slightly fractiflex, salmon-pink, 0.6–1.8 cm long; floral bracts ovate, acuminate, brown, 4–7 mm long. *Flowers c.* 1 cm across, pedicel with ovary brownish pink, sepals translucent pink to salmon-pink or pale brownish, median nerve darker red to brownish pink, petals translucent pink to salmon-pink, labellum green at base, side lobes white, mid-lobe white or dull yellow, central area red to brownish red, column pink, anther-cap white. *Pedicel with ovary* clavate, 1–2 mm long. *Sepals* and *petals* 3–nerved. *Dorsal sepal* ovate-elliptic, apiculate, somewhat carinate, curved, 7–9 × 2–3 mm. *Lateral sepals* ovate-elliptic, acute, somewhat carinate, 7–9 × 2.5–3 mm. *Petals* elliptic, acute, minutely erose towards apex, 6–8 × 2–2.8 mm. *Labellum* 3–lobed, spathulate, 5.5–7 mm long, 2 mm wide at base, 3–3.5 mm wide across mid-lobe, 3–nerved; side lobes linear-triangular, acute, minutely toothed, free apical portion 1–1.5 mm long; mid-lobe oblong-elliptic, obtuse, rather fleshy at centre, shallowly concave, margin thin, erose, 4.5–5 mm long; disc with a raised keel 1 mm wide extending from base of labellum to upper portion of mid-lobe, its base retuse, its margins raised. *Column* 4 mm long; foot prominent; apical hood elongate, irregularly shallowly 3– to 4–toothed, the median tooth longer than the outer; stelidia linear-ligulate, subacute to acute, borne from near base, longer than apical hood, 4 × 0.5–0.6 mm; stigmatic cavity quadrate; anther-cap ovate-cordate, minutely papillose.

DISTRIBUTION. Borneo.

SARAWAK. Limbang District, route from Ulu Sungai Limbang to Batu Buli (Bukit Buli), Oct. 1987, *Awa & Lee* S.50774 (K mixed collection with *D. crassilabium*!, SAR).

SABAH. Sinsuron road, Sept. 1979, *Collenette* 49/79 (E!). Mount Alab, Sinsuron road, 7 Nov. 1986, *Lamb* AL 674/86 (holotype K!). Mount Alab, 1987, *Leiden* cult. (*Vermeulen*) 26459 (K!, L!) & *Leiden* cult. (*Vermeulen*) 26521 (K!, L!). Kota Kinabalu to Tambunan road, km. 56, July 1986, *Vermeulen & Chan* 413 (L!). Sipitang District, Long Pa Sia to Long Samado trail, near crossing with Malabid River, Dec. 1986, *Vermeulen & Duistermaat* 965 (L!). Mount Alab, south ridge, 31 Oct. 1986, *de Vogel* 8646 (L!).

HABITAT. Mixed hill dipterocarp forest; lower montane mossy forest; on mossy sandstone rocks and shale banks along roadside cuttings; very low and open podsol forest. 1300–2000 m.

D. pachyphyllum is at once distinguished from all other Bornean members of section *Platyclinis* by its short, acute, thick and fleshy leaves, pendulous, filiform peduncles and brownish salmon-pink flowers.

The specific epithet is derived from the Greek *pachy*, thick or stout, and *phyllum*, leaf, referring to the fleshy leaves.

37. DENDROCHILUM INTEGRILABIUM

Dendrochilum integrilabium Carr in *Gdns. Bull. Straits Settl.* 8: 85–86 (1935). Type: Malaysia, Sarawak, Belaga District, Mount Dulit, Ulu Koyan, *c.* 1200 m, 15 Sept. 1932, *Richards* S.484 (holotype SING!, isotypes K! L!). **Fig. 72.**

DESCRIPTION. Epiphyte forming clumps up to 15 cm across, rarely terrestrial. *Rhizome* much branched, up to 10 cm long. *Roots* much branched, flexuous, minutely papillose, often elongate, 0.2–0.6 mm in diameter. *Cataphylls* 5–6, ovate-elliptic, acute to acuminate, 0.2–2 cm long, slowly becoming fibrous. *Pseudobulbs* borne 1–3 mm apart at an acute angle with the rhizome, fusiform or narrowly oblong, smooth, 0.5–1.5 × (0.25–)0.4–0.5 cm. *Leaf:* petiole sulcate, 3–6 mm long; blade narrowly linear, slightly dilated at or above middle, obtuse and minutely apiculate, 3–8 × 0.28–0.4(–0.5) cm, thin-textured, main nerves 5. *Inflorescence* many-flowered, subdense to rather lax, flowers borne 1 mm apart; peduncle filiform, 3–6 cm long; non floriferous bracts 1 or 2, ovate-elliptic, acute, 1–2 mm long, rachis narrowly quadrangular, 5–7.5 cm long; floral bracts ovate, apiculate, 1.3–2 mm long, main nerve 1. *Flowers* wide-opening, sepals and petals translucent pale green, pale yellowish green, pale greenish cream or cream, sometimes cream with palest green tips, labellum pale green or pale yellow, column pale green, orange or salmon-pink, stelidia white. *Pedicel with ovary* clavate, 1.2–1.4 mm long. *Dorsal sepal* oblong-elliptic, acute, 2.3–3 × 0.7–0.9 mm, 1–nerved. *Lateral sepals* slightly obliquely ovate-elliptic, acute, sometimes somewhat falcate, 2.3–2.8 × 0.8–0.9 mm, main nerve 1, with 2 secondary nerves. *Petals* linear, acute, often widest at middle, 2–2.2 × 0.3–0.6 mm, 1–nerved. *Labellum* entire, oblong-elliptic, apex fleshy, obtuse, recurved below the middle, very shortly clawed, 1.2–1.3 × 0.4–0.8 mm; disc with above the base 2 low keels joined at the base by a transverse ridge, extending for $^3/_4$ of length of blade. *Column* curved, 1.2–1.5 cm long; foot obscure; apical hood rounded-triangular, entire; stelidia basal, linear, subacute, 1.1–1.4 mm long; rostellum large, triangular, subacute; stigmatic cavity suborbicular, anterior margin elevate; anther-cap cucullate, minute.

DISTRIBUTION. Borneo.

SARAWAK. Belaga District, Jelini River, near junction with Nawai River, Linau-Balui divide, 31 Aug. 1978, *Burtt* B.11377 (E!). Belaga District, Jelini River, Linau-Balui divide, 31 Aug. 1978, *Lee* S.39318 (K!, L!, SAR, SING!). Batu Laga Plateau, Batang Rejang, 10 Sept. 1984, *Mohtar* S.48225 (AAU, KEP!, SAN, SAR!). Marudi District, Tama Abu Range (Mount Temabok), Upper Baram valley, 3 Nov. 1920, *Moulton* SFN 6660 (K!, SING!). Belaga District, Mount Dulit, Ulu Koyan, 15 Sept. 1932, *Richards* S.484 (holotype SING!, isotypes K!, L).

SABAH. Crocker Range, Keningau District, Kimanis road, 12 Nov. 1987, *Aningguh in Lamb* AL856/87 (K!). Sipitang district, trail from Long Pa Sia to Long Samado, *c.* 2 km from Sarawak border, 25 Oct. 1986, *de Vogel* 8579 (L!). Sipitang District, trail from Long Pa Sia to Long Samado, near Malabid River, 26 Oct. 1986, *de Vogel* 8604 (L!). Sipitang District, Ulu Long Pa Sia, 8 km. Northwest of Long Pa Sia, near Pa Sia River, 26 Oct. 1985, *Wood* 703 (K!).

KALIMANTAN BARAT. Serawai, Sungai Uut Labang, 6 Oct. 1995, *Church et al.* 2204 (AMES, BO, K!).

HABITAT. Low open kerangas forest 5–10 m high on white sand over sandstone; mixed hill dipterocarp forest on wet, sandy soil; sandy forest; on trunks of isolated trees on hill ridges and dry hill tops; lower montane forest; in moss on roadside cutting. 300–1350 m.

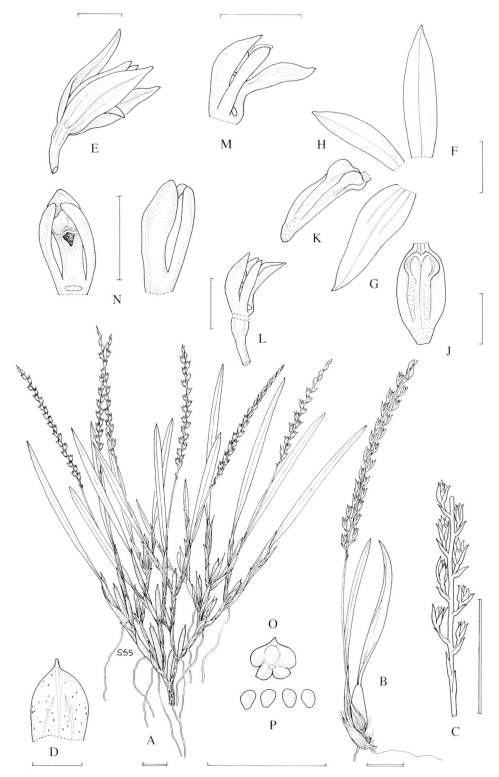

Fig. 72. *Dendrochilum integrilabium*. **A & B** habits; **C** lower portion of inflorescence; **D** floral bract; **E** flower, side view; **F** dorsal sepal, **G** lateral sepal; **H** petal; **J** labellum, front view; **K** labellum, oblique view; **L** pedicel with ovary, labellum and column, side view; **M** labellum and column, side view; **N** column, front and oblique views; **O** anther-cap, back view; **P** pollinia. **A** from *Aningguh* in *Lamb* AL 856/87, **B–P** from *Wood* 703. Scale: single bar = 1 mm; double bar = 1 cm. Drawn by Susanna Stuart-Smith.

D. *integrilabium* is a delicate species which often forms large clumps producing many small wide-opening star-shaped flowers.

The specific epithet is derived from the Latin *integri*, entire, undivided, simple, and *labium*, a lip, and refers to the simple shape of the labellum.

38. DENDROCHILUM JOHANNIS-WINKLERI

Dendrochilum johannis-winkleri J.J. Sm. in *Mitt. Inst. allg. Bot., Hamb.* 7: 36, t.v, fig. 26 (1927). Type: Indonesia, Kalimantan Barat, Tilung Hill, *Winkler* 1495 (holotype HBG). **Figs. 73 & 74.**

DESCRIPTION. Epiphyte. *Rhizome* branched, up to 3 cm long. *Roots* with a few branches, flexuous, minutely papillose, 0.5–0.6 mm in diameter. *Cataphylls* 3, ovate-elliptic, acute, *c.* 0.5–2 cm long, dark brown-spotted, becoming fibrous. *Pseudobulbs* oblong-ovoid, 0.8–1 × 0.4–0.5 cm. *Leaf*: petiole sulcate, 0.6–1.4 cm long; blade erect, linear-subspathulate, often broadest distally, apex equally shallowly retuse, rounded, 4.75–11 × 0.55–0.8 cm, main nerves 3, up to 12 or 13 in total. *Inflorescences* erect, many-flowered, lax, flowers borne 2 mm apart; peduncle filiform, 8.5–10 cm long; non-floriferous bract solitary or absent, acute, 2 mm long; rachis quadrangular, 6 cm long; floral bracts oblong-ovate, broadly obtuse, 2 × 2 mm. *Flowers* greenish white. *Pedicel with ovary* clavate, 1–1.1 mm long. *Sepals* and *petals* 3–nerved. *Dorsal sepal* oblong, subacute or acute, sometimes suberose at apex, concave, 4 × 1.5 mm. *Lateral sepals* obliquely subovate-oblong, somewhat falcate, acute, barely apiculate, concave, 3.7–3.8 × 1.5 mm. *Petals* obliquely oblong, somewhat falcate, acute, slightly concave, minutely erose, 3.2–3.3 × 1.2–1.3 mm. *Labellum* obscurely 3–lobed, clawed at base, recurved, convex, 2.2 mm long, 1 mm wide across side lobes; side lobes obscure, rounded, erose; mid-lobe oblong-trullate, subacute to acute, *c.* 1.2 × 0.7–0.8 mm; disc with 3 fleshy minutely papillose ridges joined at base, outer 2 broader, median narrow, all 3 terminating on lower half of mid-lobe. *Column* gently curved, 1.7–1.8 mm long; foot incurved, truncate, 0.4 mm long; apical hood rounded or retuse, sometimes crenate, minutely papillose; stelidia borne at the middle, obliquely linear-oblong, obtuse, slightly concave and minutely papillose distally, 2 mm long, porrect; rostellum conspicuous, convex, rounded and abruptly contracted, apiculate; anther-cap cucullate, transversely triangular, acuminate or acute, apex recurved, 0.5 mm wide.

DISTRIBUTION. Borneo.

BRUNEI. Temburong District, Mount Retak, 1987, *Leiden* cult. (*Cantley*) 26836 (K!, L!). Temburong District, Amo, Tudal Hill, 5 Oct. 1994, *Said* BRUN 15825 (BRUN, K!).

KALIMANTAN BARAT. Tilung Hill, 8 Feb. 1925, *Winkler* 1495 (holotype HBG!).

HABITAT. Lower montane forest. 800–1500 m.

The affinities of this distinctive species, known only from three collections, are uncertain. J.J. Smith (1927) suggested a possible relationship with *D. simile* Blume, a species from Peninsular Malaysia, Sumatra, Java and the Lesser Sunda Islands. The leaves are distinctive among Bornean dendrochilums in often being broadest towards the apex and subspathulate.

The specific epithet honours Professor Hans Winkler (1877–1945), a Director of Hamburg Botanic Garden in Germany, who collected the type.

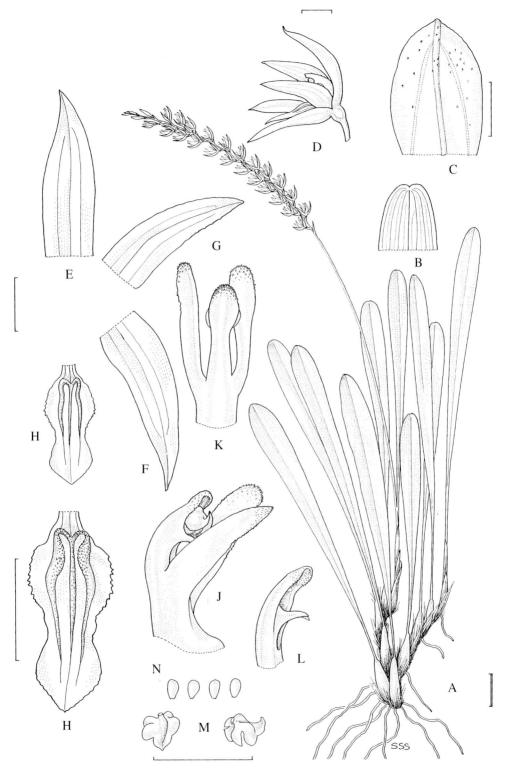

Fig. 73. *Dendrochilum johannis-winkleri*. **A** habit; **B** leaf apex; **C** floral bract; **D** flower, side view; **E** dorsal sepal; **F** lateral sepal; **G** petal; **H** labellum, front view; **J** column, side view; **K** column, back view; **L** column apex, anther-cap removed, side view; **M** anther-cap, back and side views; **N** pollinia. **A–N** from *Winkler* 1495 (holotype). Scale: single bar = 1 mm; double bar = 1 cm. Drawn by Susanna Stuart-Smith.

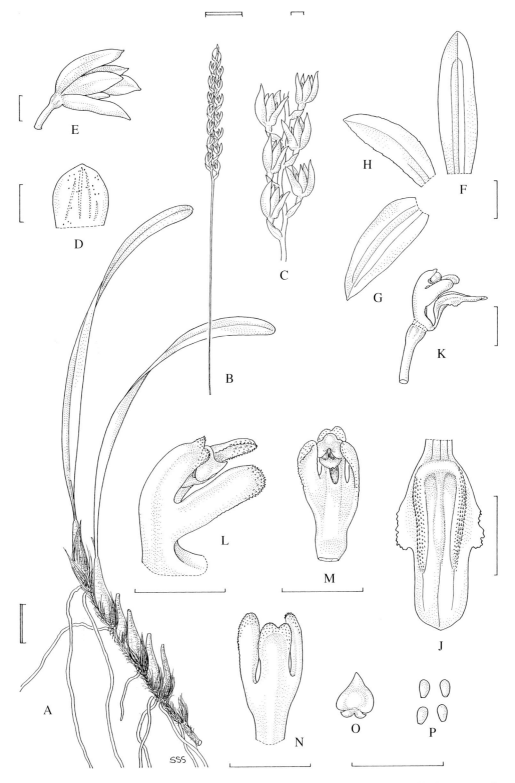

Fig. 74. *Dendrochilum johannis-winkleri*. **A** habit; **B** inflorescence; **C** lower portion of inflorescence; **D** floral bract; **E** flower, side view; **F** dorsal sepal; **G** lateral sepal; **H** petal; **J** labellum, front view; **K** pedicel with ovary, labellum and column, side view; **L** column, side view; **M** column, front view; **N** column, back view; **O** anther-cap, back view; **P** pollinia. **A–P** from *Leiden* cult. (*Cantley*) 26836. Scale: single bar = 1 mm; double bar = 1 cm. Drawn by Susanna Stuart-Smith.

39. DENDROCHILUM MULUENSE

Dendrochilum muluense J.J. Wood in *Kew Bull.* 39 (1): 80, fig. 5 (1984). Type: Malaysia, Sarawak, Marudi District, Gunung Mulu National Park, Mount Mulu, ridge at Camp 4, 27 Jan. 1978, *Nielsen* 143 (holotype AAU!, isotype K!). **Fig. 75, plate 11F & G.**

DESCRIPTION. Erect epiphyte. *Rhizome* branching, tough, clothed in fibrous sheaths, 2–18 cm long, *c.* 2 mm in diameter. *Roots* numerous, filiform, wiry, smooth, 0.2–0.3 mm in diameter. *Cataphylls* 3–4, 0.5–4.5 cm long, brown to reddish brown with darker brown spotting, becoming fibrous. *Pseudobulbs* crowded on rhizome, narrowly conical or fusiform, erect, 0.8–2(–2.5) × 0.3–0.8 cm, finely rugose when dried. *Leaf*: petiole narrow, sulcate, 1–4.5 cm long; blade linear-lanceolate to narrowly elliptic, obtuse and mucronate, (0.4–)5–15.5 × 0.4–1.3 cm, main nerves 5, with numerous small transverse nerves, thin textured, scattered calcium oxalate bodies sometimes present. *Inflorescences* 8– to 17–flowered, lax to subdense, flowers borne 1.8–2 mm apart; peduncle terete, filiform, erect, porrect or gently curving, 6–14.5 cm long, glabrous or minutely furfuraceous above; non-floriferous bracts 2–3, ovate-elliptic, acute, 1–3 mm long; rachis quadrangular, held at a sharp angle to the peduncle, 1.5–3.5 cm long, straight to slightly fractiflex, minutely furfuraceous; floral bracts ovate, obtuse, or obtuse and mucronate, 2.5–3 mm long. *Flowers* non-resupinate, sepals and petals white, cream or whitish green, labellum white or yellowish orange, with a brownish spot at the base, often yellowish at tip, column pale yellow, sometimes with a brown spot, column-foot red. *Pedicel with ovary* narrowly clavate, 1–2.5 mm long, deflexed and turning orange at anthesis. *Sepals* and *petals* 3–nerved. *Dorsal sepal* oblong to oblong-elliptic obtuse, 4–4.5 × 1.5 mm. *Lateral sepals* slightly obliquely oblong to oblong-elliptic, obtuse, 4–5 × 1.8 mm. *Petals* oblong-elliptic, obtuse, margin erose when magnified, 3.5–4.5 × 1.2–1.5 mm. *Labellum* 4–5 mm long, concave, somewhat fleshy, main nerves 5; side lobes oblong, rounded, erect, 1–1.5 × 0.8 m; mid-lobe obovate-flabellate when expanded, rounded-obtuse, attenuate at base, 1.6–2 × 3 mm; disc with 2 short basal keels extending almost to the base of the side lobes. *Column* gently curved, 2.8–3 mm long; foot distinct; apical hood ovate, truncate, entire; stelidia borne a little way below stigmatic cavity, triangular-linear, subobtuse or acute, somewhat falcate and upcurved, 0.8 mm long, reaching to and level with stigma; rostellum triangular-ovate, prominent; anther-cap cordate, cucullate, 0.8 × 0.8 mm. *Capsule* globose.

DISTRIBUTION. Borneo.

SARAWAK. Limbang District, Mount Pagon, 11 Aug. 1984, *Awa & Lee* S.47827 (AAU, K!, KEP!, L!, SAR, SING!). Limbang District, Sungai Ulu Limbang, route to Batu Lawi, Bario, 14 Aug. 1985, *Awa & Lee* S.50812 (K!, L!, SAR, SING). Limbang District, summit of a ridge connecting to Batu Lawi, Bario, 20 Aug. 1985, *Awa & Lee* S.50937 (K!, L!, NY, SAR). Lawas District, route from Ba Kelalan to Mount Murud, Camp III, 27 Sept. 1967, *Burtt & Martin* B.5244 (E!, SAR). Limbang District, trail from Bario to Batu Lawi, Batu Buli, 14 March 1998, *Leiden* cult. (*Vogel et al.*) 980071 (K!, L!) & 980142 (L!). Marudi District, Mount Mulu, 20 Oct. 1977, *Lewis* 365 (K!). Marudi District, Gunung Mulu National Park, Tutoh, Baram, 24 February 1976, *Martin* S.37059 (L!, MEL, SAR!). Marudi District, Gunung Mulu National Park, Mount Mulu, ridge at Camp 4, 27 Jan. 1978, *Nielsen* 143 (holotype AAU!, isotype K!). Marudi District, proposed Mount Murud National Park, between first and second summits, 13 Sept. 1982, *Yii* S.44616 (K!, L!, SAR).

BRUNEI. Temburong District, north ridge of Retak Hill, 28 Jan. 1989, *Wong* WKM 802 (BRUN!, K!, L!, SING!).

SABAH. Tambunan District, Crocker Range, km 55 on Kota Kinabalu to Tambunan road, 4 Sept. 1983, *Beaman* 6876 (K!, L!), & *Beaman* 8031 (K!). Ulu Apin Apin, 5 Nov. 1991,

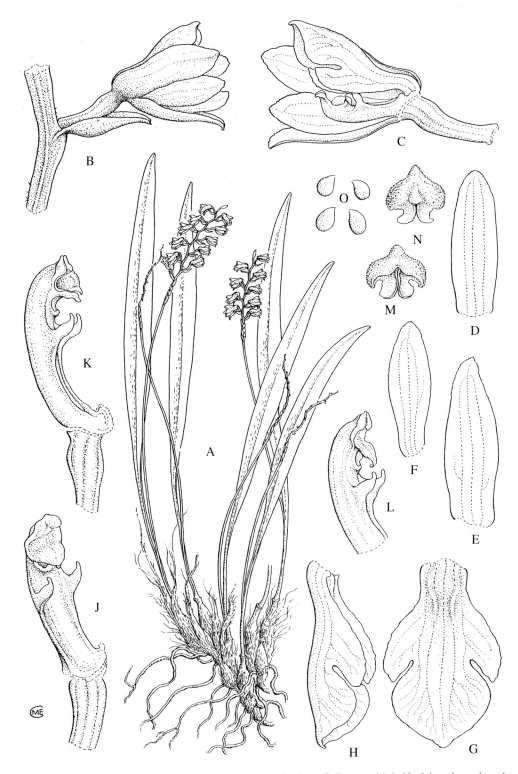

Fig. 75. *Dendrochilum muluense*. **A** habit; **B** flower, side view; **C** flower with half of dorsal sepal, and lateral sepal and petal removed, side view; **D** dorsal sepal; **E** lateral sepal; **F** petal; **G** labellum, flattened, front view; **H** labellum, side view; **J** ovary, column with anther-cap removed, oblique view; **K** ovary, column and anther-cap, side view; **L** column, distal portion, anther-cap removed, side view; **M** anther-cap, interior view; **N** anther-cap, back view; **O** pollinia. **A** x 2/3; **B** x 6; **C, D, E & F** x 8; **G & H** x 10; **J, K & L** x 14; **M, N & O** x 20. **A–O** from Nielsen 143 (holotype). Drawn by Maureen Church.

Lamb AL 1392/91 (K!). Mount Alab, 1987, *Leiden* cult. (*Vermeulen*) 26457 (L!).

KALIMANTAN TIMUR. Mount Batu Harun, north of Long Bawan, Apo Kayan, 24 July 1981, *Kato, Okamoto & Walujo* B.9758 (L!).

HABITAT. Ridge top montane forest on sandstone and shale bedrock; mossy forest; oak-laurel forest. 1400–2200 m.

A distinctive species which, since it was first described from Mount Mulu in 1984, has been recorded from several localities throughout Borneo.

40. DENDROCHILUM PLANISCAPUM

Dendrochilum planiscapum Carr in *Gdns. Bull. Straits Settl.* 8: 228–229 (1935). Type: Malaysia, Sabah, Mount Kinabalu, Tenompok, 1440 m, 29 Aug. 1933, *Carr* 3663, SFN 28020 (holotype SING!, isotypes AMES!, K!). **Fig. 76, plate 13A.**

DESCRIPTION. Epiphyte. *Rhizome* stout, branching, shortly creeping, to 4 cm long. *Roots* elongate, flexuous, much-branched, very minutely papillose, 1–2 mm in diameter. *Cataphylls* 3, ovate to lanceolate, acute, 1–7 cm long, pinkish cream. *Pseudobulbs* crowded, clump-forming, ovoid, smooth to wrinkled, 1.5–2.7 × 1–1.5 cm. *Leaf*: petiole sulcate, slender, 2–7 cm long; blade linear-ligulate, apex conduplicate, acute, sulcate above, rigid, (13–)17–26(–30) × 0.5–0.85 cm, main nerves 5, mid nerve prominent on abaxial surface. *Inflorescences* emerging with almost fully expanded leaf, erect to curving, *c.* 20– to 50 or more-flowered, lax, flowers borne 4–6 mm apart; peduncle laterally compressed, 9–16 cm long, *c.* 1.5 mm in diameter, pale yellowish green; non-floriferous bract solitary, ovate, acuminate, subulate, 0.9–1.5 cm long; rachis quadrangular, (9–)18–24 cm long, pale green; floral bracts subulate, involute, 0.5–1 cm long, brown. *Flowers* fragrant, pedicel with ovary pinkish green, sepals and petals bright yellow green, lemon-yellow (citron), or pale greenish cream, sepals often suffused red at base, labellum pale flesh-coloured to dark brown or maroon-purple, column pale green above, brownish pink below, stelidia pink, or pale yellow, hood whitish, stelidia whitish suffused pale salmon-pink, or entirely white, anther-cap yellowish green. *Pedicel with ovary* narrowly clavate, 5–7 mm long. *Sepals* and *petals* rather rigid, papillose, 3–nerved, mid nerve prominent on abaxial surface, lateral nerves obscure. *Dorsal sepal* narrowly oblong-ligulate, subacute to acute, margin slightly revolute, *c.* 8.5–11 × 1.8–2.1 mm. *Lateral sepals* narrowly linear-elliptic to ligulate, acute, margin revolute, 7–10.5 × 1.8–2.1 mm. *Petals* oblong to oblong-elliptic, acute, sometimes falcate, margin strongly revolute, 6.8–10 × 1.7–2 mm. *Labellum* coiled-up apically, subentire, *c.* 4.8–7 × 2 mm, obscurely papillose; side lobes very obscure, scarcely rounded, erect; mid-lobe narrowly triangular-ovate or triangular-elliptic, acute or obtuse; disc with 2 separate keels extending from near base, fading near apex as raised nerves, median nerve prominent on mid-lobe. *Column* 2.8–3 mm long; foot distinct; apical hood oblong-triangular, obtuse, or 2– to 3–toothed; stelidia borne on either side of stigmatic cavity, broadly triangular, acuminate, longer than apical hood, *c.* 2 mm long; rostellum ovate, obtuse; stigmatic cavity oblong, with elevated brown margins; anther-cap cordate, cucullate.

DISTRIBUTION. Borneo.

SABAH. Mount Kinabalu, Mamut, 19 Nov. 1988, *Amin et al.* SAN 129373 (K!, SAN!). Mount Kinabalu, Bukit Ular, *Barkman* 260 (Sabah Parks Herbarium, Kinabalu Park). Mount Kinabalu, Menteki Ridge, 6 Jan. 1995, *Beaman* 11188 (K). Mount Kinabalu, Tenompok, 29 Aug. 1933, *Carr* 3663, SFN 28020 (holotype SING!, isotypes AMES!, K!). Mount Kinabalu, Mesilau Trail, 23 Sept. 1972, *Chow & Leopold* SAN 74511 (K!, SAN!). Crocker Range,

Fig. 76. *Dendrochilum planiscapum.* **A** habit; **B** junction of rachis with peduncle showing a solitary non-floriferous bract and three floral bracts; **C** transverse section through peduncle; **D** floral bract; **E & F** flowers, side view; **G** dorsal sepal; **H** lateral sepal; **J** petal; **K** labellum, front view; **L** basal portion of labellum showing keels, oblique view; **M** labellum, side views; **N** pedicel with ovary, labellum and column, side view ;**O** column, oblique view; **P** column, side view;**Q** column, back view; **R** anther-cap, back and interior views; **S** pollinia. **A, D & L** from *Meijer* 48111, **B & C, E–K & M–R** from *Giles & Woolliams* s.n. Scale: single bar = 1 mm; double bar = 1 cm. Drawn by Susanna Stuart-Smith.

Sinsuron road, Sept. 1979, *Collenette* 46/79 (E!). Locality unknown, probably Tenom District, *Giles & Woolliams* 467 (K!). Crocker Range, Sinsuron road, 29 July 1989, *Lamb* s.n., Chan drawing 258 (K!). Mount Alab, 6 Sept. 1980, *Lamb* SAN 92253 (K, sketch only!). Sipitang District, Long Pa Sia area, *Lamb* (colour photograph K!). Mount Kinabalu, Mesilau Cave Trail, 21 Feb. 1965, *Meijer* SAN 48111 (K!, SAN!). Pinosuk Plateau, near Kundasang, 3 Oct. 1986, *de Vogel* 8032 (L!) & 8033 (L!).

HABITAT. Lower montane forest; upper montane forest; mossy *Agathis* forest; Fagaceae/*Dacrydium*/*Leptospermum* forest *c*. 35 m high with little undergrowth on quarternary gravels with sandstone, ultramafic and dioritic rock; recorded as an epiphyte on *Castanopsis*. 1300–2400 m.

A distinctive plant unlikely to be confused with any other Bornean species on account of the flattened peduncle and coiled labellum. The coiled labellum recalls the Sumatran *D. mirabile* J.J. Wood, belonging to the monospecific section *Longicolumna*. It is probably most closely allied, however, to *D. odoratum* (Ridl.) J.J. Sm. from Peninsular Malaysia and *D. simile* Blume from Peninsular Malaysia, Sumatra, Java and the Lesser Sunda Islands.

The specific epithet is derived from the Latin *planus*, even, flat, and *scapus*, leafless floral axis or peduncle, in reference to the flattened peduncle.

41. DENDROCHILUM SUBLOBATUM

Dendrochilum sublobatum Carr in *Gdns. Bull. Straits Settl.* 8: 86–87 (1935). Type: Malaysia, Sarawak, Marudi District, Mount Dulit, Dulit Ridge, 5 Sept. 1932, *Synge* S.406 (holotype SING!, isotypes K!, L!). **Fig. 77.**

DESCRIPTION. Epiphyte. *Rhizome* abbreviated, up to 2 cm long, 1 mm in diameter. *Roots* slender, flexuous, smooth. *Cataphylls* 3, 1–2.5 cm long, brown, with scattered darker spotting, becoming fibrous. *Pseudobulbs* cylindrical, wrinkled, 0.6–1.5 × 0.3–0.4 cm. *Leaf*: petiole sulcate, 1–2.8 cm long; blade narrowly linear-oblanceolate, attenuated below, obtuse and minutely apiculate, 8–16.5 × 0.3–0.5 cm, main nerves 3, thinly coriaceous, containing numerous crystalline calcium oxalate bodies. *Inflorescences* from the apex of the mature pseudobulb, many-flowered, erect, subdense, flowers borne 1–2 mm apart; peduncle terete, 1.2–5 cm long; non-floriferous bracts absent; rachis quadrangular, 8–17.5 cm long; floral bracts ovate, acute, 1 mm long, 1–nerved. *Flowers* greenish yellow, tip of column brown, with a very slight sweet scent. *Pedicel with ovary* narrow, 2 mm long. *Sepals* 3–nerved. *Dorsal sepal* oblong-lanceolate, acute, 2.5 × 0.9 mm. *Lateral sepals* oblong-ovate, acute or subacute, slightly falcate, 2 × 1 mm. *Petals* oblanceolate-ligulate, acute, slightly falcate, 2.2 × 0.6 mm, 1– to 3–nerved. *Labellum* slightly elastically attached, obscurely subquadrate in the lower half, shortly clawed below, narrowed about the middle into a triangular acuminate apex, 1.5 × 0.9 mm when flattened; disc with a conspicuous basal horseshoe-shaped keel. *Column* 1.5 mm long; foot very small; apical hood oblong to triangular, obtuse, entire; stelidia basal, linear, obtuse, slightly shorter than apical hood; rostellum large, broadly rounded; anther-cap cucullate, glabrous.

DISTRIBUTION. Borneo.

SARAWAK. Marudi District, Mount Dulit, 18 Oct. 1932, *Ford* 2290 (K!). Marudi District, Mount Murud, Oct. 1922, *Mjöberg* 46 (AMES!). Marudi District, Mount Dulit, Dulit Ridge, 5 Sept. 1932, *Synge* S.406 (holotype SING!, isotypes K!, L!).

HABITAT. Mossy forest. 1230– above 1900 m.

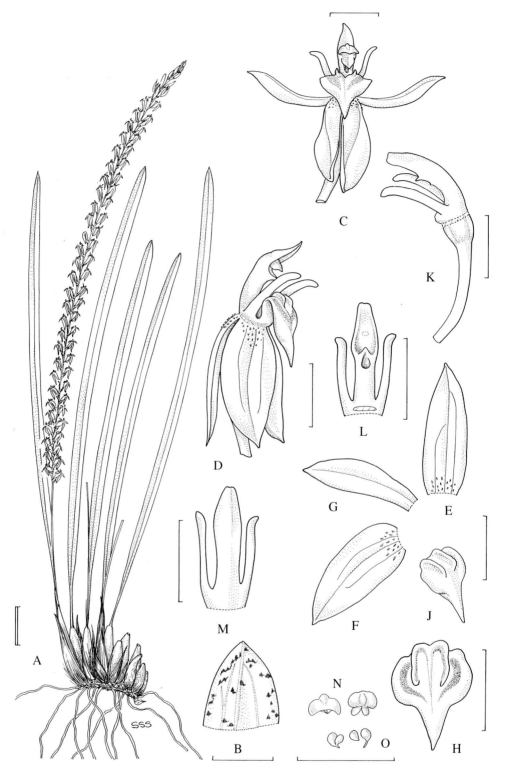

Fig. 77. *Dendrochilum sublobatum.* **A** habit; **B** floral bract; **C** flower, front view; **D** flower, side view; **E** dorsal sepal; **F** lateral sepal; **G** petal; **H** labellum, front view; **J** labellum, oblique view; **K** pedicel with ovary and column, anther-cap removed, side view; **L** column, anther-cap removed, front view; **M** column, back view; **N** anther-cap, front and back views; **O** pollinia. **A–O** from *Synge* S.406 (holotype). Scale: single bar = 1 mm; double bar = 1 cm. Drawn by Susanna Stuart-Smith.

Carr (1935) considered this little-collected species to be allied to *D. dolichobrachium, D. gibbsiae* and *D. haslamii* (section *Cruciformia*). Despite obvious similarities, I have provisionally included it within section *Platyclinis* on account of the presence of a small column-foot. Molecular analysis may help to clarify its true affinities and perhaps future inclusion within section *Cruciformia*.

The specific epithet is derived from the Latin *sub*, somewhat, not completely, and *lobus*, a lobe, and refers to the labellum which is obscurely subquadrate in the lower half.

42. DENDROCHILUM SURATII

Dendrochilum suratii J.J. Wood in *Lindleyana* 7 (2): 77–79, fig. 3 (1992). Type: Malaysia, Sabah, Tambunan District, Mount Trus Madi, *c.* 2490 m, 15 June 1988, *Wood* 905 (holotype K!, isotype Tenom Orchid Centre!). **Fig. 78.**

DESCRIPTION. Epiphyte. *Rhizome* up to 15 cm or more long, slender. *Roots* slender, wiry, filiform, with a few branches, smooth. *Cataphylls* 3–4, 1–3 cm long, pale brown with darker brown speckles, becoming fibrous. *Pseudobulbs* borne 6–8 mm apart, ovate to ovate-elliptic, smooth, 1–1.2 × 0.4–0.6 cm, plum-purple. *Leaf*: petiole sulcate, 0.5–0.7 cm long; blade linear-elliptic, ligulate, obtuse, slightly constricted 0.8–1.3 cm below apex, 3–5.2 × 0.3–0.8 cm, thin-textured, main nerves 5. *Inflorescences* laxly 5– to 8–flowered, flowers opening from the top, borne 2 mm apart; peduncle erect, filiform, 2.7–4 cm long; non-floriferous bract solitary, ovate-elliptic, acute, 3 mm long; rachis curved, quadrangular, 1.6–2.2 cm long; floral bracts ovate-elliptic, rather truncate, with a small central mucro, 4 × 1.5 mm. *Flowers* yellow, flushed plum purple. *Sepals* obscurely 3–nerved. *Dorsal sepal* ovate-elliptic, acute, 4 × 1.1 mm. *Lateral sepals* ovate-elliptic, slightly oblique, acute, 4 × 1.5 mm. *Petals* narrowly elliptic, acute, 4 × 1 mm. *Labellum* oblong-ovate, acute, margin minutely erose, curved, deflexed distally, 3.5–3.6 mm long, 1.9–2 mm wide near base; disc with 3 prominent nerves, the outermost developed into low basal keels and becoming obscure distally, the median a prominent line. *Column* gently curved, slender, 2.6 × 0.6 mm; foot small; apical hood ovate, entire; stelidia basal, oblong, obtuse, 1 mm long; rostellum truncate; anther-cap ovate, cucullate, smooth.

DISTRIBUTION. Borneo.

SABAH. Tambunan District, Mount Trus Madi, 15 June 1988, *Wood* 905 (holotype K!, isotype Tenom Orchid Centre!).

HABITAT. Upper montane mossy ericaceous forest, growing on exposed branches and twigs associated with mosses, lichens and *Coelogyne plicatissima* Ames & C. Schweinf. 2490 m.

A distinctive, yellow-flowered species known only from the type collection. Only a few plants were seen in flower during June and full flowering probably takes place later in the year.

The specific epithet honours Aningguh (Andi) Surat, employed at the Agricultural Park, Tenom, Sabah, who has accompanied me on several field trips in Sabah including the ascent of Mount Trus Madi in 1988.

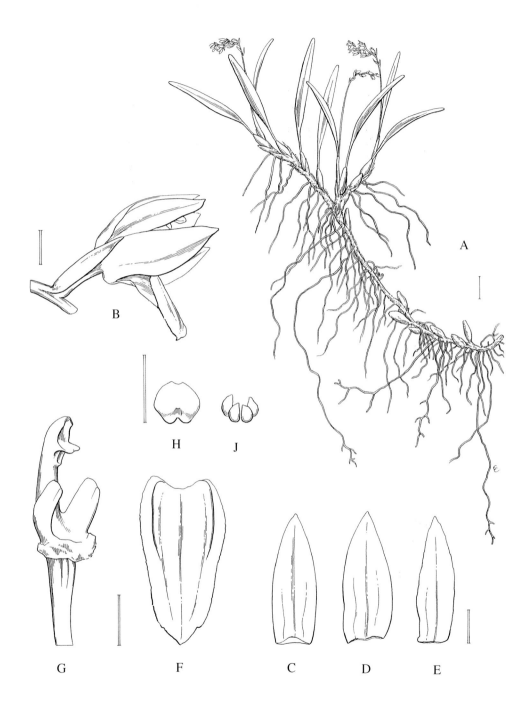

Fig. 78. *Dendrochilum suratii*. **A** habit; **B** flower, side view; **C** dorsal sepal; **D** lateral sepal; **E** petal; **F** labellum, front view; **G** pedicel with ovary and column, anther-cap removed, oblique view; **H** anther-cap, back view; **J** pollinia. **A–J** from *Wood* 905 (holotype). Scale: single bar = 1 cm; double bar = 1 mm. Drawn by Eleanor Catherine.

43. DENDROCHILUM GRACILIPES

Dendrochilum gracilipes Carr in *Gdns. Bull. Straits Settl.* 8: 81–83 (1935). Type: Malaysia, Sarawak, Marudi District, Mount Dulit, Dulit Ridge, *c.* 1290 m, 6 Sept. 1932, *Synge* S.418 (holotype SING!, isotype K!). **Fig. 79.**

DESCRIPTION. Epiphyte, rarely terrestrial. *Rhizome* 8–10 cm long. *Roots* wiry, much-branched, minutely papillose, elongate, 1.2–1.8 mm in diameter. *Cataphylls* 3–4(–6), ovate-elliptic, obtuse or subacute, (0.5–)2–8(–12.5) cm, speckled brown. *Pseudobulbs* crowded or to 2.5 cm apart, clump-forming, narrowly cylindrical, very slender, (2–)13–16(–22) cm long, 0.3–0.5 cm wide near base, 0.18–0.3 cm wide near apex, finely shallowly longitudinally furrowed, older bulbs minutely wrinkled. *Leaf*: petiole sulcate, 1–3.5 cm long; blade narrowly elliptic, elliptic or oblanceolate, acute, sometimes constricted below apex, cuneate at base, thinly coriaceous, (6.5–)16–21 × 1.2–3(–3.8) cm, main nerves 7, with numerous small transverse secondary nerves. *Inflorescences* erect, many-flowered, lax to subdense, flowers borne 3–6 mm apart; peduncle stout, wiry, 8.5–14 cm long; non-floriferous bract solitary, oblong-ovate, obtuse to subacute, 3–3.5 mm long; rachis quadrangular, (11–)14–17(–25) cm long; floral bracts ovate, obtuse to subacute, 3.5–5 × 2.5 mm. *Flowers* mostly all open together, unscented, slightly scented or with a strong anise-like scent, described as very pale greenish, very pale yellowish, greenish orange or cream, labellum pale ochre. *Pedicel with ovary* clavate, 2–3 mm long. *Sepals* and *petals* spreading, with sparse basal hairs, 3–nerved. *Dorsal sepal* narrowly elliptic, subacute to acute, 5.8–7 × 1.5–2 mm. *Lateral sepals* narrowly ovate-elliptic, slightly oblique, acuminate, 6.5–7 × 1.8–2 mm. *Petals* narrowly elliptic or narrowly oblong-elliptic, shortly acuminate, 6.5–7 × 1.5–1.8 mm. *Labellum* strongly recurved above base, obscurely 3-lobed, joined to column-foot by a short claw, 5–5.1 mm long when flattened; side lobes rudimentary, triangular, tooth-like, margin erose, 2 mm wide across both; mid-lobe ovate or spathulate-elliptic, obtuse or subacute, margins sometimes decurved, minutely erose or minutely ciliolate, papillose, 3.5–4 × 2.9–3.5 mm, 3–nerved; disc provided between side lobes with 2 short tall fleshy keels joined at base and a little introrse in the upper half and terminating above base of mid-lobe, median nerve shortly elevate. *Column* curved, 3–3.5 mm long; foot distinct; apical hood oblong, obtuse or subacute, entire; stelidia borne just above base, linear-ligulate, obtuse or subacute, sometimes with a tiny tooth-like lacinule near the apex, nearly as long as hood, 2–2.1 mm long; anther-cap ovate, cucullate.

DISTRIBUTION. Borneo.

SARAWAK. Marudi District, Tama Abu Range, Bario, 6 Sept. 1985, *Awa & Lee* S.51102 (AAU, K!, KEP, L, MEL, SING). Belaga District, Bukit Dema, 28 Aug. 1978, *Burtt* B.11325 (E!, SAR!). Marudi District, Mount Mulu, 15 June 1962, *Burtt & Woods* B.2123 (E!, K!). Marudi District, Mount Dulit, Dulit Ridge, 5 Sept. 1932, *Harrisson* S.405 (K!), & 17 Sept. 1932, *Hartley* S.512 (K!). Marudi District, trail to Mount Mulu, 21 March 1998, *Leiden* cult. (*Vogel et al.*) 980415 (K!, L!). Marudi District, Tama Abu Range (Mount Temabok), Upper Baram Valley, 5 Nov. 1920, *Moulton* SFN 6663 (K!, SING!). Belaga District, Mount Dulit, Ulu Koyan, 15 Sept. 1932, *Richards* S.482 (K!). Belaga District, Mount Dulit, Dulit Ridge, 5 Sept. 1932, *Synge* S.418 (holotype SING!, isotype K!). Kapit District, northern Hose Mountains, base of ridge leading to Bukit Batu, 2 Nov. 1991, *de Vogel* 1116 (L!) & *de Vogel* 1116A (K, L). Murudi District, Gunung Mulu National Park, Mount Mulu, 10 March 1990, *Yii & Talib* S.58429 (KEP, SAR!).

BRUNEI. Mount Pagon, Temburong River, April 1958, *Ashton* A.211 (K!).

SABAH. Sipitang District, ridge between Maga River and Malabid River headwaters, Dec. 1986, *Vermeulen & Duistermaat* 1020 (K!, L!).

Fig. 79. *Dendrochilum gracilipes.* **A &B** habits; **C** mature pseudobulb and leaf; **D** floral bract; **E** flower, oblique view; **F** dorsal sepal; **G** lateral sepal; **H** petal; **J** base of labellum showing callus, back view; **K** pedicel with ovary, labellum and column, side view; **L** column, oblique view; **M** column, side view; **N** column, back view; **O** anther-cap, back and side views; **P** pollinia. **A** from *Yii & Talib* S. 58429, **B** from *Burtt* B. 11325, **C** from *Awa & Lee* S. 51102, **D–P** from *de Vogel* 1116. Scale: single bar = 1 mm; double bar = 1 cm. Drawn by Susanna Stuart-Smith.

HABITAT. Montane elfin forest *c.* 25 metres high on a very narrow sandstone ridge; open, mossy low forest; submontane forest; mossy forest; on the ground among moss in sand forest; epiphytic low on trees. 900–1600 m.

An attractive species with very narrow, pencil-like pseudobulbs up to 22 cm in length and flowers mostly all opening simultaneously on the inflorescence. The margin of the labellum mid-lobe is often somewhat decurved.

The specific epithet is derived from the Latin *gracilis*, thin, slender, and refers to the narrow pseudobulbs.

44. DENDROCHILUM LANCILABIUM

Dendrochilum lancilabium Ames, *Orchidaceae* 6: 58–59, pl. 83 (bottom) (1920). Type: Malaysia, Sabah, Mount Kinabalu, Marai Parai Spur, 23 Nov. 1915, *J. Clemens* 280 (holotype AMES!, isotypes BM!, K!, NY!, SING!). **Fig. 80, plate 11A & C.**

DESCRIPTION. Epiphyte, occasionally lithophytic. *Rhizome* branching, 12–20 cm long, 0.3–0.5 cm in diameter. *Roots* wiry, flexuous, much-branched, forming a tangled mat, very minutely papillose, elongate, up to 1.8 mm in diameter. *Cataphylls* 3, ovate-elliptic, acute, 0.5–6 cm long, finely speckled. *Pseudobulbs* crowded or up to 2 cm apart, narrowly cylindrical or narrowly fusiform, 1.5–7(–9.5) cm long, 0.5 cm wide near base, 0.3–0.4 cm wide near apex, finely wrinkled when dry, green. *Leaf*: petiole sulcate, 0.7–3.5 cm long; blade linear-ligulate or narrowly elliptic, obtuse and mucronate, sometimes constricted below apex, coriaceous, (6–)10–18.5(–23) × 0.6–1.5 cm, main nerves 5–6, dark green, almost metallic, sometimes rather yellowish. *Inflorescences* erect, many-flowered, subdense, flowers borne 2–3 mm apart; peduncle 5–11 cm long; non-floriferous bract solitary, ovate-elliptic, acute, 3–5 mm long; rachis quadrangular, 15–18(–26) cm long; floral bracts ovate, subacute, 2–4 mm long. *Flowers* sweetly scented, greenish yellow, pale lemon or pale green. *Pedicel with ovary* clavate, 3 mm long. *Sepals* 3–nerved, sparsely hairy at base. *Dorsal sepal* narrowly elliptic, acuminate, slightly concave, 4.1–6 × 1.4–1.5 mm. *Lateral sepals* slightly obliquely narrowly elliptic, acute to acuminate, 4.1–5 × 1.4–1.5 mm. *Petals* narrowly elliptic, acuminate, 4–4.25 × 0.9–1 mm, margin sometimes minutely denticulate, 1–nerved. *Labellum* entire, shortly clawed, narrowly elliptic, acuminate, apex decurved, margin minutely denticulate to erose, except towards base, 3–4 × 1.5–1.6 mm, very versatile, 3–nerved; disc with 2 prominent fleshy basal keels joined at base, median nerve elevate and prominent from base to about halfway along labellum. *Column* very slender, slightly curved, 2.5 mm long; foot small; apical hood elongate, entire, subquadrate or minutely 3–lobed, acute; stelidia basal, linear-ligulate, obtuse, 2 mm long; rostellum acute; anther-cap minute, cucullate. *Capsule* narrowly ovoid, 3.2 × 2.1 mm.

DISTRIBUTION. Borneo.

SARAWAK. Lawas District, route from Ba Kelalan to Mount Murud, 30 Sept. 1967, *Burtt & Martin* 5319 (E!, mixed collection with *D. crassifolium*, SAR). Murudi District, Gunung Mulu National Park, Mount Mulu, path from sub-camp 3 to sub-camp 4, 19 Oct. 1977, *Lewis* 343 (K!, SAR!). Marudi District, Mount Murud, Oct. 1922, *Mjöberg* 52 (AMES!). Marudi District, Mount Murud, Lawas, *Paie* S.26455 (E!, K!, L!, SAR, SING).

SABAH. Mount Kinabalu, Pig Hill, *Barkman* 49 (Sabah Parks Herbarium, Kinabalu Park). Mount Kinabalu, Pig Hill, 25 May 1984, *Beaman* 9888 (K!). Mount Kinabalu, Menteki Ridge, 6 Jan. 1995, *Beaman* 11196 (K!). Kemburongoh, Oct. 1933, *Carr* 3752 (SING!). Paka-paka

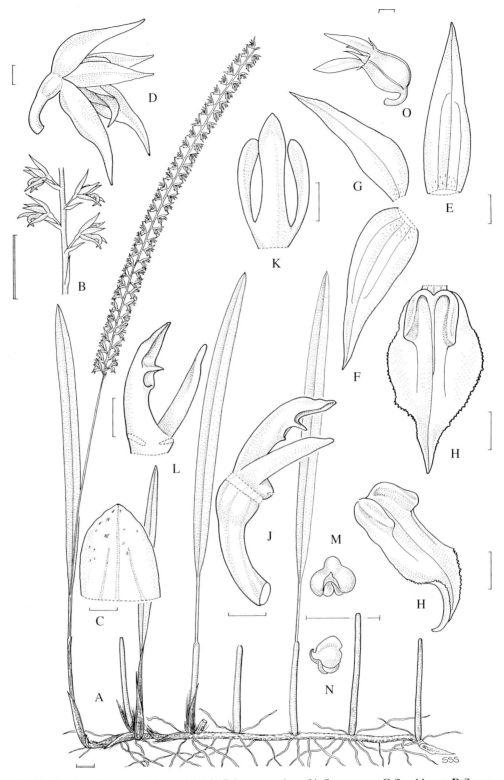

Fig. 80. *Dendrochilum lancilabium*. **A** habit; **B** lower portion of inflorescence; **C** floral bract; **D** flower, side view; **E** dorsal sepal; **F** lateral sepal; **G** petal; **H** labellum, front and oblique views; **J** pedicel with ovary and column, stelidium and anther-cap removed, side view; **K** column, back view; **L** column, stelidium and anther-cap removed, oblique view; **M** anther-cap, front view; **N** anther-cap, side view; **O** fruit capsule. **A** from *Leopold* SAN 71928, **B** from *Paie* S. 26455, **C–N** from *Lewis* 343, **O** from *Burtt & Martin* B. 5319. Scale: single bar = 1 mm; double bar = 1 cm. Drawn by Susanna Stuart-Smith.

Cave, *J. Clemens* 114 (AMES). Marai Parai Spur, *J. Clemens* 224A (AMES), *J. Clemens 242* (AMES), 23 Nov. 1915, *J. Clemens* 280 (holotype AMES!, isotypes BM!, K!, NY!, SING!) & 3 Dec. 1915, *J.Clemens* 386 (BM!, BO!, E!, K!, SING!). Upper Kinabalu, 12 Nov. 1931, *J. & M.S. Clemens* 27147 (AMES!, BO!, HBG!, K!, L!, SING!). Mesilau River, 26 Dec. 1933, *J. & M.S. Clemens* 51626 (BM!). Lumu-Lumu, 10 Jan. 1932, *J. & M.S. Clemens* s.n. (BM!). Mount Kinabalu, locality unknown, July-Aug. 1916, *Haslam* s.n. (AMES, BM!, E!, K!, SING!), & Sept. 1981, *Sato* 2178 (UKMS!). Pig Hill, 8 June 1993, *SNP* 5526 (Sabah Parks Herbarium, Kinabalu Park). Mount Trus Madi, 4 miles up from kampong Sinua, 4 Nov. 1971, *Leopold* SAN 71928 (K!, L!, SAN!, SING!).

KALIMANTAN TIMUR. West Kutai (Koetai), Mount Kemul (Kemoel), 14 Oct. 1925, *Endert* 4129 (L!).

HABITAT. Lower montane forest, especially on exposed ridges; low mossy and xerophyllous scrub forest on extreme ultramafic rock; on tree trunks and roots; on boulders; thrives as a 'terrestrial' in mossy hummocks, usually creeping within mats of bog moss, sometimes in full sunlight. 1200–2550 m.

A distinct species probably most closely related to *D. subintegrum* Ames, but with, among other things, much narrower leaves and a broader labellum. Ames (1920) comments that the terminal wing, i.e. the apical hood, of the column is very variable, that of the type being three-lobed, three-toothed or almost entire.

The specific epithet is derived from the Latin *lanci*, lanceolate or spear-shaped, and *labium*, a lip, referring to the shape of the labellum.

45. DENDROCHILUM LONGIPES

Dendrochilum longipes J.J. Sm. in *Bull. Jard. Bot. Buitenz.*, ser. 2, 3: 55 (1912). Type: Malaysia, Sarawak, Limbang District, Mount Batu Lawi, Ulu Limbang, flowering May 1911, *Moulton* 15 (holotype possibly destroyed BO, isotypes L!, SAR!). **Fig. 81.**

D. mantis J.J. Sm. in Bull. Jard. Bot. Buitenz., ser. 3, 11: 108 (1931). Type: Indonesia, Kalimantan Timur, West Kutai (Koetai), summit of Mount Kemul (Kemoel), 1800 m, 13 Oct. 1925, *Endert* 3991 (holotype L!).

DESCRIPTION. Epiphyte or terrestrial, sometimes lithophytic. *Rhizome* branching, elongate, internodes 0.9–3 cm long, 0.3–0.5 cm in diameter. *Roots* tough, flexuous, branching, smooth, elongate, 0.5–2 mm in diameter. *Cataphylls* on rhizome tubular, ovate, obtuse, 1–2 cm long; at base of pseudobulbs 3–4, ovate-elliptic, obtuse or subacute, accrescent, 0.5–9 cm long, speckled. *Pseudobulbs* 1.5–6.5 cm apart, often borne at an acute angle to rhizome, narrowly cylindrical, cauliform, 5–13 × 0.3–0.4 cm. *Leaf*: petiole sulcate, 0.4–2 cm long; blade narrowly elliptic, acute or shortly obtuse and somewhat conduplicate, often constricted below apex, stiff-textured, coriaceous, 5.5–18 × 1–3.5 cm, main nerves 5–9, with numerous small transverse nerves, pale brownish green or pale green. *Inflorescences* erect, rigid, densely many-flowered, flowers borne 2–3.2 mm apart; peduncle 1–4 cm long, greenish brown or pinkish; non-floriferous bract solitary, or rarely 2, ovate, obtuse, adpressed, 2–3.5 mm long; rachis quadrangular, 17–28 cm long; floral bracts broadly ovate to semi-orbicular, obtuse, erose, spreading, 2–3.7 × 2.7–2.8 mm. *Flowers* fragrant, variously described as yellowish green, tinged brownish, labellum light salmon-pink, or sepals and petals translucent greenish-yellow, light brown, glistening flesh-pink or very pale green, flushed pink, labellum pink, dull

185

Fig. 81. *Dendrochilum longipes*. **A–C** habits; **D** lower portion of inflorescence; **E** floral bract; **F** flower, side view; **G** dorsal sepal; **H** lateral sepal; **J** petal; **K** labellum, front view; **L** labellum, side view; **M** labellum, front view; **N** labellum, oblique view; **O** basal portion of labellum showing keels, front view; **P** pedicel with ovary, labellum and column, side view; **Q** column, oblique view; **R** column, back view; **S** column, side view, **T** anther-cap, back view; **U** pollinia; **V** fruit capsule. **A, K, L & S** from *Endert* 3991 (holotype of *D. mantis*), **B** from *Yii* S. 44432, **C** from *Kato et al.* B11071, **D–J, M–R & T–V** from *Lewis* 345. Scale: single bar = 1 mm; double bar = 1 cm. Drawn by Susanna Stuart-Smith.

deep red, peach, pale orange or with red mid-lobe, yellow side lobes, calli yellow or creamy white, column yellow or pale flesh-pink. *Pedicel with ovary* clavate, 3 mm long. *Sepals* 3–nerved. *Dorsal sepal* ovate-oblong to oblong-elliptic, shortly acute to acuminate, concave, 3–3.7 × 1.2–1.5 mm. *Lateral sepals* obliquely oblong-ovate, shortly acute to acuminate, concave, subcarinate, 3–3.6 × 1.4–1.7 mm. *Petals* obliquely narrowly elliptic, sometimes somewhat falcate, acute, concave, 2.8–3.5 × 0.8–1 mm, 1– to obscurely 3–nerved. *Labellum* subentire, strongly decurved at middle, 3–nerved, 2–2.6 mm long, 1.4 mm wide across side lobes when flattened; side lobes obscure, erect, quadrangular or semi-orbicular and rounded, slightly undulate, 1 mm long; mid-lobe broadly ovate-triangular, rounded at base, acute, undulate, papillose, 1.5–1.6 × 1.5–1.8 mm; disc with 2 prominent, tall, semi-orbicular keels adnate above base to margin of side lobes and terminating at or just beyond junction of side lobes and mid-lobe, appearing as a flange each side running to margin of side lobes and forming a concave area. *Column* curved, 2 mm long; foot distinct; apical hood obtuse and rounded or obscurely 3–lobed; stelidia basal, linear-ligulate, dilated and obliquely obtuse, spathulate-truncate and convex at apex, often subfalcate, longer than column, 2.2–3.1 mm long; rostellum shortly ovate-triangular, acute, convex; stigmatic cavity semi-orbicular, lower margin elevate; anther-cap cucullate, broadly ovate-triangular, apex narrowly truncate, 0.5–0.6 mm wide.

DISTRIBUTION. Borneo

SARAWAK. Limbang District, lower slope of Mount Batu Lawi, Bario, 17 Aug. 1985, *Awa & Lee* S.50900 (K!, L, SAR). Lawas District, route from Ba Kelalan to Mount Murud, near Camp IV, 5 Oct. 1967, *Burtt & Martin* B.5399 (E!). Marudi District, Gunung Mulu National Park, Ulu Sungai Tutoh, ridge path to summit of Mount Mulu, 13 June 1975, *Chai* S.35896 (K!, KEP!, L!, MO, SAR, SING!). Marudi District, Gunung Mulu National Park, path from sub-camp 3 to sub-camp 4, 19 Oct. 1977, *Lewis* 345 (K!). Marudi District, Gunung Mulu National Park, Mount Mulu summit, 20 Oct. 1977, *Lewis* 364 (K!). Marudi District, Mount Murud, Oct. 1922, *Mjöberg* 66 (AMES!). Kapit District, Batu Tiban Hill, 1920's, *Mjöberg* s.n. (AMES!) & *Mjöberg* s.n. (AMES!). Limbang District, Mount Batu Lawi, Ulu Limbang, May 1911, *Moulton* 15 (holotype of *D. longipes* possibly destroyed BO, isotypes L! SAR!). Marudi District, Kelabit Highlands, eastern path to summit of Mount Murud, 5 April 1970, *Nooteboom & Chai* 01954 (L!). Marudi District, Kelabit Highlands, summit of Mount Batu Lawi, 28 April 1970, *Nooteboom & Chai* 02278 (K!, L!, SAR!). Marudi District, Mount Murud, second sandstone summit, 12 Sept. 1982, *Yii* S.44432 (K!, L!, SAR, SING!).

KALIMANTAN BARAT. Serawai, summit of Raya Hill (Bukit Raya), 21 Oct. 1995, *Church et al.* 2610 (AMES, BO, K!).

KALIMANTAN TENGAH. Raya Hill (Bukit Raya) and upper Katingan (Mendawai) River area, south-east side of Raya Hill (Bukit Raya), Upper Samba River, 5 Dec. 1982, *Mogea* 3963 (BO, L!).

KALIMANTAN TIMUR. West Kutai (Koetai), summit of Mount Kemul (Kemoel), 13 Oct. 1925, *Endert* 3991 (holotype of *D. mantis* L!). Mount Buduk Rakik, north of Long Bawan, Apo Kayan, 10 Aug. 1981, *Kato et al.* B11071 (L!).

HABITAT. Lower and upper montane ridge forest; mossy forest; exposed places in cloud forest associated with *Nepenthes* spp., *Rhododendron* spp. and *Vaccinium* spp.; montane ericaceous scrub; on boulders. 1400–2400 m.

Closely allied to *D. subintegrum* Ames but distinguished primarily by the narrower leaves, broader labellum mid-lobe, prominent flange-like keels adnate to the margin of the side lobes and stelidia which exceed the apical hood of the column and are dilated at the apex.

J.J. Smith described *D. mantis* from Mount Kemul as late as 1931 stating that it "must be a near ally of *D. subintegrum* Ames." From examination of the type it is clear that *D. mantis* is conspecific with *D. longipes* which Smith had described much earlier in 1912. He apparently either overlooked or had forgotten *D. longipes* when studying the material from Mount Kemul.

The specific epithet is derived from the Latin *longus*, long, elongate and refers to the elongated rhizomes and pseudobulbs.

46. DENDROCHILUM SUBINTEGRUM

Dendrochilum subintegrum Ames, *Orchidaceae* 6: 65–66 (1920). Type: Malaysia, Sabah, Mt Kinabalu, Lubang (Lobong) Cave, Nov. 1915, *J.Clemens* 285 (holotype AMES!). **Fig. 82.**

DESCRIPTION. Epiphyte or terrestrial. *Rhizome* branching, elongate, internodes 2.8–4.5 cm long, up to 0.6 cm in diameter. *Roots* tough, flexuous, branching mostly distally, very minutely papillose, 0.3–1 mm in diameter. *Cataphylls* on rhizome tubular, ovate, obtuse, 1–3 cm long, apical margins dark green to brown; at base of pseudobulbs 3, ovate-elliptic, obtuse to acute, accrescent, 3–7 cm long, dark green to brown distally. *Pseudobulbs* 4–12 cm apart, often borne at an acute angle to rhizome, narrowly cylindrical, cauliform, (4–)8–11 × 0.4–0.5 cm. *Leaf*: petiole sulcate, 1–2.5 cm long; blade broadly elliptic or oblong-elliptic, obtuse and shortly mucronate, sometimes constricted below apex, coriaceous, (7.5–)16–18.5 × 1.5–5 cm, main nerves 7–9. *Inflorescences* erect to gently curving, densely many-flowered, flowers borne 1–2 mm apart; peduncle slender, 8–14 cm long; non-floriferous bracts 1 or 2, ovate-elliptic, acute, 1.5–4 mm long; rachis quadrangular, 12–16 cm long; floral bracts oblong to ovate-oblong, obtuse, 2–2.5 mm long. *Flowers* sweetly fragrant, cream, yellowish or very pale greenish, labellum white or pale greenish, pale orange or ochre in centre. *Pedicel with ovary* very narrowly clavate, curved, 1.9–2 mm long. *Sepals* and *petals* 3–nerved. *Dorsal sepal* oblong-lanceolate, acute, 3–4.1 × 1 mm. *Lateral sepals* slightly obliquely oblong-lanceolate, almost subfalcate, acute, 3–4 × 1 mm. *Petals* oblong-lanceolate, subfalcate, acute, minutely erose, 2.5–3.5 × 0.9–1 mm. *Labellum* shortly clawed, versatile, narrowly trullate-lanceolate, subentire, acute, 3–nerved, 2.25–2.9 × 0.8–0.9 mm; side lobes minute, rounded or triangular, irregularly toothed, mid-lobe margin entire; disc with a low M-shaped basal callus terminating at about junction of side lobes and mid-lobe. *Column* curved, 1.5–1.6 mm long; foot distinct; apical hood obscurely triangular, obtuse; stelidia basal, ligulate, subfalcate, obtuse or subacute, 1.2–1.6 mm long; rostellum obtuse; stigmatic cavity narrowly oblong; anther-cap cucullate, ovate-triangular.

DISTRIBUTION. Borneo.

SARAWAK. Belaga District, Batu Laga, Kapit, 29 Aug. 1984, *Mohtar* S.48086 (AAU, K!, L!, MEL, SING). Marudi District, Tama Abu Range (Mount Temabok), Upper Baram Valley, 5 Nov. 1920, *Moulton* SFN 6670 (K!, SING!).

SABAH. Mount Kinabalu, Lubang (Lobong) Cave, Nov. 1915, *J. Clemens* 285 (holotype AMES!). Sipitang District, about halfway along Maligan to Long Pa Sia trail, Dec. 1986, *Vermeulen & Duistermaat* 931 (K!, L!). Sipitang District, along Long Pa Sia to Long Samado trail, near crossing with Malabid River, Dec. 1986, *Vermeulen & Duistermaat* 959 (K!, L!). Sipitang District, Rurun River headwaters, Dec. 1986, *Vermeulen & Duistermaat* 1042 (K!, L!).

HABITAT. Lower montane forest; low open kerangas forest, often with *Pandanus* spp. and rattan understorey. 1200–1500 m.

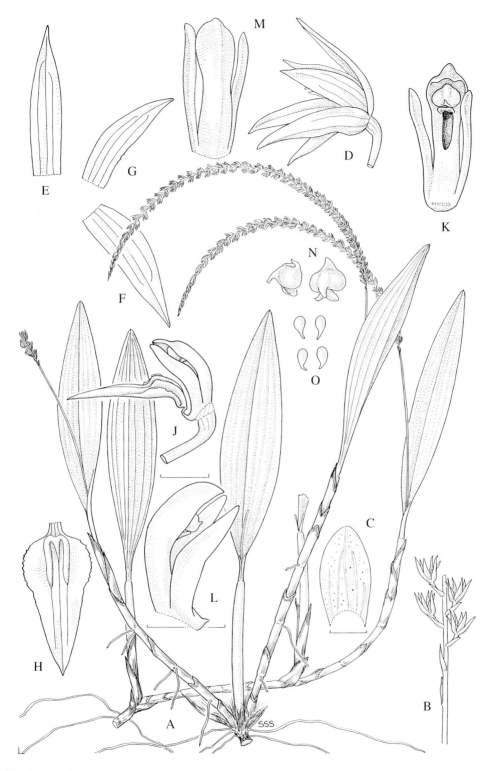

Fig. 82. *Dendrochilum subintegrum.* **A** habit; **B** lower portion of inflorescence; **C** floral bract; **D** flower, side view; **E** dorsal sepal; **F** lateral sepal; **G** petal; **H** labellum, front view; **J** pedicel with ovary, labellum and column, side view; **K** column, front view; **L** column, side view; **M** column, back view; **N** anther-cap, side and back views; **O** pollinia. **A & B** from *Vermeulen & Duistermaat* 959, **C–O** from *Moulton* SFN 6670. Scale: single bar = 1 mm; double bar = 1 cm. Drawn by Susanna Stuart-Smith.

Ames, in his original description, states that the pseudobulbs are clustered at the growing point. This is a misinterpretation based on an incomplete type specimen. Two of the 'clustered pseudobulbs' described by Ames are actually branches of the rhizome borne at a node bearing a solitary mature pseudobulb. This habit of growth is identical to *D. longipes*.

The specific epithet is derived from the Latin *sub-*, somewhat, not completely, a little, and *integri-*, entire, whole, and refers to the very obscurely lobed, subentire labellum.

47. DENDROCHILUM GLUMACEUM

Dendrochilum glumaceum Lindl. in *Edward's Bot. Reg.* 27: 23 (1841). Type: Philippines, sine loc., *Cuming*/cult. *Loddiges* s.n. (holotype K!). **Fig. 83, plate 5A.**

Platyclinis glumacea (Lindl.) Benth. ex Hemsl. in *Gdnrs. Chron.* 2, 16: 656 (1881).
Platyclinis glumacea (Lindl.) Benth. ex Hemsl. in *J. Linn. Soc., Botany* 18: 295 (1891), *comb. inval.*
Platyclinis glumacea (Lindl.) Benth. ex Hemsl. var. *valida* Rolfe in *Orchid Rev.* 1: 115 (1893).
    Syntypes (sine loc:) sine coll./cult. Hort. Bot. *Glasnevin* s.n., April 1891 (syn. K!); sine
    coll./cult. *Hunter* s.n. (not located); sine coll./cult. *Peeters* s.n., Feb. 1893 (syn. K!).
Acoridium glumaceum (Lindl.) Rolfe in *Orchid Rev.* 12: 220 (1904).
Dendrochilum glumaceum Lindl. var. *validum* (Rolfe) Pfitzer & Kraenzl., in Engler,
    Pflanzenreich 4, 50, 2 B 7: 105 (1907).
D. uncatum auct., non Rchb. f.: Graf 1963: a fig. s.n. on p. 1256 (excl. text on p. 1594);
    Greatwood 1973: fig. 65 (excl. text on p. 116).

DESCRIPTION. Epiphyte. *Rhizome* abbreviated. *Roots* fleshy, with a few branches, minutely papillose, elongate, up to 3 mm in diameter. *Cataphylls* 3–6, almost perfectly tubular, somewhat inflated, rounded to acute, 0.5–12 cm long, soon disintegrating into persistent fibres. *Pseudobulbs* crowded, fusiform to obpyriform, 1.1–5.2(–6.2) × 0.3–1.3(–2) cm. *Leaf*: petiole sulcate, 1.8–12(–14) cm long; blade linear-lanceolate, lanceolate or elliptic, obtuse to subacute, sometimes constricted 1/4 to 1/3 below apex, thin-textured, 7.6–41.5 × (0.7–)1.9–4.8(–6) cm, main nerves 5–7(–9). *Inflorescences* suberect to curving, densely 13– to many-flowered, flowers borne 2–3.5(–4) mm apart; peduncle slender to stout, terete to subquadrate, 9–37 cm long; non-floriferous bract solitary, sometimes absent, oblong, obtuse and apiculate to acute, 0.5–1 cm long; rachis nodding, sometimes slightly flexuose towards apex, slightly furrowed (5.3–)7–21.5(–25) cm long; floral bracts elliptic (-oblong) or ovate, obtuse to acute, often finely mucronate, entire, 3.6–10.6 × 2.8–5.7 mm. *Flowers* opening from proximal or central part of rachis, sweetly fragrant, sepals and petals white to cream, labellum yellow to orange, rarely white or brown. *Pedicel with ovary* conical to terete, straight, 1.8–4 mm long. *Sepals* and *petals* wide-spreading, 3–nerved. *Dorsal sepal* linear to lanceolate, acuminate, often finely mucronate, 5.3–12.4 × 1.2–2.4 mm. *Lateral sepals* linear to lanceolate, sometimes slightly oblique and subfalcate, acuminate, often finely mucronate, 5.1–13.7 × 1–2.9 mm. *Petals* lanceolate-elliptic, acute to acuminate, often finely mucronate, finely erose-dentate to nearly entire, 3–11 × 1–2.5 mm. *Labellum* versatile, pendent, sessile, 3–lobed, 3–nerved, finely papillose, 2.3–3.7 × (1.2–)1.5–2.7 mm; side lobes erect to spreading, narrowly triangular or obliquely triangular-oblong, obtuse or acute, often serrate-dentate, free portion 1.1–1.2 mm long; mid-lobe broadly elliptic to suborbicular, obtuse, rounded or subacute, margin irregular to finely crenate; disc with 2 fleshy, entire to serrate-dentate keels located on lateral nerves, running from near base to about middle of labellum or shorter,

190

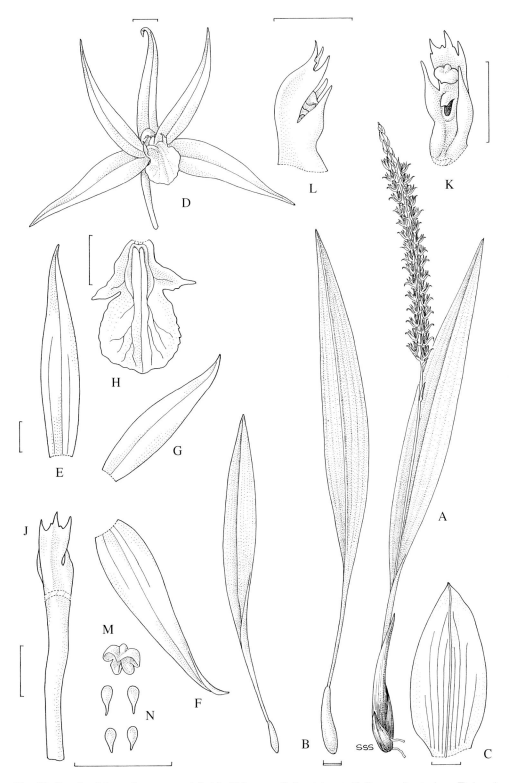

Fig. 83. *Dendrochilum glumaceum*. **A** habit; **B** leaves; **C** floral bract; **D** flower, front view; **E** dorsal sepal; **F** lateral sepal; **G** petal; **H** labellum, side view; **J** pedicel with ovary and column, back view; **K** column, oblique view; **L** column, side view; **M** anther-cap, interior view; **N** pollinia. **A–N** from *Bogor* cult., no. 928.11.46. Scale: single bar = 1 mm; double bar = 1 cm. Drawn by Susanna Stuart-Smith.

usually free at base, occasionally united. *Column* suberect, slightly incurved, 1.5–2.7 mm long; foot short; apical hood laciniate, irregularly (2–) 3– to 5–toothed; stelidia subbasal, suberect, triangular-oblong to ligulate-falcate, obtuse to acute, slightly shorter than apical hood; rostellum barely protruding, subtriangular, shortly truncate; stigmatic cavity subcircular, margins slightly elevate; anther-cap ovate-cordate, with a distinct conical apical wart. *Capsule* ellipsoid, somewhat adpressed, extant column forming a short beak, 1.8–1.9 × 1–1.2 cm.

DISTRIBUTION. Borneo and the Philippines.

KALIMANTAN TENGAH. Kotawaringin, cult. *Bogor*, no. 928. 11.46, flowered March 1929 (BO!).

HABITAT. Not recorded for Borneo, but in the Philippines it is usually found in mossy forest between 500 and 2330 m elevation.

Popular in cultivation, *D. glumaceum* is a common species in the Philippines where populations exhibit great variation in stature and leaf-width.

The specific epithet is derived from the Latin *gluma*, a husk of corn, resembling the chaffy two-ranked bract-like organs subtending the spikelets of grasses and related plants which the prominent floral bracts resemble.

48. DENDROCHILUM HAVILANDII

Dendrochilum havilandii Pfitzer in Engler, *Pflanzenreich* 4, 50, 2 B 7: 107–108, fig. 35L (1907). Type: Malaysia, Sarawak, *Haviland* 2346 (holotype K!; isotype SING!).

D. hewittii J.J. Sm. in *Bull Dép. Agric. Indes Néerl.* 22: 14–15 (1909). Type: Malaysia, Sarawak, Kuap (Quop), *Hewitt* s.n. (holotype BO!). **Fig. 84.**

DESCRIPTION. Epiphyte or lithophyte. *Rhizome* clothed in fibrous remains of cataphylls, to *c.* 6 cm long. *Roots* branching, flexuose, smooth, 0.3–1 mm in diameter. *Cataphylls* 3–4(–5), lanceolate, acuminate, smooth and shiny, 1.3–15 cm long, disintegrating into persistent fibres, pale brown. *Pseudobulbs* crowded, approximate, cylindrical, terete to narrowly ovate, 4.5–9 × 0.5–1 cm. *Leaf*: petiole conduplicate, 3.5–7.5 cm long; blade narrowly elliptic to lanceolate-obovate, obtuse to subacute, cuneate, attenuate below, often constricted 0.4–1.2 cm below apex, tough, coriaceous, 11–28(–32) × 2–3.5(–3.8) cm, shiny above, mat beneath, main nerves 5, with numerous irregular transverse nerves. *Inflorescences* erect, densely many-flowered, flowers borne 2–2.5(–3) mm apart; peduncle terete, 12.5–28 cm long, *c.* 1.5 mm in diameter; non-floriferous bract solitary, ovate-elliptic, acuminate, 4–9 mm long; rachis quadrangular, 10.5–16 cm long; floral bracts broadly ovate, acute to acuminate, involute, spreading, 5–8 × 4 mm, sometimes dorsally punctate, white. *Flowers* white or cream fading to pale lemon-yellow or pale greenish. *Pedicel with ovary* clavate, 2 mm long. *Sepals* and *petals* 3–nerved. *Dorsal sepal* linear-lanceolate, apex conical-acute, reflexed distally, 7.3–8 × 1.5–1.7 mm. *Lateral sepals* slightly obliquely linear-lanceolate, often somewhat sigmoid, apex incurved, conical-apiculate, 7–7.1 × 1.2–1.7 mm. *Petals* linear-lanceolate, acute, somewhat concave, apex often reflexed, 6–7 × 1.1–1.4 mm. *Labellum* versatile, sessile, sigmoid, strongly decurved below middle, slightly concave at base, 3–lobed, 3–nerved, minutely finely papillose, (4–)4.6–4.7 × 1.4–1.5 mm; side lobes tiny, subulate-falcate, acute, tooth-like, erect or decurved; mid-lobe oblong-elliptic or oblong-rhomboid, obtuse to subacute, concave; disc with 2 short, erect, parallel keels extending from above base and terminating just beyond side lobes, median nerve

Fig. 84. *Dendrochilum havilandii.* **A** habit; **B** leaf apex; **C** floral bract; **D** flower, side view; **E** dorsal sepal; **F** lateral sepal; **G** petal; **H** labellum, front view; **J** pedicel with ovary, labellum and column, side view; **K** pedicel with ovary and column, side view; **L** column, front view; **M** column, back view. **N** anther-cap, back and interior views; **O** pollinia. **A** from *Yii* S. 39541, **B** from *Paie* S. 28076, **C–O** from *Cribb* 89/15. Scale: single bar = 1 mm; double bar = 1 cm. Drawn by Susanna Stuart-Smith.

raised and prominent, especially towards base. *Column* gently incurved, 2.5 mm long; foot prominent; apical hood irregularly toothed, somewhat truncate; stelidia borne centrally each side of stigmatic cavity, linear-ligulate, acute to acuminate, equalling apical hood, 1.8–1.9 mm long; rostellum ovate-triangular, obtuse; stigmatic cavity semi-ovate; anther-cap cucullate, conical, with a conical apical wart.

DISTRIBUTION. Borneo.

SARAWAK. Bau District, Bau/Seburan, 20 Oct. 1958, *J.A.R. Anderson* S.11099 (K!, L!, SAR!). Samarahan District, Kuap (Quop), 25 Aug. 1912, *J.W. Anderson* s.n. (SING!). Bau District, Bidi Cave, 19 Oct. 1929, *J. & M.S. Clemens* 20729 (AMES!, BO!, K!, L!, SAR!, SING!). Bau District, Bau, 11 Nov. 1989, *Cribb* 89/15 (K!). Locality unknown, 1893, *Haviland* 2346 (holotype of *D. havilandii* K!, isotype SING!). Samarahan District, Kuap (Quop), *Hewitt* s.n. (holotype of *D. hewittii* BO!). Samarahan District, Kuap (Quop), Nov. 1906, *Hewitt* 9 (SING!). Samarahan District, Kuap (Quop), Dec. 1906, *Hewitt* s.n. (K!, SAR!). Samarahan District, Kuap (Quop), 17 Dec. 1907, *Hewitt* 1 (SING!). Samarahan District, Kuap (Quop), Oct. 1908, *Hewitt* s.n. (K!). Serian District, Selabor Hill, Lobang Mawang, Tebekang road, 27 Sept. 1968, *Paie* S. 28076 (E!, K!, L!, SAR, SING!). Marudi District, Gunung Mulu National Park, summit of Berar Hill, 28 Sept. 1977, *Yii* S.39541 (K!, L!, MEL, SAR). Miri District, Baram, *Unknown collector* (probably *Hose*) 25 (W!).

HABITAT. Lowland and hill forest on limestone; limestone rocks and cliffs in light shade. Sea level– 400 m.

Four Hewitt collections are extant, each annotated with a different date. These are deposited at K, SAR and SING. A collection at BO, without a date, labelled as the type of *D. hewittii*, has been remounted together with a Herb. Hort. Bot. Bog. label annotated in J.J. Smith's hand. Unfortunately, Smith never specified which of the Hewitt collections was the holotype and the original Hewitt label providing the date of collection has been lost.

The specific epithet honours George Darby Haviland (1857–1901), surgeon and naturalist and a Curator of the Sarawak Museum in Kuching, who collected the type.

49. DENDROCHILUM ANGUSTILOBUM

Dendrochilum angustilobum Carr in *Gdns. Bull. Straits Settl.* 8: 222–223 (1935). Type: Malaysia, Sabah, Mount Kinabalu, Tenompok, *c.* 1500 m, April 1933, *Carr* 3233, SFN 26874 (holotype SING!; isotypes AMES!, K!). **Fig. 85, plate 1C & D.**

DESCRIPTION. Epiphyte. *Rhizome* abbreviated, branching. *Roots* numerous, branching, flexuose, minutely papillose, 0.3–1.1 mm in diameter. *Cataphylls* 2–4, lanceolate or ovate-elliptic, obtuse to acute, smooth and shiny, 0.6–3 cm long, disintegrating into persistent fibres, chestnut-brown, sometimes speckled. *Pseudobulbs* crowded, mostly approximate, narrowly ovoid or fusiform, smooth, 1–5 × 0.4–1 cm, shiny grey-green, more or less suffused red. *Leaf*: petiole sulcate, 2–7 mm long; blade ovate, oblong-elliptic or oblong-oblanceolate, obtuse to subacute, sometimes constricted a little way below apex, thin-textured, (3.5–)5.5–8(–10) × (0.9)2.5–3.1 cm, green above, with narrow red margins, or suffused red, grey-green beneath, often suffused red, main nerves 7–9, with numerous tiny transverse nerves. *Inflorescences* pendulous, subdensely many-flowered, often 80 or more flowers per inflorescence, flowers

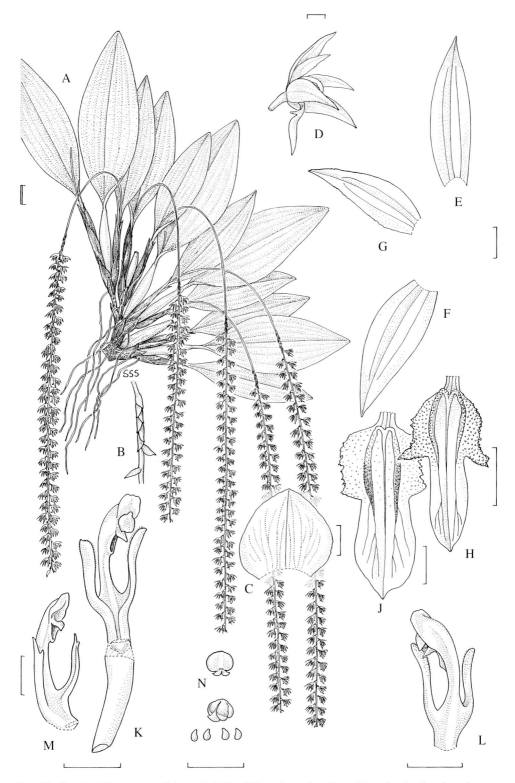

Fig. 85. *Dendrochilum angustilobum*. **A** habit; **B** junction of rachis with peduncle showing six non-floriferous bracts and two floral bracts; **C** floral bract; **D** flower, side view; **E** dorsal sepal; **F** lateral sepal; **G** petal; **H**, labellum, front view; **J** labellum, front view; **K** pedicel with ovary and column, front view; **L** column, back view; **M** column, oblique view; **N** anther-cap, side and back views, and pollinia. **A–H, K–L & N** from *Carr* C. 3233, SFN 26874 (holotype), **J & M** from *Beaman* 7309. Scale: single bar = 1 mm; double bar = 1 cm. Drawn by Susanna Stuart-Smith.

borne 3–4 mm apart; peduncle filiform, slightly dilated distally, 2–5.5 cm long; non-floriferous bracts 3–8, ovate, obtuse, erose, imbricate, 2–5 mm long; rachis quadrangular, 18–35 cm long, pale dull red; floral bracts ovate, obtuse to acute, 2.5–3(–4) mm long, bright flesh-coloured. *Flowers* often first opening near centre of inflorescence, sepals and petals bright yellowish green, sepals with a darker keel outside, labellum pale olive, side lobes pale greenish white, mid-lobe yellowish green with 2 converging brown streaks, column and stelidia greenish, tipped darker. *Pedicel with ovary* narrowly clavate, 2–3 mm long. *Sepals* and *petals* 3–nerved, a little incurved about the middle, margins recurved. *Dorsal sepal* oblong-elliptic, acute, *c.* 5.3 × 1.8 mm. *Lateral sepals* slightly obliquely ovate-elliptic, acute, subfalcate, *c.* 5 × 2 mm. *Petals* oblong-elliptic, acute, margins minutely erose, subfalcate, *c.* 4.8 × 1.7 mm. *Labellum* stipitate to column-foot by a narrowly oblong claw, 3–lobed, 3–nerved, *c.* 3.7 mm long, *c.* 2 mm wide across side lobes when flattened, often provided with 1 or more minute teeth in the sinus of the lobes; side lobes rounded, forming with the blade an ovate or suborbicular hypochile, free apical portion narrowly triangular, posterior margin minutely erose, surface minutely papillose; mid-lobe oblong-oblanceolate or oblong from a cuneate base, rounded and shortly apiculate, margins sometimes very minutely erose distally, *c.* 2.3 × 0.9 mm; disc with 2 papillose keels incurved and contiguous at base, extending from apex of claw and terminating at junction of side lobes and mid-lobe, or a little beyond, median nerve thickened and elevate, especially towards base. *Column* gently incurved, *c.* 2.5 mm long; foot short; apical hood ovate, truncate, obscurely toothed; stelidia arising above the base, linear-ligulate, obtuse, sometimes bifid at apex, minutely papillose distally or glabrous, reaching to just beyond rostellum or to middle of apical hood; rostellum ovate-triangular, prominent; stigmatic cavity narrowly oblong, anther-cap ovate, cucullate.

DISTRIBUTION. Borneo.

Kota Kinabalu to Sinsuron (Sensuron) road, Mile 27, bought from roadside stall, Dec. 1971, *Bacon* 183 (E!). Tambunan District, Crocker Range, Km 55.4 on Kota Kinabalu to Tambunan road, 30 Oct. 1983, *Beaman* 7309 (K!). Tambunan District, Crocker Range, Km 59.5 on Kota Kinabalu to Tambunan road, 2 Nov. 1983, *Beaman* 7373 (K!). Mount Kinabalu, Tenompok, April 1933, *Carr* 3233, SFN 26874 (holotype SING!, isotypes AMES!, K!). Mount Kinabalu, Tenompok, 4 April 1932, *J. & M.S. Clemens* 28949 (BM!), 7 April 1932, *J. & M.S. Clemens* 29295 (BM!) & 22 April 1932, *J. & M.S. Clemens* 29361 (BM!, E!, HBG!, K!, L!). Mount Kinabalu, Tenompok/Tomis, 2 May 1932, *J. & M.S. Clemens* 29412 (BM!, E!). Mount Kinabalu, Kinateki River Head, 16 Jan. 1933, *J. & M.S. Clemens* s.n. (BM!). Mount Alab, 1987, *Leiden* cult. (*Vermeulen*) 26514 (K!, L!). Mount Kinabalu, Park Headquarters, May 1967, *Price* 217 (K!).

HABITAT. Lower montane forest on sandstone and shale bedrock; oak-laurel forest. 1400–2000 m.

Beaman 7309 and 7373 differ in having a broader labellum mid-lobe up to 1.8–1.9 mm wide, while 7309 also has bifid stelidia.

The specific epithet is derived from the Latin *angustatus*, narrowed, and *lobus*, a lobe, referring to the unusually narrow labellum mid-lobe.

50. DENDROCHILUM LONGIRACHIS

Dendrochilum longirachis Ames, *Orchidaceae* 6; 60–62 (1920). Type: Malaysia, Sabah, Mount Kinabalu, Kiau, 29 Nov. 1915, *J. Clemens* 332 (holotype AMES; isotypes BO!, K!). **Fig. 86, plate 11B.**

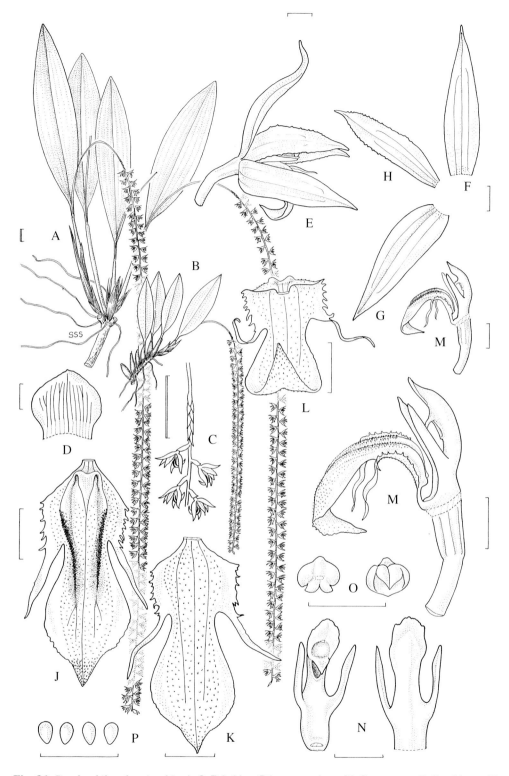

Fig. 86. *Dendrochilum longirachis*. **A & B** habits; **C** lower portion of inflorescence; **D** floral bract; **E** flower, side view; **F** dorsal sepal; **G** lateral sepal; **H** petal; **J** labellum, flattened, front view; **K** labellum, flattened, back view; **L** labellum, natural position, back view; **M** pedicel with ovary, labellum and column, side view; **N** column, anther-cap removed, front and back views; **O** anther-cap, back and interior views. **P** pollinia. **A** from *Nooteboom & Chai* 01890, **B** from *Richards* S. 483, **C–P** from *Wood* 694. Scale: single bar = 1 mm; double bar = 1 cm. Drawn by Susanna Stuart-Smith.

DESCRIPTION. Epiphyte. *Rhizome* abbreviated, branching, to 3 cm long. *Roots* numerous, forming a dense mass, branching, flexuose, smooth, 0.2–1.5 mm in diameter. *Cataphylls* 3–4, ovate-elliptic, obtuse to acute, 0.5–5 cm long, speckled. *Pseudobulbs* crowded, approximate, cylindrical to narrowly fusiform, (1.5–)5–9.5 × 0.5–0.7 cm, broadest proximally, shiny green. *Leaf*: petiole sulcate, (2–)6–11 mm long; blade oblong-elliptic to elliptic, rounded to acute, thin-textured, (4–)8–15(–19) × (1.5–)3–4.3 cm, main nerves (7–)9, with numerous tiny transverse nerves. *Inflorescences* pendulous, laxly up to 130 or more-flowered, flowers borne (2.5–)4 mm apart; peduncle filiform, slightly dilated distally, 3–8.5 cm long, brownish pink; non-floriferous bracts (3–)5–9, ovate, obtuse, erose, 1.5–2 mm long, pinkish brown; rachis quadrangular, 20–45 cm long, 1 mm in diameter, brownish pink; floral bracts broadly ovate, rounded, somewhat erose, 2–2.5 mm long, salmon-pink. *Flowers* slightly sweet-scented, sepals and petals greenish yellow, palest green, translucent yellow, dull greenish cream or white, labellum yellow or pale green, side lobes often white, keels brown, column pale yellow. *Pedicel with ovary* narrowly clavate, 2–2.2 mm long. *Sepals* and *petals* 3–nerved. *Dorsal sepal* oblong-elliptic, acute, 4.5–5.5 × 1.5 mm. *Lateral sepals* slightly obliquely oblong-elliptic, acute, 4.5–5.6 × 1.6–1.7 mm. *Petals* linear-elliptic, acute, entire or minutely erose, 4–5.5 × 1 mm. *Labellum* stipitate to column-foot by a narrowly oblong claw 0.1–0.2 mm long, 3–lobed, 3–nerved, curved, minutely papillose, (3.5–)4.5–4.6 mm long, 1.8–1.9 mm wide across side lobes; side lobes narrowly triangular with a narrow setaceous free apical portion 1.5–1.6 mm long, reaching to middle of mid-lobe, outer margin irregularly serrate; mid-lobe cuneate-obovate, apiculate, margin entire or minutely erose, 2.5–2.9 × 1.5–2 mm; disc with 2 erect-spreading, fleshy, partially papillose keels joined at the base with the thickened or elevate, median nerve, extending from apex of claw and terminating on proximal portion of mid-lobe, elevate portion of median nerve terminating about halfway along disc. *Column* gently incurved, 2.5–2.6 mm long; foot short; apical hood oblong, rounded to truncate, entire or obscurely erose; stelidia median, arising just below level of stigmatic cavity, linear-subulate, acute, slightly exceeding anther-cap, 1.4–1.5 mm long; rostellum ovate-triangular, rostrate; stigmatic cavity narrowly oblong-elliptic, lower margin elevate; anther-cap cucullate, minutely papillose.

DISTRIBUTION. Borneo.

SARAWAK. Marudi District, Kelabit Highlands, Mount Murud east, Kelapang (Belapan) River, 3 April 1970, *Nooteboom & Chai* 01890 (L!, SAR!). Belaga District, Mount Dulit, Ulu Koyan, 15 Sept. 1932, *Richards* S.483 (K!, SING!).

SABAH. Mount Kinabalu, below Kemburongoh, *Barkman* 264 (Sabah Parks Herbarium, Kinabalu Park.) Mount Kinabalu, Kiau, 29 Nov. 1915, *J. Clemens* 332 (holotype AMES, isotypes BO!, K!). Mount Kinabalu, Marai Parai Spur, 2 Dec. 1915, *J. Clemens* 377 (AMES, BM!, K!, SING!). Mount Kinabalu, Penibukan, 30 Dec. 1932, *J. & M.S. Clemens* 30601 (BM!, BO!, E!, L!) & 16 Jan. 1933, *J. & M.S. Clemens* 31002 (BM!, BO!). Mount Kinabalu, locality unknown, 1933, *J. & M.S. Clemens* 35205 (BM!, BO!). Mount Kinabalu, Gurulau Spur, Nov. 1933, *J. & M.S. Clemens* 51080 (BM!, K, mixed sheet with *D. grandiflorum*!). Mount Kinabalu, Tahubang River, 4 Jan. 1933, *J. & M.S. Clemens* s.n. (BM!). Crocker Range, Sinsuron road, Sept. 1979, *Lamb* SAN 89674 (K, sketch only!, SAN). Mount Kinabalu, Mamut River, 27/28 Oct. 1981, *Sato et al.* 1558 (UKMS!). Sipitang District, ridge between Maga River and Pa Sia River, 18 Oct. 1986, *de Vogel* 8375 (L!). Sipitang District, Ulu Long Pa Sia, 8 km Northwest of Long Pa Sia, along Pa Sia River, 26 Oct. 1985, *Wood* 694 (K!).

KALIMANTAN TIMUR. Apo Kayan, en route from Pa Panik to Pa Pelinitan, north of Long Bawan, 29 July 1981, *Kato et al.* B 10110 (BO!, L!).

HABITAT. Forest on sandy soil; riverine lower montane forest with *Rhododendron* understorey; rather dense primary forest to 30 metres high dominated by *Agathis* spp. and *Lithocarpus* spp. on poor sandy soil over sandstone. 1100–1500(–2000) m.

The specific epithet is derived from the Latin *longus*, long, and *rhachis*, the rachis, i.e. the axis of the inflorescence above the peduncle, which is very elongate in this species.

51. DENDROCHILUM IMITATOR

Dendrochilum imitator J.J. Wood in Wood & Cribb, *Checklist orch. Borneo*: 179–180, fig. 22G (1994). Type: Malaysia, Sabah, Sipitang District, trail from Long Pa Sia to Long Samado, *c.* 4 km Southwest of Long Pa Sia, 1300 m, 22 Oct. 1986, *de Vogel* 8451 (holotype L!; isotype K!). **Fig. 87.**

DESCRIPTION. Pendulous epiphyte. *Rhizome* branching, to 18 cm long. *Roots* branching, flexuose, densely minutely papillose, 0.2–1 mm in diameter. *Cataphylls* 2–3(–4), ovate-elliptic, acute, 1–6 cm long, speckled brown. *Pseudobulbs* crowded, approximate, narrowly cylindrical or narrowly fusiform, 1.2–3.5(–4.5) × 0.2–0.3 cm. *Leaf*: petiole sulcate, 3–6 mm long; blade narrowly elliptic, acute, thin-textured, 4–10 × 0.8–1.4 cm, main nerves 5, with numerous tiny transverse nerves. *Inflorescences* pendulous, laxly 6– to 20 or more-flowered, flowers borne 3–5 mm apart; peduncle filiform, 4.5–9 cm long; non-floriferous bracts 1–2, ovate, acute, 2–5 mm long; rachis quadrangular, 2.5–10.5 cm long; floral bracts ovate, acute, 2–2.5 mm long. *Flowers* pale green, dull pale greenish brown or reddish tan, labellum white with a large central brownish blotch, or with yellow brown keels, median keel pale green, column cream, stelidia speckled brownish inside. *Pedicel with ovary* narrowly clavate, slightly curved, 4 mm long. *Sepals* 3–nerved. *Dorsal sepal* oblong-elliptic, concave, acute, often reflexed, 4.5–5 × 1–1.2 mm. *Lateral sepals* obliquely ovate-elliptic, acute, reflexed, 4–4.5 × 1.5–1.9 mm. *Petals* ligulate, subacute, slightly falcate, 4–5 × 0.9–1 mm. *Labellum* stipitate to column-foot by a narrowly oblong claw, 3–lobed, 3–nerved, minutely papillose, 3.2–4 mm long, 1.6–2.1 mm wide across mid-lobe; side lobes triangular, acute, outer margin sometimes irregularly erose-denticulate; mid-lobe obovate, rounded to subtruncate, sometimes with a tiny apical mucro; disc with 3 fleshy papillose keels, the outer contiguous at base and extending to at or just below base of mid-lobe, median keel broadest at base, narrowing at the middle and usually becoming pronounced again distally, extending almost to the apex of the mid-lobe. *Column* curved, 3.5 mm long; foot 0.5–0.8 mm long; apical hood ovate-flabellate, cucullate, margin sometimes reflexed and often slightly uneven; stelidia arising just above base, ligulate or narrowly oblong, apex bifid, 1.5 mm long; stigmatic cavity narrowly oblong; rostellum triangular-ovate, acute, prominent; anther-cap cucullate, with an apical wart.

DISTRIBUTION. Borneo.

SARAWAK. Marudi District, Gunung Mulu National park, Northwest ridge of Mount Tamacu, 8 May 1978, *Argent & Coppins* 1202 (SAR!).

SABAH. Ranau District, Mamut Copper Mine, ridge above quarry at guard house on road from Lohan, 28 June 1984, *Beaman* 10347 (K!). Sipitang District, Maga River, at confluence with Pa Sia River, 19 Oct. 1986, *de Vogel* 8426 (L!). Sipitang District, along trail from Long Pa Sia to Long Samado, *c.* 4 km Southwest of Long Pa Sia, 22 Oct. 1986, *de Vogel* 8451 (holotype L!, isotype K!). Sipitang District, along trail from Long Pa Sia to Long Samado, near crossing with Malabid River, Dec. 1986, *Vermeulen & Duistermaat* 962 (L!).

Fig. 87. *Dendrochilum imitator*. **A–C** habits; **D** floral bract; **E** flower, side view; **F** dorsal sepal; **G** lateral sepal; **H** petal; **J** labellum, front view; **K** pedicel with ovary and column, anther-cap removed, oblique view; **L** column, back view; **M** column apex, anther-cap removed, front view; **N** anther-cap, back view; **O** pollinia. A and **D–O** from *de Vogel* 8451 (holotype), **B** from *Beaman* 10347, **C** from *de Vogel* 8426. Scale: single bar = 1 mm; double bar = 1 cm. Drawn by Susanna Stuart-Smith.

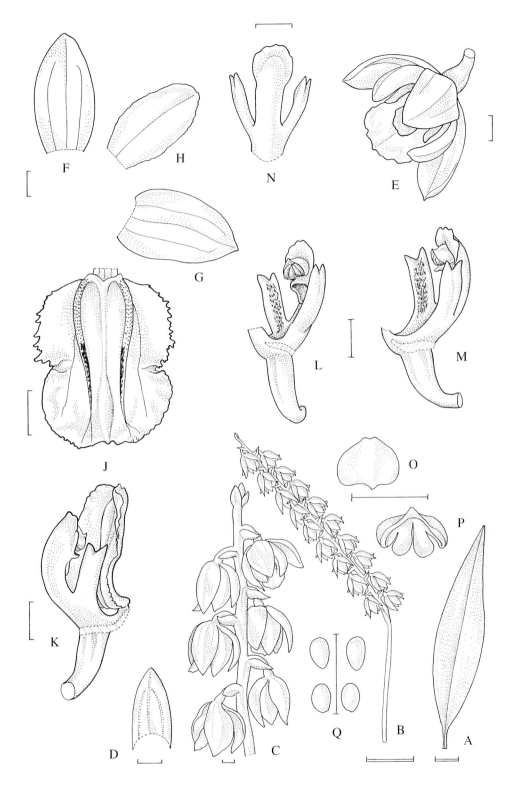

Fig. 88. *Dendrochilum* sp. aff. *imitator*. **A** leaf; **B** inflorescence; **C** upper portion of inflorescence; **D** floral bract; **E** flower, from above; **F** dorsal sepal; **G** lateral sepal; **H** petal; **J** labellum, front view; **K** pedicel with ovary, labellum and column, side view; **L** pedicel with ovary and column, oblique view; **M** pedicel with ovary and column, side view; **N** column, back view; **O** anther-cap, back view; **P** anther-cap, interior view; **Q** pollinia. **A–Q** from *Leiden* cult. (*de Vogel*) 911135. Scale: single bar = 1 mm; double bar = 1 cm. Drawn by Susanna Stuart-Smith.

HABITAT. Epiphytic on tree trunks in dry, rather low and open ridge forest on soil probably derived from sandstone and shales, with small open patches of grass and *Gleichenia* spp.; epiphytic on branches overhanging water in dense primary riverine forest up to 30 metres high on soil derived from sandstone; open podsol forest; low stature forest on ultramafic substrate. 1300–1500 m.

D. imitator is distinguished from *D. lacteum* Carr by the more pronounced spotted cataphylls, floral bracts shorter than the pedicel with ovary, smaller flowers with a longer pedicel with ovary, shorter and a little narrower, acute rather than acuminate sepals and petals, reflexed lateral sepals, shorter labellum with a much more pronounced median keel, and bifid stelidia.

Leiden cult. (*de Vogel*) 911135 (originally cited by Wood & Cribb 1994 as *de Vogel* 2079) (K, L, spirit material only, see fig. 88), collected at 1300 metres on Mount Pagon in Brunei, is similar to *D. imitator* but has minute hairs at the base of broader sepals and petals, non-reflexed lateral sepals as in *D. lacteum*, and a broader labellum. The pronounced median keel on the labellum and the bifid stelidia are typical, however, of *D. imitator*. Further collections are necessary before its status can be clarified.

The Greek specific epithet *imitator*, to imitate, is in reference to the close resemblance of this species to *D. lacteum*.

52. DENDROCHILUM LACTEUM

Dendrochilum lacteum Carr in *Gdns. Bull. Straits Settl.* 8: 223–224 (1935). Type: Malaysia, Sabah, Mount Kinabalu, above Lumu Lumu, July 1933, *Carr* C. 3608, SFN 27892 (holotype SING!; isotypes AMES!, K!, LAE). **Fig. 89, plate 10C.**

DESCRIPTION. Erect to pendulous epiphyte. *Rhizome* branching, to *c.* 4 cm long. *Roots* branching, forming a dense mass, flexuose, densely minutely papillose, 0.2–1mm in diameter. *Cataphylls* 3, ovate-elliptic, acute, 1–4 cm long, speckled brown. *Pseudobulbs* crowded, approximate, narrowly ovoid, fusiform or ovoid, 1.2–5.5 × 0.4–1.2 cm. *Leaf*: petiole sulcate, 0.5–2 cm long; blade narrowly elliptic to elliptic, obtuse to acute, margin often gently undulate, thin-textured, 2.3–18.5 × 1–3.3 cm, main nerves 7–9, with numerous tiny transverse nerves, often flushed red, particularly when young. *Inflorescences* pendulous, laxly *c.* 16- to many-flowered, flowers borne 4–5 mm apart; peduncle suberect at first, filiform, 2–9(–11.5) cm long; non-floriferous bracts 1–6, ovate, acute, 3–5 mm long; rachis quadrangular, 8–20 cm long, pinkish; floral bracts ovate to ovate-elliptic, acute or minutely apiculate, 4–4.5 × 3.7 mm, pale flesh-coloured. *Flowers* with yellowish cream or transparent cream sepals and petals, often suffused pale salmon-pink down the middle and at the base, labellum cream, mid-lobe with a brownish pink centre or median line, base and keels pale salmon-pink, column pale salmon-pink, stelidia similar or yellowish cream, apical hood pale yellow, ovary pink. *Pedicel with ovary* clavate, 1.5–1.8 mm long. *Sepals* 3–nerved, carinate, not reflexed. *Dorsal sepal* narrowly elliptic, acute, 5–7 × 2–2.3 mm. *Lateral sepals* oblong-elliptic, slightly oblique, acute, 5.5–7 × 2.2–2.7 mm. *Petals* oblong-elliptic, narrowed at base, acute, margins minutely erose, 4.5–6.7 × 2–2.3 mm. *Labellum* shortly stipitate to column-foot by a narrowly oblong claw, 3–lobed, 3–nerved, minutely papillose, particularly on side lobes, 4.5–*c.* 6 mm long, 3.9–5.2 mm wide across side lobes when flattened; side lobes triangular, acute, outer margin irregularly erose-denticulate, free apical portion 0.8–1 mm long; mid-lobe obovate or

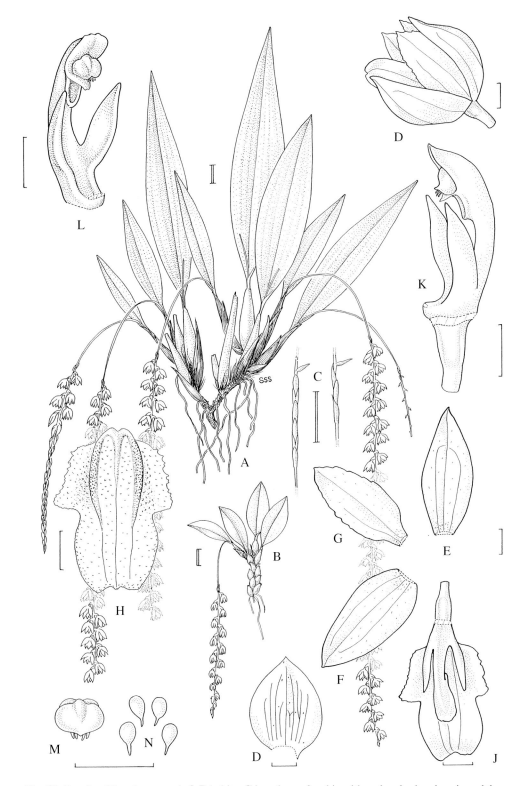

Fig. 89. *Dendrochilum lacteum*. **A & B** habits; **C** junctions of rachis with peduncle showing six and three non-floriferous bracts and one floral bract; **D** flower, side view; **E** dorsal sepal; **F** lateral sepal; **G** petal; **H** labellum, front view; **J** pedicel with ovary, labellum and column, viewed from above; **K** pedicel with ovary and column, side view; **L** column, oblique view; **M** anther-cap, back view; **N** pollinia. **A & C** from *de Vogel* 8036, **B** from *Carr* C. 3608, SFN 27892 (holotype), **D–N** from *Bailes & Cribb* s.n. Scale: single bar = 1 mm; double bar = 1 cm. Drawn by Susanna Stuart-Smith.

suborbicular, obtuse, often retuse and apiculate, 2.1–3.4 × 2.5–3.4 mm; disc with 2 papillose incurved keels joined at base with elevate median nerve to form an M-shape, extending to junction of side lobes and mid-lobe, elevate median nerve extending to about middle of side lobes. *Column* gently incurved, dorsally somewhat carinate, 3.5–4.3 mm long; foot small; apical hood oblong, obtuse, margin often uneven; stelidia arising from just above base, oblong-elliptic or narrowly elliptic, acute, reaching to rostellum, 1.8–2 mm long; stigmatic cavity ovate, lower margin elevate; rostellum triangular-ovate, acute; anther-cap cucullate.

DISTRIBUTION. Borneo.

SARAWAK. Lundu District, Mount Pueh, 24 Sept. 1955, *Purseglove & Shah* P. 4747 (SAR!, SING!).

SABAH. Locality unknown, cult. RBG Kew, accession number 229–83. 02953, *Bailes & Cribb* s.n. (K!). Mount Kinabalu, off road to Power Station, *Barkman* 235 (Sabah Parks Herbarium, Kinabalu Park). Mount Kinabalu, above Lumu Lumu, July 1933, *Carr* 3608, SFN 27892 (holotype SING!, isotypes AMES!; K!; LAE). Locality unknown, cult. RBG Kew, ex *M.Held*, Swiss Orchid Society (K!). Mount Kinabalu, Park Headquarters, 15 May 1967, *Price* 145 (K!). Mount Kinabalu, locality unknown, 22 Sept. 1966, *Togashi* s.n. (TI!). Mount Kinabalu, Pinosuk Plateau, 3 Oct. 1986, *de Vogel* 8036 (L!). Crocker Range, Mount Alab, south ridge, 31 Oct. 1986, *de Vogel* 8632 (L!).

HABITAT. Lower montane forest; Fagaceae/*Dacrydium*/*Leptospermum* forest *c*. 35 metres high with little undergrowth on sloping gravels with sandstone, ultramafic and dioritic rock; mossy forest with little undergrowth on west facing slopes and ridges on sandstone and shale. 1400–2000 m.

The specific epithet is derived from the Latin *lacteus*, milky-white, and refers to the flower colour.

53. DENDROCHILUM DULITENSE

Dendrochilum dulitense Carr in *Gdns. Bull. Straits Settl.* 8: 84–85 (1935). Type: Malaysia, Sarawak, Marudi District, Mount Dulit, Dulit Ridge, *c*. 1230 m, 7 Sept. 1932, *Synge* S. 435 (holotype SING!; isotype K!). **Fig. 90.**

DESCRIPTION. Epiphyte. *Rhizome* abbreviated, up to *c*. 3 cm long. *Roots* branching, forming a dense mass, flexuose, densely minutely papillose, 0.2–0.4 mm in diameter. *Cataphylls* 3–4, ovate, obtuse, 0.2–1.5 cm long. *Pseudobulbs* approximate, ovoid or subglobose, 0.3–0.6 × 0.3 cm, dull green. *Leaf*: petiole sulcate, 2–5 mm long; blade linear, ligulate, obtuse, coriaceous, margins often revolute in dried material, 1.8–3.4 × 0.2–0.3 cm, main nerves 5, mid-green. *Inflorescences* erect or nodding, emerging with the unexpanded leaf, laxly 6- to 8–flowered, flowers borne 4–7 mm apart; peduncle filiform, 1.2–2.5(–3) cm long; non floriferous bracts absent; rachis quadrangular, 2.5–3 cm long, pale pinkish; floral bracts oblong-ovate, very shortly acuminate, acute, 3–nerved, 1.8–3 × 1.5–1.6 mm. *Flowers* unscented, translucent pale salmon-pink or orange ochre, labellum pale or dull yellow, or orange ochre with creamy margin, column dull peach or orange ochre, stelidia yellow ochre; also described as apricot. *Pedicel with ovary* narrowly clavate, 3 mm long. *Sepals* spreading, 3–nerved, papillose at base. *Dorsal sepal* narrowly ovate-elliptic, acute to acuminate, 5–5.5 × 1.5–2 mm. *Lateral sepals* narrowly ovate-elliptic to subfalcate, acuminate, 5.5–6.2 × 1.8–2.5 mm. *Petals* narrowly ovate, long acuminate, margin minutely uneven to erose, 1–nerved,

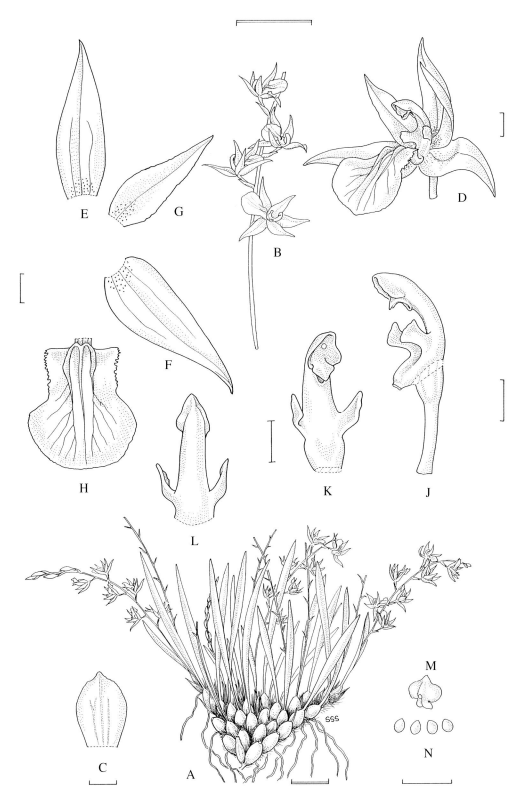

Fig. 90. *Dendrochilum dulitense*. **A** habit; **B** inflorescence; **C** floral bract; **D** flower, oblique view; **E** dorsal sepal; **F** lateral sepal; **G** petal; **H** labellum, front view; **J** pedicel with ovary and column, side view; **K** column, front view; **L** column, back view; **M** anther-cap, back view; **N** pollinia. A–N from *Burtt & Martin* B. 4894. Scale: single bar = 1 mm; double bar = 1 cm. Drawn by Susanna Stuart-Smith.

4.7–5 × 1.3–1.8 mm. *Labellum* shortly stipitate to column-foot by a narrowly oblong claw, obscurely 3–lobed, 3–nerved, smooth, 4.3–5 mm long, 2.9–3 mm wide at base, 3.7–4 mm wide across mid-lobe; side lobes triangular, subacute, often ill-defined, somewhat retrorse, outer margin minutely erose to irregularly serrulate; mid-lobe transversely elliptic, rounded, broader than long; disc with 2 short keels joined at base to the elevate median nerve to form an M-shape, extending to junction of side lobes and mid-lobe, or just beyond. *Column* incurved, 2.8–3.1 mm long; foot prominent, 0.8–0.9 × 0.8–0.9 mm; apical hood ovate-triangular, obtuse, entire; stelidia arising just above base, oblong-spathulate, cuneate, rounded to subtruncate, reaching to just below stigmatic cavity, c. 0.9 mm long; stigmatic cavity transversely oblong, lower margin elevate; rostellum prominent, ovate, subacute, porrect; anther-cap cucullate, with a conical dorsal wart.

DISTRIBUTION. Borneo.

SARAWAK. Kapit District, southeast end of Hose Mountains, Pantoh Hill due east of Nibong, 10 Aug. 1967, *Burtt & Martin* B. 4894 (E!, SAR!). Marudi District, Mount Dulit, Dulit Ridge, 7 Sept. 1932, *Synge* S. 435 (holotype SING!, isotype K!).

BRUNEI. Amo Subdistrict, ridge to northeast of Mount Retak, 11 March 1991, *Sands* 5335 (BRUN, K!). Temburong, Mount Retak, 1989, *Leiden* cult. (*Wong*) 27853 (K, sketch only!, L).

HABITAT. Lower montane mossy forest; epiphytic amongst moss in *Rhododendron/Cyathea* community 2–4 metres high on sandstone ridges. 1200–1400 m.

An attractive dwarf species easily distinguished by its salmon to ochre-coloured flowers with oblong-spathulate and somewhat truncate stelidia on the column.

The specific epithet refers to Mount Dulit (1369 m) in Sarawak, the type locality.

54. DENDROCHILUM OCHROLABIUM

Dendrochilum ochrolabium J.J. Wood in Wood & Cribb, *Checklist orch. Borneo*: 187–189, fig. 24 J & K (1994). Type: Malaysia, Sabah, Sipitang District, ridge east of Maga River, c. 1.5 km south of confluence with Pa Sia River, 1450 m, 17 Oct. 1986, *de Vogel* 8351 (holotype L!; isotype K!). **Figs. 91 & 92, plate 11E.**

DESCRIPTION. Epiphyte. *Rhizome* branching, up to 4 cm long. *Roots* branching, forming a dense mass, flexuose, very minutely papillose, 0.1–0.5 mm in diameter. *Cataphylls* 3–4, ovate-elliptic, acute, apiculate, 0.4–1.5 cm long. *Pseudobulbs* ovoid to elliptic, or oblong-cylindrical, 0.5–1.2 × 0.2–0.3 cm. *Leaf*: petiole sulcate, 1–4 mm long; blade narrowly elliptic to linear-ligulate, obtuse, usually minutely mucronate, margin minutely papillose, surface very minutely black punctate-ramentaceous, particularly abaxially, 1.2–5.9 × 0.3–0.4 cm, main nerves 9, median and 2 outermost marginal being raised and prominent. *Inflorescences* erect to curving, laxly 5– to 13–flowered, flowers borne 2–3 mm apart; peduncle filiform, 1.2–4 cm long; non-floriferous bracts absent; rachis quadrangular, 1.5–4 cm long; pale reddish green to salmon-pink; floral bracts oblong-elliptic, concave, acute, 2–2.5 × 1.1–1.2 mm. *Flowers* slightly scented, sepals and petals yellowish green or pale green, labellum pale green, orange or ochre, reddish brown to brownish centrally or with two dark reddish brown basal bands, reverse green with two longitudinal brown stripes, or orange with pale green side lobes, column pale green, brownish in front below stigma, anther-cap cream. *Pedicel with ovary* cylindrical to narrowly clavate, outer wall appearing fleshy and translucent, 1.9–2 mm long.

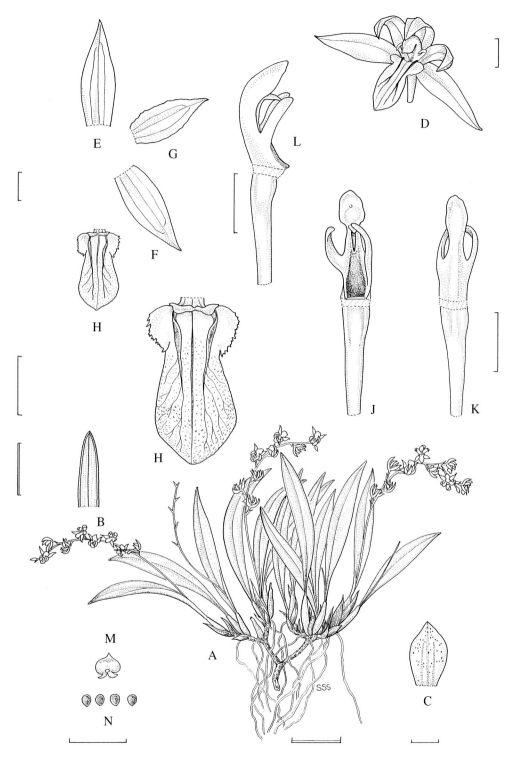

Fig. 91. *Dendrochilum ochrolabium*. **A** habit; **B** leaf apex; **C** floral bract; **D** flower, oblique view; **E** dorsal sepal; **F** lateral sepal; **G** petal; **H** labellum, front view; **J** pedicel with ovary and column, anther-cap removed, front view; **K** pedicel with ovary and column, back view; **L** pedicel with ovary and column, anther-cap removed, side view; **M** anther-cap, back view; **N** pollinia. **A–N** from *de Vogel* 8351 (holotype). Scale: single bar = 1 mm; double bar = 1 cm. Drawn by Susanna Stuart-Smith.

Sepals and *petals* 3–nerved. *Dorsal sepal* narrowly ovate-elliptic, acute, 3.9–4 × 1.1 mm. *Lateral sepals* ovate-elliptic, acute, 3.9–4 × 1.2–1.3 mm. *Petals* ovate-elliptic, slightly erose, acute, 3 × 1.1 mm. *Labellum* shortly stipitate to column-foot by a short, narrowly oblong claw, 3–lobed, 3–nerved, minutely papillose, 2.8 mm long, 1.2–1.3 mm wide across side lobes, 1.4–1.5 mm wide across mid-lobe; side lobes auriculate, irregularly toothed; mid-lobe obovate, obtuse, somewhat convex; disc with 2 somewhat papillose keels joined at the base to a narrow, transverse flange, each terminating a little above base of mid-lobe, median nerve slightly elevate. *Column* incurved, *c.* 1.8 mm long; foot short; apical hood ovate to oblong, rounded, sometimes very obscurely 3–lobed; stelidia arising from at or near the middle, ligulate, subacute, equalling or a little longer than rostellum, *c.* 0.8 mm long; stigmatic cavity narrowly oblong; rostellum triangular-ovate, acute; anther-cap cucullate.

DISTRIBUTION. Borneo.

SARAWAK. Kapit District, Mengiong/Balleh Rivers, 26 July 1987, *Lee S.* 54797 (SAR!). Kuching District, Mount Penrissen, near Padawan, Feb. 1993, *Leiden* cult. (*Schuiteman, Mulder & Vogel*) 933108 (K!, L!).

SABAH. Crocker Range, Sinsuron road, *Lamb* (colour slide only!). Sipitang District, Long Pa Sia to Long Samado trail near crossing with Malabid River, Dec. 1986, *Vermeulen & Duistermaat* 967 (L!). Sipitang District, ridge east of Maga River, *c.* 1.5 km south of confluence with Pa Sia River, 17 Oct. 1986, *de Vogel* 8351 (holotype L!, isotype K!).

HABITAT. Open, low, dry stunted forest 5 to 10 metres high on sandstone ridges, with a dense undergrowth of terrestrial orchids and other herbs; very low and open podsol forest;

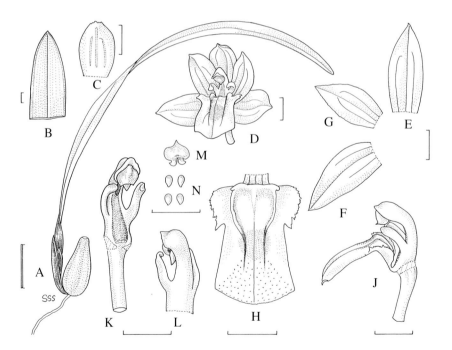

Fig. 92. *Dendrochilum ochrolabium.* **A** habit; **B** leaf apex; **C** floral bract; **D** flower, front view; **E** dorsal sepal; **F** lateral sepal; **G** petal; **H** labellum, front view; **J** pedicel with ovary, labellum and column, side view, **K** pedicel with ovary and column, oblique view; **L** column, back view; **M** anther-cap, back view; **N** pollinia. A–N from *Leiden* cult. (*Schuiteman et al.*) 933108. Scale: single bar = 1 mm; double bar = 1 cm. Drawn by Susanna Stuart-Smith.

primary submontane forest; recorded as an epiphyte on a tree trunk near a stream. 800–1500 m.

D. ochrolabium appears to be related to *D. dulitense* Carr from Sarawak but differs in having slightly broader, narrowly elliptic, minutely black punctate-ramentaceous leaves, pale green flowers with an orange to ochre lip, slightly smaller sepals and petals lacking basal papillae, a distinctly smaller, narrower labellum, indistinctly elevate median nerve, and ligulate, acute stelidia. The leaves are also notable for having prominent outermost, marginal nerves (see Fig. 14B).

The specific epithet is derived from the Latin *ochraceus*, ochre-yellow or yellowish brown, and *labium*, lip, referring to the colour of the labellum.

55. DENDROCHILUM GALBANUM

Dendrochilum galbanum J.J. Wood in *Orchid Rev.* 102 (1197): 147, fig. 80 (1994). Type: Malaysia, Sarawak, Lawas District, route from Ba Kelalan to Mount Murud, Southwest of Camp III, 1800 m, 30 Sept. 1967, *Burtt & Martin* B. 5328 (holotype E!; isotypes K!, SAR!). **Fig. 93.**

DESCRIPTION. *Rhizome* abbreviated, up to *c.* 3 cm long. *Roots* branching, forming a dense mass, flexuose, smooth or with a few minute papillae, 0.3–1 mm in diameter. *Cataphylls* disintegrated in Bornean material examined, (5–6 in Sumatran material), ovate, obtuse, 0.4–3.5 cm long. *Pseudobulbs* ovoid to ovate-oblong, 0.5–1 × 0.4–0.5 cm, greenish yellow. *Leaf*: petiole sulcate, 2–5 mm long; blade erect, linear-ligulate, obtuse and minutely mucronate, 4–7 × 0.3–0.6 cm, main nerves 5. *Inflorescences* erect, laxly 12– to 20–flowered, flowers borne 3–5 mm apart; peduncle wiry, 0.3–1.8 cm long; non-floriferous bracts 1–3, ovate-elliptic, acute, 3–4 mm long; rachis quadrangular, 4–9.5 cm long, orange-coloured; floral bracts ovate-elliptic, acute to acuminate, 3.5–4 mm long, pale brown. *Flowers* greenish yellow or bright yellow, labellum yellow, brown at the middle, keels red, column green, anther-cap yellow. *Pedicel with ovary* narrowly clavate, 3 mm long. *Sepals* and *petals* spreading, 3–nerved. *Dorsal sepal* narrowly elliptic, acute, 7.5–8 × 2 mm. *Lateral sepals* narrowly ovate-elliptic, acute, 7.9–8 × 2 mm. *Petals* narrowly elliptic, subacute to acute, 7 × 2 mm. *Labellum* sessile, entire, 3–nerved, elliptic, acute, slightly rounded and erose towards base; disc with 2 fleshy keels and a prominent elevate median nerve joined at base, terminating a little above erose portion. *Column* gently incurved, slender, 3 mm long; foot short; apical hood ovate, entire, rounded; stelidia arising a short distance above base, linear, obtuse to acute, not quite reaching stigmatic cavity, 1 mm long; stigmatic cavity narrowly oblong; rostellum prominent, ovate-triangular; anther-cap ovate, cucullate.

DISTRIBUTION. Sumatra and Borneo.

SARAWAK. Lawas District, route from Ba Kelalan to Mount Murud, Southwest of Camp III, 20 Sept. 1967, *Burtt & Martin* B. 5328 (holotype E!, isotypes K!, SAR!). Lawas District, proposed Mount Murud National Park, second sandstone summit, 12 Sept. 1982, *Yii* S. 44430 (K!, L!, SAR!).

HABITAT. Mossy forest on sandstone. 1800–2250 m.

An attractive species closely related to *D. graminoides* Carr, described from Mount Kinabalu in Sabah, but distinguished by the much shorter, broader leaves, erect inflorescences

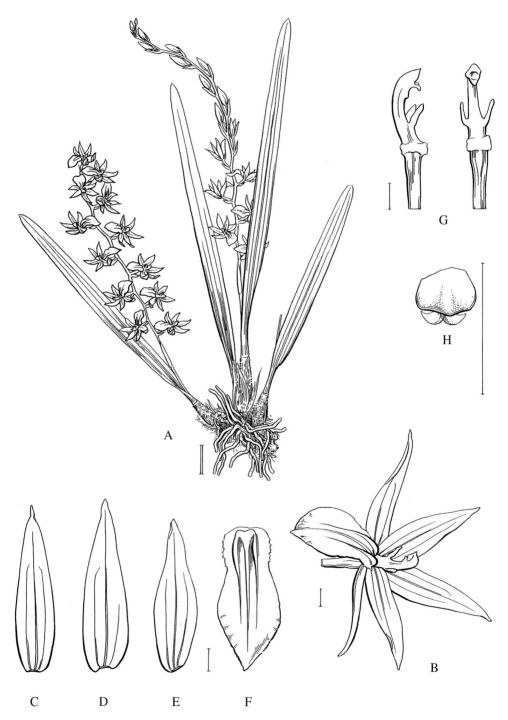

Fig. 93. *Dendrochilum galbanum*. **A** habit; **B** flower, front view; **C** dorsal sepal; **D** lateral sepal; **E** petal; **F** labellum, front view; **G** pedicel with ovary and column, anther-cap removed, front and side views; **H** anther-cap, back view. Scale: single bar = 1 mm; double bar = 1 cm. **A–H** from *Burtt & Martin* B.5328 (holotype). Drawn by Eleanor Catherine.

and larger flowers with a longer, broader labellum and longer column with slightly shorter stelidia.

Populations collected at 2250 m on Mount Ketambe situated in Mount Leuser Nature Reserve in Aceh Province, Sumatra, are more robust in habit, have slightly broader leaves and longer inflorescences. Their floral morphology, however, differs little from Sarawak collections.

The specific epithet is derived from the Latin *galbanus*, the colour of gum galbanum, which is greenish yellow, and refers to the flower colour.

56.DENDROCHILUM GRAMINOIDES

Dendrochilum graminoides *Carr* in *Gdns. Bull. Straits Settl.* 8: 229–230 (1935). Type: Malaysia, Sabah, Mount Kinabalu, near Pinansak, *c.* 900 m., Nov. 1933, *Carr* C. 3680, SFN 28006 (holotype SING!, isotypes AMES!, K!). **Fig. 94, plate 5D & E.**

DESCRIPTION. *Rhizome* branching, up to 5 cm long. *Roots* with a few branches, forming a dense mass, flexuose, smooth or with many minute papillae, 0.6–1 mm in diameter. *Cataphylls* 3–5, ovate-elliptic, acute, 0.5–3 cm long, speckled, soon disintegrating. *Pseudobulbs* grouped closely together, approximate, ovoid, or oblong-ovoid, minutely wrinkled, 0.8–1.4 × 0.4–0.6 cm, pale orange-brown or reddish-orange. *Leaf:* petiole slender, sulcate, rigid, 1.2–3.5 cm long; blade narrowly linear, obliquely obtuse to acute, thin-textured, 4.5–13.5 × 0.1–0.3 cm, main nerves 3. *Inflorescences* gently curving, produced with the leaf nearly fully expanded, subdensely 16– to *c.* 26–flowered, flowers borne 2–3 mm apart; peduncle filiform, 5–9.5 cm long; non-floriferous bracts 3–4, ovate-elliptic, acute, 2–4.5 mm long; rachis nodding, quadrangular, 5–7(–9) cm long; floral bracts narrowly ovate-elliptic, subacute to acute, 2–3 mm long, pale brown. *Flowers* scented or unscented, pedicel with ovary greenish brown, sepals and petals deep yellow, bright lemon-yellow or greenish yellow, salmon-pink or brownish orange at base, labellum cream or yellow, flushed green centrally, salmon-pink at base, with a brown median line, keels green at base, column greenish white, brown at base. *Pedicel with ovary* narrowly clavate, 1.8–2 mm long. *Sepals* and *petals* spreading, 3–nerved. *Dorsal sepal* narrowly oblong-elliptic, acute, 4–4.3 × 1–1.1 mm. *Lateral sepals* narrowly elliptic, acute, often somewhat falcate, (4–)4.4–4.5 × 1–1.1 mm. *Petals* narrowly elliptic, acute, minutely erose, often somewhat falcate, 3.7–4 × 1–1.2 mm. *Labellum* stipitate to column-foot by a small claw, inconspicuously 3–lobed, 3–nerved, 3–3.1 mm long, 1–1.5 mm wide across side lobes, 1.7–1.8 mm wide across mid-lobe; side lobes rounded, irregularly erose-serrate; mid-lobe elliptic, subacute to acute, minutely erose, concave; disc with 2 short, erect keels united at the base, extending to just beyond base of mid-lobe, median nerve thickened and elevate for a short distance, whole structure M-shaped. *Column* incurved, 1.5–2 mm long; foot short; apical hood cuneate, rounded to triangular, obtuse; stelidia arising below stigmatic cavity, slender, subulate, reaching rostellum, 0.5–0.8 mm long; stigmatic cavity narrowly ovate, margins elevate; rostellum large, ovate, obtuse; anther-cap cucullate. *Capsule* obovoid, retaining persistent flower remains, *c.* 4.5 × 2.7 mm.

DISTRIBUTION. Borneo.

SARAWAK. Lawas District, Mount Murud, 1967, *Burtt & Martin* s.n., cult. RBG Edinburgh C. 5468 (E!).

BRUNEI. Temburong District, Mount Pagon, Temburong River, April 1958, *Ashton* 205 (K! L!). Temburong District, Mount Pagon, 8–11 Jan. 1992, *Leiden* cult. (*de Vogel*) 914922

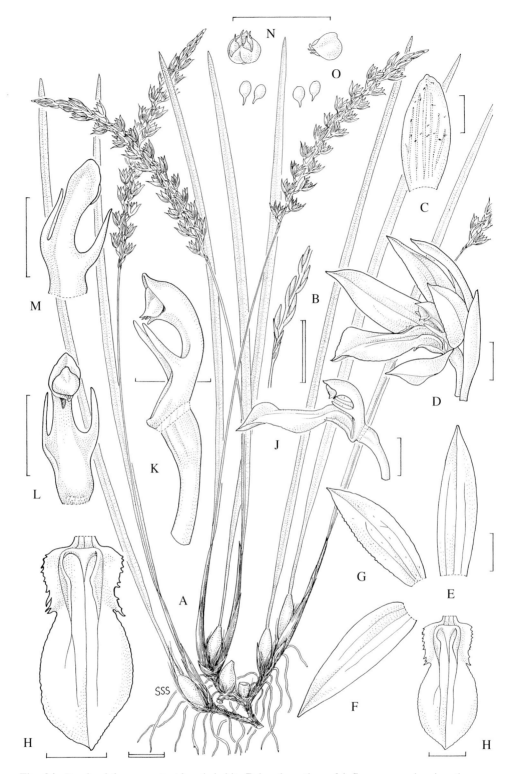

Fig. 94. *Dendrochilum graminoides.* **A** habit; **B** basal portion of inflorescence showing three non-floriferous bracts, five floral bracts and five buds; **C** floral bract; **D** flower with floral bract, side view; **E** dorsal sepal; **F** lateral sepal; **G** petal; **H** labellum, front view; **J** pedicel with ovary, labellum and column, side view, **K** pedicel with ovary and column, side view; **L** column, front view; **M** column, back view; **N** anther-cap, interior and side views; **O** pollinia. **A & B** from *Chan & Lamb* SAN 87477, **C–O** from *Leiden* cult. (*de Vogel*) 911236. Scale: single bar = 1 mm; double bar = 1 cm. Drawn by Susanna Stuart-Smith.

(L!). Temburong District, Mount Pagon, 8–11 Jan. 1992, *Leiden* cult. (*de Vogel*) 911214 (L!), 911227 (L!) & 911236 (L!), 911245 (L!). Belait District, Badas Forest Reserve, 17 Jan. 1992, *Leiden* cult. (*de Vogel*) 911272 (L!).

SABAH. Crocker Range, Keningau to Kimanis road, near RTM station, *Barkman* 227 (Sabah Parks Herbarium, Kinabalu Road). Mount Kinabalu, near Pinansak, Nov. 1933, *Carr* C. 3680, SFN 28006 (holotype SING!, isotypes AMES!, K!). Crocker Range, top of Keningau to Kimanis road, 21 Nov. 1977, *Chan & Lamb* SAN 87477 (SAN!). Mount Lotung, edge of escarpment, April 1982, *Lamb* LMC 2303, SAN 93489 (K!). Crocker Range, Keningau to Kimanis road, Dec. 1986, *Vermeulen & Duistermaat* 665 (K!, L!). Sandakan Zone, Mount Tawai, Nov. 1986, *Vermeulen & Lamb* 706 (K!, L!).

HABITAT. Hill and lower montane forest, sometimes on ultramafic substrate; mossy forest with many *Casuarina* spp., undergrowth of climbing bamboo; montane ridge forest 5–20 metres high on sandstone; lowland kerangas forest *c.* 30 metres high, with *Agathis* spp. and undergrowth of pole trees, on pure white sand; recorded as a branch epiphyte. Near sea level —1650 m.

This pretty yellow-flowered species epitomises the graceful habit of many dendrochilums. Although generally a plant of cool, montane mossy forest, a collection by de Vogel is said to have been collected in kerangas near sea level in Brunei. This disparity in habitat is unusual in the genus. However, de Vogel (pers. comm.) states that the provenance may have been cited in error and its true origin is probably Mount Pagon.

The specific epithet is derived from the Latin *gramineus*, grassy, referring to the grass-like leaves.

57. DENDROCHILUM MAGAENSE

Dendrochilum magaense J.J. Wood in *Orchid Rev.* 102 (1197): 147–149, fig. 81 (1994). Type: Malaysia, Sabah, Sipitang District, Ulu Long Pa Sia, 8 km. Northwest of Long Pa Sia, above Maga River, *c.* 1400 m, 24 Oct. 1985, *Wood* 657 (holotype K!). **Fig. 95.**

DESCRIPTION. Epiphyte. *Rhizome* abbreviated, to *c.* 2 cm long. *Roots* branching, flexuose, minutely papillose, 0.3–0.7 mm in diameter. *Cataphylls* 4–5, ovate-elliptic, acute, 0.4–1.8 cm long. *Pseudobulbs* grouped close together, approximate, ovoid-elliptic, 0.8–1.2 × 0.2–0.5 cm, olive-green. *Leaf:* petiole slender, sulcate, 0.6–1.8 cm long; blade ligulate, obtuse, apiculate, margin very minutely papillose at apex, thin-textured, 3–7.5 × 0.3–0.4 cm, main nerves 3. *Inflorescences* gently curving to pendulous, produced with the leaf nearly fully expanded, laxly 4– to 8–flowered, flowers borne 3–4 mm apart; peduncle filiform, 3–5 cm long; non-floriferous bracts 2–3, ovate, acuminate to setose, 3–5 mm long; rachis quadrangular, 2–3.8 cm long; floral bracts ovate to ovate-elliptic, apiculate, 4–5 mm long, pale brown. *Flowers* wide-opening, salmon-pink to buff; or cream with orange to pale ochre centre to sepals and petals and pale orange to ochre centre to labellum. *Pedicel with ovary* clavate, 2.5–2.8 mm long. *Sepals* and *petals* 3–nerved. *Dorsal sepal* oblong-elliptic, acute, 7.2–7.3 × 2.2–2.3 mm. *Lateral sepals* oblong-elliptic, acute, 7.5 × 2.4–2.5 mm. *Petals* irregularly elliptic, obtuse and mucronate, margin very minutely erose, 6.5 × 2.5–2.6 mm. *Labellum* stipitate to column-foot by a short claw, 3–lobed, 3–nerved, 4.9–5 mm long, 2 mm wide across base of side lobes; side lobes triangular, acuminate, irregularly serrate to lacerate, free apical portion *c.* 1 mm long;

Fig. 95. *Dendrochilum magaense*. **A** habit; **B** flower, oblique view; **C** dorsal sepal; **D** lateral sepal; **E** petal; **F** labellum, front view; **G** pedicel with ovary and column, anther-cap removed, oblique view; **H** anther-cap, back view; **J** pollinia. Scale: single bar = 1 mm; double bar = 1 cm. **A–J** from *Wood* 657 (holotype). Drawn by Eleanor Catherine.

mid-lobe broadly elliptic from a cuneate base, rounded, margin slightly irregular, 2.5–2.6 × 2.8–2.9 mm; disc thickened, with 2 low fleshy ridges joined at base and terminating at base of mid-lobe, median nerve not elevate. *Column* incurved, 3. 5 mm long; foot 0.5–0.6 mm long; apical hood elongate, oblong, bifid, recurved; stelidia basal, acicular, acute, reaching base of apical hood, 2 mm long; rostellum ovate-triangular; stigmatic cavity ovate, lower margin elevate; anther-cap ovate, cucullate, 0.9 × 0.9 mm. *Capsule* ellipsoid, flower remains persistent, 1.1 × 0.8 cm.

DISTRIBUTION. Borneo.

SABAH. Sipitang District, ridge east of Maga River, *c.* 1.5 km south of confluence with Pa Sia River, 17 Oct. 1986, *de Vogel* 8350 (L!). Sipitang District, Ulu Long Pa Sia, 8 km Northwest of Long Pa Sia, above Maga River, 24 Oct. 1985, *Wood* 657 (holotype K!).

HABITAT. Oak/chestnut ridge forest with *Agathis borneensis* Warb.; open, low, dry stunted forest 5–10 metres high, with a dense field layer of terrestrial orchids and other herbs, on narrow sandstone ridge. 1400–1500 m.

I would expect *D. magaense* to occur in the mountains of neighbouring Sarawak and Brunei.

The specific epithet refers to the Maga River in Southwest Sabah, above which the type was collected.

58. DENDROCHILUM SUBULIBRACHIUM

Dendrochilum subulibrachium J.J. Sm. in *Bull. Jard. Bot. Buitenz.* ser. 3, 11: 109–110 (1931). Type: Indonesia, Kalimantan Timur, Long Petak, 800 m, 12 Sept. 1925, *Endert* 3221 (holotype L, not located). **Fig. 96.**

DESCRIPTION. Epiphyte. *Rhizome* abbreviated, to *c.* 3 cm long. *Roots* elongate, with very few branches, flexuose, minutely papillose, 0.2–0.5 mm in diameter. *Cataphylls* 3, ovate-elliptic, obtuse to acute, 0.4–2 cm long, speckled. *Pseudobulbs* grouped close together, approximate, oblong-ovoid, 0.5–1.1 × 0.4–0.5 cm. *Leaf*: petiole slender, sulcate, 0.3–1 cm long; blade linear, obtuse, rigid, fleshy, 2.5–6.5 × 0.1–0.4 cm, main nerves 3. *Inflorescences* gently curving, produced with the leaf nearly fully expanded, laxly 9– to *c.* 28–flowered, flowers borne 2.5–3.5 mm apart; peduncle filiform, 5–9 cm long; non-floriferous bracts absent or rarely solitary, 3 mm long; rachis quadrangular, often flexuose, 3–9 cm long; floral bracts oblong, obtuse, erose distally, 2(–3.5) × 1 mm, brown. *Flowers* pale greenish to yellow, with 2 pale brown longitudinal central streaks on labellum. *Pedicel with ovary* narrowly clavate, 1.5–2 mm long. *Sepals* 3–nerved, incurved. *Dorsal sepal* oblong-ovate, obtuse to acute, 3.7–4.1 × 1.4–1.5 mm. *Lateral sepals* obliquely oblong-elliptic or subovate oblong, acute, 3.7–4 × 1.5 mm. *Petals* narrowly ovate-elliptic, acute to acuminate, sometimes slightly erose, 3.5–3.9 × 1 mm, incurved, 1– to 2–nerved. *Labellum* stipitate to column-foot by a short claw, 3–lobed, 3–nerved, gently undulate, 3.5–4 mm long; side lobes spreading, transversely quadrangular, truncate at base, apex acuminate, outer margin irregularly lacinulate, *c.* 1.4 × 2–2.3 mm; mid-lobe obovate-truncate from a narrow cuneate base, minutely erose distally, 2–2.5 × 2.7–3 mm; disc with 2 prominent, smooth keels united at base, terminating above base of mid-lobe, median nerve slightly elevate. *Column* incurved, fleshy at base, slender above, 2–2.5 mm long; foot prominent; apical hood entire, obtuse, recurved; stelidia borne above the

215

Fig. 96. *Dendrochilum subulibrachium.* **A & B** habits; **C** floral bract; **D** flower, side view; **E** dorsal sepal; **F** lateral sepal; **G** petal; **H** labellum, front view; **J** pedicel with ovary and column, anther-cap removed, side view; **K** column, oblique view; **L** column, back view; **M** anther-cap, back view; **N** pollinia. **A & C–N** from *Lewis* 349, **B** from *Awa & Lee* S. 50756. Scale: single bar = 1 mm; double bar = 1 cm. Drawn by Susanna Stuart-Smith.

base, linear-subulate, acute, reaching to base of stigmatic cavity, 1 mm long; rostellum ovate-triangular, obtuse, convex, obscurely carinate; stigmatic cavity oblong; anther-cap cucullate, with an apical wart.

DISTRIBUTION. Borneo.

SARAWAK. Limbang District, Bario, Pa Mario River, Ulu Limbang, route to Batu Lawi, 9 Aug. 1985, *Awa & Lee* S.50756 (K!, L!, SAR, SING). Marudi District, Gunung Mulu National Park, Mount Mulu, path from sub-camp 3 to sub-camp 4, 100 yards from camp 4, 19 Oct. 1977, *Lewis* 349 (K!).

KALIMANTAN TIMUR. Long Petak, 12 Sept. 1925, *Endert* 3221 (holotype L, not located).

HABITAT. Lower montane ridge forest; hill forest; 800–1730 m.

The specific epithet is derived from the Latin *subula*, a fine sharp point, and *brachium*, an arm, in reference to the subulate stelidia on the column.

59. DENDROCHILUM ANGUSTIPETALUM

Dendrochilum angustipetalum Ames, *Orchidaceae* 6: 47–48, pl. 83, centre (1920). Type: Malaysia, Sabah, Mount Kinabalu, Marai Parai Spur, 22 Nov. 1915, *J. Clemens* 270 (holotype AMES!). **Fig. 97.**

DESCRIPTION. Epiphyte. *Rhizome* to *c.* 8 cm long. *Roots* elongate, branching, flexuose, very minutely papillose, to 1 mm in diameter. *Cataphylls* 3–4, ovate-elliptic, obtuse to acute, 0.5–6 cm long, speckled. *Pseudobulbs* grouped close together, approximate, fusiform, slender distally, finely rugulose when dry, 2–4 × 0.3–0.5 cm. *Leaf*: petiole slender, sulcate, 2–5.5 cm long; blade narrowly oblong, ligulate, subacute to acute, minutely mucronate, thin-textured, 11–22 × 0.6–1.2 cm, main nerves 5. *Inflorescences* gently curving, produced with the leaf nearly fully expanded, subdensely up to 40–flowered, flowers borne 2 mm apart; peduncle filiform, 8–16.5 cm long; non-floriferous bracts 1–2, ovate-elliptic, obtuse or acute, 2–3.5 mm long; rachis quadrangular, 5–8(–9) cm long; floral bracts ovate, obtuse to subacute, 2.5–3 mm long. *Flowers* strongly sweetly-scented, sepals and petals pale yellow, labellum pale yellow with whitish margins, keels and elevate mid-nerve bright salmon-pink, column salmon-pink, apical hood and stelidia whitish, anther-cap salmon-pink, also described as greenish yellow with an orange and white labellum, flesh or salmon with two purple spots on labellum, cream and pink, cream and brown. *Pedicel with ovary* narrowly clavate, 2 mm long. *Sepals* 3–nerved. *Dorsal sepal* oblong to oblong-elliptic, acute, 3.5–4.5 × 0.9–1 mm. *Lateral sepals* oblong-elliptic, acute, sometimes slightly falcate, concave at base, obscurely carinate, 3.5–4.5 × 1–1.1 mm. *Petals* linear-oblong, acute, 4.5 × 5.7 mm, main nerves 1. *Labellum* 3–lobed, 3–nerved, 2.2–3 mm long; side lobes small, triangular, irregularly serrate or erose, acute, antrorse, *c.* 0.8 mm across; mid-lobe triangular-ovate or elliptic, acute, entire or minutely erose, 2 × 1.2–1.5 mm; disc with 2 tall keels united at base with the elevate mid-nerve forming an M-shape, terminating on lower portion of mid-lobe. *Column* incurved, 2 mm long; foot prominent; apical hood rounded, minutely erose; stelidia arising just below stigmatic cavity, oblong, subobtuse, equalling apical hood; rostellum ovate, obtuse; stigmatic cavity narrowly triangular; anther-cap ovate, cucullate, with an apical wart.

DISTRIBUTION. Borneo.

SABAH. Mount Kinabalu, Marai Parai Spur, 29 Aug. 1933, *Carr* 3684, SFN 28019 (K!, SING!); Kinateki River, 21 Sept. 1958, *Collenette* 104 (BM!); Marai Parai Spur, 22 Nov.

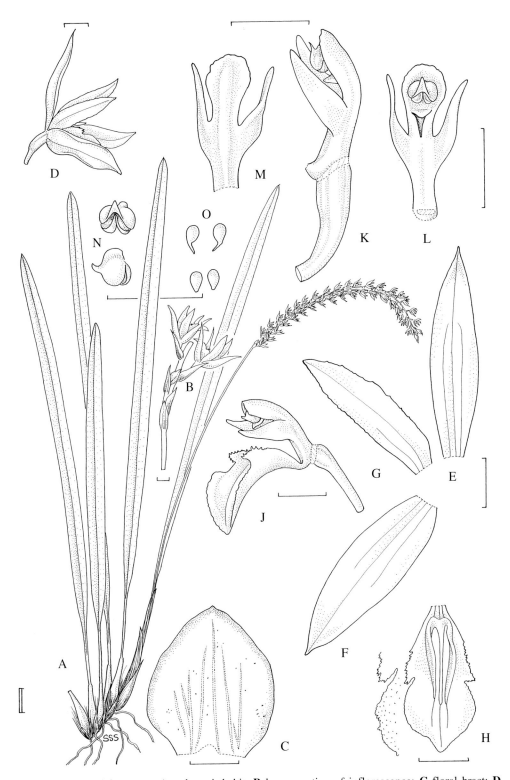

Fig. 97. *Dendrochilum angustipetalum.* **A** habit; **B** lower portion of inflorescence; **C** floral bract; **D** flower, side view; **E** dorsal sepal; **F** lateral sepal; **G** petal; **H** labellum, front view, with close-up of marginal zone; **J** pedicel with ovary, labellum and column, side view; **K** pedicel with ovary and column, side view; **L** column; front view; **M** column, back view; **N** anther-cap, front and side views; **O** pollinia. **A–O** from *J. & M.S. Clemens* 51754. Scale: single bar = 1 mm; double bar = 1 cm. Drawn by Susanna Stuart-Smith.

1915, *J. Clemens* 270 (holotype AMES!); Marai Parai Spur, 5 April 1933, *J. & M.S. Clemens* 32617 (E!); Penibukan, 8 Sept. 1933, *J. & M.S. Clemens* 40255 (BM!), 26 Sept. 1933, *J. & M.S. Clemens* 40467 (AMES!, BM!, K!) & 3 Oct. 1933, *J. & M.S. Clemens* 51714 (AMES!, BM!, K!, L!). Kampung Melangkap Tomis, 250 m ke rumah Gunung Doa Melangkap Tomis, 8 Oct. 1995, *Lugas* PEK 1048 (K!, Sabah Parks Herbarium, Kinabalu Park). Kampung Melangkap Tomis, Kawasan Taman bukit Dengiranuk, 1 Mar. 1996, *Lugas* PEK 1818 (K!, Sabah Parks Herbarium, Kinabalu Park). Mount Tembuyuken, near summit, 10 May 1990, *Nais et al.* SNP 4724 (Sabah Parks Herbarium, Kinabalu Park!).

HABITAT. Lower montane forest on ultramafic substrate; Collenette notes it growing one metre up a moss-covered tree in semi-mossy forest on top of a ridge. 1200–2000 m.

The specific epithet is derived from the Latin *angustatus*, narrowed, and *petalum*, petal, in reference to the rather narrow petals of this species.

60. DENDROCHILUM DEWINDTIANUM

Dendrochilum dewindtianum W.W. Sm. in *Notes R. Bot. Gdn. Edinb.* 8: 321–322 (1915). Type: Malaysia, Sabah, Mount Kinabalu, 25 Aug. 1913, *native collector* 68 (holotype E!). **Figs. 98 & 99, plates 3A & B; 9F.**

D. lobongense Ames, *Orchidaceae* 6: 59–60 (1920). Type: Malaysia, Sabah, Mount Kinabalu, Lubang, 17 Nov. 1915, *J. Clemens* 116 (holotype AMES!).

D. perspicabile Ames, *Orchidaceae* 6: 62–63, pl. 82, IV, 5 (1920). Type: Malaysia, Sabah, Mount Kinabalu, 1500–3000 m, 15 Nov. 1915, *J. Clemens* 202 (holotype AMES!, isotypes BM!, BO!, E!, K!, SING!).

D. furfuraceum J.J. Sm. in *Bull. Jard. Bot. Buitenz.*, ser. 3, 5: 55–56 (1922). Type: Indonesia, Sumatra, Sumatera Barat, Brani, 900 m, *Bünnemeijer* 3333 (holotype BO).

D. dewindtianum W.W. Sm. var. *sarawakense* Carr in *Gard. Bull. Straits Settl.* 8: 83–84 (1935). Types: Malaysia, Sarawak, Marudi District, Mount Dulit, 13 Aug. 1932, *Shackleton* S. 186 (lectotype SING!, isolectotype K!); Dulit Ridge, 7 Oct. 1932, *native collector in Synge* S.558 (syntype SING!, isosyntype K!), **syn. nov.**

D. sp.: Sato, *Flowers and Plants of Mount Kinabalu*: 38, top (1991).

DESCRIPTION. Epiphyte or lithophyte. *Rhizome* branching, abbreviated, up to *c.* 5 cm long. *Roots* elongate, branching, flexuose, forming a dense mass, smooth, 0.5–2 mm in diameter. *Cataphylls* (3–)4–5, ovate-elliptic or oblong-ovate, acute to acuminate, 0.5–6 cm long. *Pseudobulbs* crowded together to up to 0.5 cm apart, approximate, ovoid-oblong to globose-ovoid when mature, 0.8–4.5 × 0.5–1.5 cm, smooth, becoming wrinkled, yellowish when dry. *Leaf*: petiole rigid, sulcate, 0.4–2.5 cm long; blade linear to oblong or oblong-elliptic, apex often conduplicate, obtuse and mucronate to acute, cuneate at base, thinly coriaceous to rigid and coriaceous, 3–12 × 0.6–2 cm, main nerves 7–8. *Inflorescences* erect to gently curving, produced with the young folded leaf or nearly fully expanded leaf, laxly to subdensely *c.* 10- to many-flowered, to 2 cm in diameter, flowers borne 0.3–1 cm apart; peduncle often rather stout, 3–15 cm long, up to *c.* 3 mm in diameter; non-floriferous bracts absent, or 1 or 2, ovate, acuminate, 5–6 mm long; rachis quadrangular, becoming sulcate when fruiting, (4.5–)10–22 (–28) cm long; floral bracts oblong-ovate or ovate-elliptic, obtuse to acuminate, 3–5 × 2–2.4 mm. *Flowers* strongly sweet-scented, comparatively large, 0.8–1 cm in diameter, sepals and

219

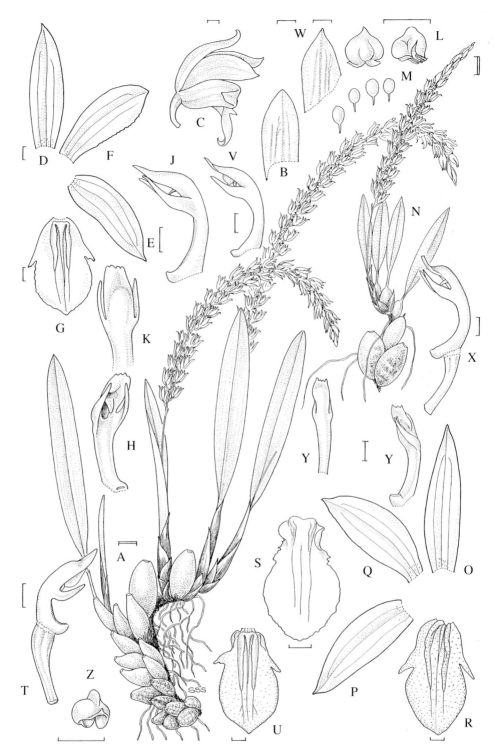

Fig. 98. *Dendrochilum dewindtianum*. **A** habit; **B** floral bract; **C** flower, side view; **D** dorsal sepal; **E** lateral sepal; **F** petal; **G** labellum, front view; **H** column, oblique view; **J** column, side view; **K** column, back view; **L** anther-cap, back and side views; **M** pollinia; **O** dorsal sepal; **P** lateral sepal; **Q** petal; **R** labellum, front view; **S** labellum, front view; **T** pedicel with ovary and column, side view; **U** labellum, front view; **V** column, side view; **W** floral bract; **X** pedicel with ovary and column, side view, **Y** column, back and oblique views; **Z** anther-cap, front view. **A–M** from *Native collector* 68 (holotype), **N–R** from *Holttum* s.n., SFN 36566, **S & T** from *J. Clemens* 202 (isotype of *D. perspicabile*), **U & V** from *J. & M.S. Clemens* 50774, **W–Z** from *J. Clemens* 116 (holotype of *D. lobongense*). Scale: single bar = 1 mm; double bar = 1 cm. Drawn by Susanna Stuart-Smith.

petals lemon-yellow, yellowish green or rarely creamy white, sometimes tipped brown, labellum lemon-yellow suffused bright green down centre, keels bright green, column pale green, apical hood and stelidia whitish, anther-cap white. *Pedicel with ovary* clavate, 3–5 mm long. *Sepals* and *petals* spreading, 3–nerved, sparsely hirsute at base. *Dorsal sepal* oblong-elliptic, acute, somewhat carinate, 6–8 × 1.8–2.5 mm. *Lateral sepals* oblong-elliptic, subfalcate, acute, 6–8 × 2–3 mm. *Petals* oblong-elliptic or oblong-ovate, obtuse to shortly acuminate, sometimes inconspicuously clawed at base, margin irregularly denticulate or almost entire, (5.7–)6–8 × (1.8–)2–2.5 mm. *Labellum* 3–lobed, 3–nerved, 5–7 mm long, (2–)3–4.5 mm wide across mid-lobe; side lobes narrowly triangular to subulate, often somewhat falcate, acute or obtuse, posterior margin entire, with a few irregular teeth, or minutely erose, equalling base or lower portion of mid-lobe, free apical portion 1 mm or less long; mid-lobe broadly ovate or obovate, rounded, obtuse, margin minutely crenulate to erose, (2.6–)3–5 mm long; disc tricarinate, keels smooth, united at base by a transverse ridge to form an elongate M-shape, lateral keels one on each lateral nerve, prominent, rounded, extending to lower portion or middle of mid-lobe and sometimes with a minute extrorse basal tooth each side, median nerve thickened and elevate to form a third keel, usually shorter, occasionally longer, sometimes not elevate. *Column* slender, incurved, (3.3–)4–5 mm long; foot prominent, sometimes minute; apical hood elongate, oblong-elliptic, entire, acute, or oblong, truncate and variously shortly toothed; stelidia arising at or just below middle of column, usually opposite or just below lower portion of stigmatic cavity, narrowly triangular, or linear, falcate, acute or obtuse, equalling middle of apical hood, *c.* 2 mm long; rostellum broadly triangular-ovate, acute; stigmatic cavity oblong to narrowly oblong, lower margin often elevate; anther-cap ovate, cucullate, dorsally keeled. *Capsule* oblong-ellipsoid, to *c.* 2.3 cm long.

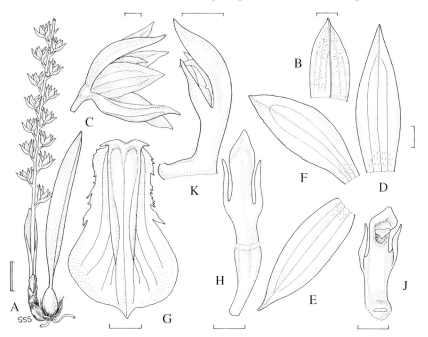

Fig. 99. *Dendrochilum dewindtianum.* **A** habit; **B** floral bract; **C** flower, side view; **D** dorsal sepal; **E** lateral sepal; **F** petal; **G** labellum, front view; **H** pedicel with ovary and column, back view; **J** column, front view; **K**, column, side view. **A–K** from *Shackleton* S.186 (lectotype of *D. dewindtianum* var. *sarawakense*). Scale: single bar = 1 mm; double bar = 1 cm. Drawn by Susanna Stuart-Smith.

DISTRIBUTION. Sumatra and Borneo.

SARAWAK. Marudi District, summit ridge of Mount Murud, 14 April 1995, *Beaman* 11448 (K!). Marudi District, Mount Dulit, 13 Aug. 1932, *Shackleton* S.186 (lectotype of *D. dewindtianum* var. *sarawakense* SING!, isolectotype K!). Marudi District, Dulit Ridge, 7 Oct. 1932, *native collector* in *Synge* S. 558 (syntype SING!, isosyntype K!).

SABAH. Mount Kinabalu: above Layang Layang, *Barkman* 1 (K!, Sabah Parks Herbarium, Kinabalu Park); above Kemburongoh, *Barkman* 267 (Sabah Parks Herbarium, Kinabalu Park); Lumu-Lumu, June 1933, *Carr* 3533, SFN 27562 (K!, SING!); Marai Parai Spur, 10 June 1933, *Carr* 3534 (SING!); below Kemburongoh, 28 Oct. 1933, *Carr* 3717 (SING!) & *Carr* 3730 (SING!); Lubang, 17 Nov. 1915, *J. Clemens* 116 (holotype of *D. lobongense* AMES!); 1500–3000 m, 15 Nov. 1915, *J. Clemens* 202 (holotype of *D. perspicabile* AMES!, isotypes BM!, BO!, E!, K!, SING!); 2400 m, 15 Nov. 1931, *J. & M.S. Clemens* 27140 (BM!, E!, HBG!, K!, L!, SING!); 2700 m, 15 Nov. 1931, *J. & M.S. Clemens* 27145 (BM!, E!, K!, L!); Upper Kinabalu, 12 Nov. 1931, *J. & M.S. Clemens* 27147 (E!, mixed collection with *D. lancilabium*); 2300 m, 12 Nov. 1931, *J. & M.S. Clemens* 27148 (BM!, E!, K!, L!); above Lumu-Lumu, 15 Nov. 1931, *J. & M.S. Clemens* 27168 (BO!); 2400 m, Jan. 1932, *J. & M.S. Clemens* 30173 (E!); Gurulau Spur, 6 Dec. 1933, *J. & M.S. Clemens* 50774 (AMES!, BM!, E!, K!, L!); Mount Kinabalu, locality unknown, July-Aug. 1916, *Haslam* s.n. (BM!, E!, K!); Kemburongoh, 16 Nov. 1931, *Holttum* s.n., SFN 36566 (AMES!, SING!); Mount Trus Madi, summit, 6 March 1986, *Kitayama* K.2086 (UKMS!) & 30 Feb. 1992, *Kitayama* 11–13 (Sabah Parks Herbarium, Kinabalu Park!). Locality unknown, 1986, *Lamb* AL 687/86 (K!). Mount Kinabalu, 25 Aug. 1913, *native collector* 68 (holotype of *D. dewindtianum* E!) & 28 Aug. 1913, *native collector* 99 (paratype of *D. dewindtianum* E!, K!). Mount Kinabalu, locality unknown, 12 Sept. 1981, *Sato et al.* 040 (UKMS!). Mount Kinabalu, Carson's Camp, 25 July 1981, *Sato et al.* 0703 (UKMS!), & *Sato et al.* 1404 (UKMS!); East Mesilau, 11 May 1986, *SNP* 2890 (Sabah Parks Herbarium, Kinabalu Park!); Kemburongoh, 25 Sept. 1966, *Togashi* s.n. (TI!). Mount Alab, south ridge leading towards summit, 31 Oct. 1986, *de Vogel* 8667 (L!).

HABITAT. Lower montane forest; upper montane forest and scrub, mostly on ultramafic substrate; *Leptospermum/Dacrydium* forest about 8 metres high on steep east-facing slopes on sandstone and shale; rocky places. 1500–3000 m.

This attractive species is abundant above about 2600 m on Mount Kinabalu, particularly in forest and scrub developed on ultramafic substrate. Leaf and flower dimensions and labellum shape are very variable throughout its range. Some individuals, e.g., the type collection and *Clemens* 50774 are showy plants having among the largest flowers of any dendrochilum in Borneo. Others, e.g., *Clemens* 27140 and 27168, are more slender in habit and smaller flowered. An attractive creamy white-flowered form occurs at about 2200 m around Kemburongoh on Mount Kinabalu. The mid elevation form is rather intermediate in vegetative and floral morphology between the higher elevation forms and *D. tenompokense*. A collection from the Crocker Range, *Wood* 583, illustrated as *D. dewindtianum* by Wood, Beaman & Beaman (1993, fig. 29 M–U) is referable to *D. tenompokense* Carr var. *papillilabium* J.J. Wood.

A small plant, referred by Carr under *D. dewindtianum* as var. *sarawakense*, is very similar to plants refered to *D. perspicabile* by Ames and may represent a dwarf ecotype. The lateral keels on the labellum sometimes have small retrorse basal teeth similar to those found, for example, in *D. crassifolium*.

The specific epithet honours the Ranee of Sarawak, Margaret Lili Alice de Windt, sister of the explorer Henry de Windt.

222

61. DENDROCHILUM CRASSIFOLIUM

Dendrochilum crassifolium Ames, *Orchidaceae* 6: 49–50, pl. 84a (1920). Type: Malaysia, Sabah, Mount Kinabalu, July-Aug. 1916, *Haslam* s.n. (holotype AMES!). **Figs. 100 & 101, plate 2A.**

DESCRIPTION. Epiphyte, occasionally terrestrial. *Rhizome* abbreviated. *Roots* branching, flexuose, forming a dense mass, minutely papillose or smooth, 1–3 mm in diameter. *Cataphylls* 3–4, ovate-elliptic or oblong-ovate, obtuse to acute, rigid, 1–7 cm long. *Pseudobulbs* crowded to *c.* 0.5 cm apart, approximate, ovoid, ovoid-elliptic, or conical, 0.8–3.5 × 0.5–1.2 cm, smooth, becoming wrinkled, yellowish. *Leaf:* petiole sulcate, slender to stout and fleshy, 2–6 cm long, up to 4 mm in diameter; blade linear to linear-ligulate, or oblong to oblong-elliptic, obtuse to shortly acute, or subacute and mucronate, sometimes slightly cucullate, gradually attenuated and cuneate below, thick and fleshy, rigid, or toughly coriaceous, 6–20.5 × 0.5–2.5 cm, main nerves 5–7 visible in dried material. *Inflorescences* gently curving to pendulous, produced with the young folded leaf or nearly fully expanded leaf, much longer than leaves, subdensely many-flowered, flowers borne 3–5 mm apart, lowermost up to 1 cm apart; peduncle wiry, (3.5–)12–18(–20) cm long; non-floriferous bracts absent, or 2–4, oblong-ovate, obtuse and mucronate, 3.5–7 mm long; rachis quadrangular, (4.5–)15–36 cm long; floral bracts ovate, apiculate, subacute or obtuse, 3–5 × 2.1–2.4 mm. *Flowers* unscented, sepals and petals green, greenish cream, greenish yellow or pale yellow with a darker median nerve on exterior, labellum similar, green between side lobes and with 2 darker green, pale yellowish brown or orange streaks from middle of side lobes to about middle of mid-lobe, column pale yellow, stelidia whitish. *Pedicel with ovary* clavate, 2–3 mm long. *Sepals* and *petals* 3–nerved, with or without sparse hairs at base. *Dorsal sepal* oblong-elliptic to narrowly elliptic, acute to shortly acuminate, dorsally carinate, 6–7 × 1.7–2.5 mm. *Lateral sepals* oblong-elliptic, acute to shortly acuminate, dorsally carinate, 5.8–7 × 1.5–2.5 mm. *Petals* oblong-elliptic or narrowly elliptic, acute to shortly acuminate, erose to irregularly denticulate, sometimes entire, 5.7–7 × 1.8–3 mm. *Labellum* 3–lobed, 3–nerved, shortly stipitate to column-foot, 5–5.5 mm long, 1.1–1.2 mm wide above base, 2–3 mm wide across mid-lobe; side lobes divaricate, free apical portion narrowly triangular, subulate, setaceous distally, 1–1.1 mm long, posterior margin minutely erose to denticulate; mid-lobe obovate, ovate-elliptic, broadly obtrullate or subcordate, acute to acuminate, or rounded and mucronate, uneven to minutely erose, 2.6–3.5 mm long; disc with 2 smooth to papillose keels terminating towards distal portion of mid-lobe, joined at base by a short transverse ridge which sometimes bears a minute extrorse tooth on each side, or sometimes developed into 2 erose basal flanges, median nerve elevate to a varying degree. *Column* gently incurved, 2.5–4 mm long; foot distinct; apical hood triangular-elliptic, obtuse to acute, entire, or irregularly tridentate; stelidia arising just above base and below stigmatic cavity, linear-ligulate or narrowly triangular, falcate, subacute to acuminate, reaching or exceeding rostellum, 1.8–2 mm long; rostellum triangular, acute; stigmatic cavity narrowly oblong, lower margin a little elevate; anther-cap ovate, cucullate.

DISTRIBUTION. Borneo.

SARAWAK. Lawas District, Kelabit Highlands, route from Ba Kelalan to Mount Murud, Camp III, 27 Sept. 1967, *Burtt & Martin* B.5241 (E!, SAR!); Marudi District, Mount Murud, 7 April 1970, *Nooteboom & Chai* 01995A (L!).

SABAH. Mount Kinabalu, Tenompok, Aug. 1933, *Carr* 3671, SFN 28047 (K!, SING!); Tenompok Orchid Garden, 17 Nov. 1933, *J. & M.S. Clemens* 40434 (BM!, K!); Tinekuk Falls, 27 Oct. 1933, *J. & M.S. Clemens* 40931 (BM!); Tenompok, 9 Nov. 1933, *J. & M.S. Clemens*

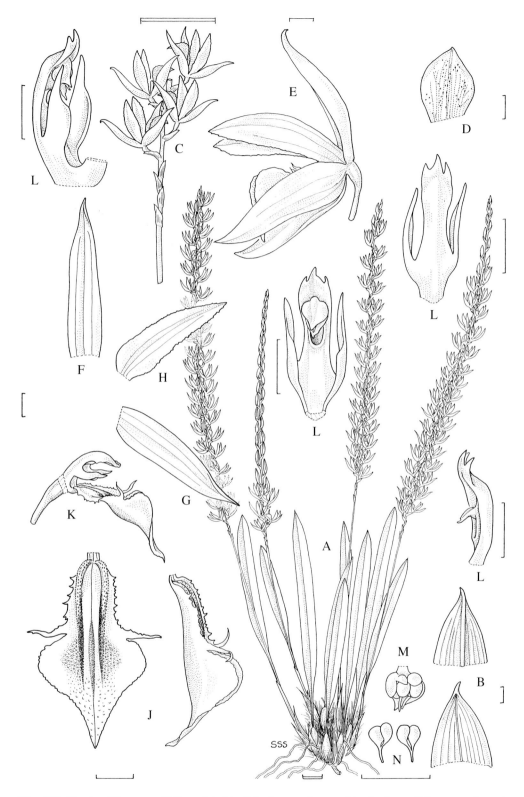

Fig. 100. *Dendrochilum crassifolium*. **A** habit; **B** leaf apex, front and back views; **C** lower portion of inflorescence; **D** floral bract; **E** flower, side view; **F** dorsal sepal; **G** lateral sepal; **H** petal; **J** labellum, front and side views; **K** pedicel with ovary, labellum and column, side view; **L** column, front, oblique, side and back views; **M** anther cap with pollinia, interior view; **N** pollinia. **A–N** from *Burtt & Martin* B. 5241. Scale: single bar = 1 mm; double bar = 1 cm. Drawn by Susanna Stuart-Smith.

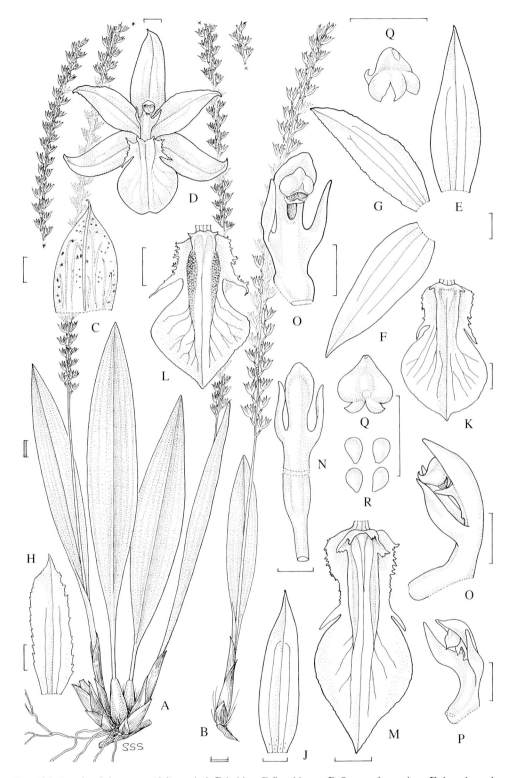

Fig. 101. *Dendrochilum crassifolium*. **A** & **B** habits; **C** floral bract; **D** flower, front view; **E** dorsal sepal; **F** lateral sepal; **G, H** & **J** petals; **K–M**, labelli, front views; **N** pedicel with ovary and column, back view; **O** & **P**, columns, front, oblique and side views; **Q** anther-cap, back and oblique views; **R** pollinia. **A** from *Lamb* AL876/87, **B, H** & **L** from *Haslam* s.n. (holotype), **C–G, K, O, Q** & **R** from *Vermeulen & Duistermaat* 686, **J, M** & **P** from *Vermeulen & Duistermaat* 1043. Scale: single bar = 1 mm; double bar = 1 cm. Drawn by Susanna Stuart-Smith.

50234 (SING!); Tenompok Orchid Garden, 9 Nov. 1933, *J. & M.S. Clemens* 50235 (AMES, mixed collection with *D. exasperatum* and *D. kamborangense!*, BM!) & 9 Nov. 1933, *J. & M.S. Clemens* 50258 (AMES!, BM!, K!); Mount Kinabalu, locality unknown, July-Aug. 1916, *Haslam* s.n. (holotype AMES!). Mount Alab, Sept. or Nov. 1987, *Lamb* AL 876/87 (K!). Crocker Range, Keningau to Kimanis road, Dec. 1986, *Vermeulen & Duistermaat* 686 (L!). Sipitang District, Rurun River headwaters, Dec. 1986, *Vermeulen & Duistermaat* 1043 (K!, L!).

HABITAT. Lower montane forest; low, rather open, wet, somewhat podsolic forest, with a very dense undergrowth of *Pandanus* spp. and rattans; cleared, steeply sloping roadsides on sandstone and shale outcrops partially covered with grass and small bushes, in full sun. 1300–2400 m.

D. crassifolium appears to be closely related to *D. dewindtianum* and is equally variable. It is sometimes difficult to distinguish *D. crassifolium* from some forms of *D. dewindtianum* in the herbarium, although the tough fleshy leaves of the former are distinctive. *Burtt & Martin* B5241 differs in having no extrorse tooth at the base of the keels, a more narrow, acuminate labellum mid-lobe and an irregularly toothed column apical hood. *Vermeulen & Duistermaat* 1043 has entire petals and smooth keels with two curious erose flanges at the base. *Vermeulen & Duistermaat* 686 also has smooth keels.

The specific epithet is derived from the Latin *crassus*, thick, and *folium*, leaf, referring to the tough fleshy leaves.

62. DENDROCHILUM GEESINKII

Dendrochilum geesinkii J.J. Wood in Wood & Cribb, *Checklist orch. Borneo*: 175–176, fig. 17 (1994). Type: Indonesia, Kalimantan Timur, Apo Kayan, base of Mount Tapa Sia, between Long Bawan & Panado, 1400 m, 22 July 1981, *Geesink* 9180 (holotype L!, isotypes BO, K!, KYO). **Fig. 102.**

DESCRIPTION. Epiphyte. *Rhizome* up to 6 cm long, 0.5 cm in diameter. *Roots* branching, flexuose, smooth, *c*. 0.5–1 mm in diameter. *Cataphylls* 3, ovate-elliptic, acute, 1–8 cm long. *Pseudobulbs* crowded to 0.5–0.6 cm apart, cylindrical or narrowly oblong, 6–9 × 0.6–1 cm, wrinkled when dry. *Leaf*: petiole sulcate to conduplicate, 1.5–2.5 cm long; blade oblong-elliptic, ligulate, obtuse and mucronate to subacute, narrowed and cuneate at base, coriaceous, 11.5–22 × 1.4–2.2 cm, main nerves 7. *Inflorescences* erect to gently curving, produced with the nearly fully expanded leaf, densely many-flowered, flowers borne 1.8–2 mm apart, lowermost up to 3 mm apart; peduncle 13–17 cm long; non-floriferous bracts solitary, ovate-elliptic, acuminate, 4.5–5.5 mm long; rachis quadrangular, 20–23 cm long; floral bracts narrowly triangular-ovate, subulate, acute to acuminate, 5–7 × 1–1.1 mm. *Flowers* pale green. *Pedicel with ovary* narrowly clavate, 2 mm long. *Sepals* and *petals* usually 3–nerved, sometimes lateral nerves very short. *Dorsal sepal* narrowly triangular, acuminate, 6 × 1 mm. *Lateral sepals* narrowly elliptic, subfalcate, acute, 5.5 × 1 mm. *Petals* linear to narrowly elliptic, subfalcate, acute, minutely erose, 5 × 0.8 mm. *Labellum* obscurely 3–lobed, 3–nerved, oblong-elliptic in outline, obtuse, 3.8–4 × 2 mm; side lobes auriculate, tooth-like, acute; mid-lobe minutely papillose, margin rather irregular, disc 3–keeled, united at base by a transverse ridge, the third keel formed by the elevate median nerve, all grading into prominent nerves which terminate beyond middle of mid-lobe. *Column* slightly incurved, 2 mm long; foot prominent; apical hood oblong, obscurely irregularly trilobed; stelidia borne opposite lower

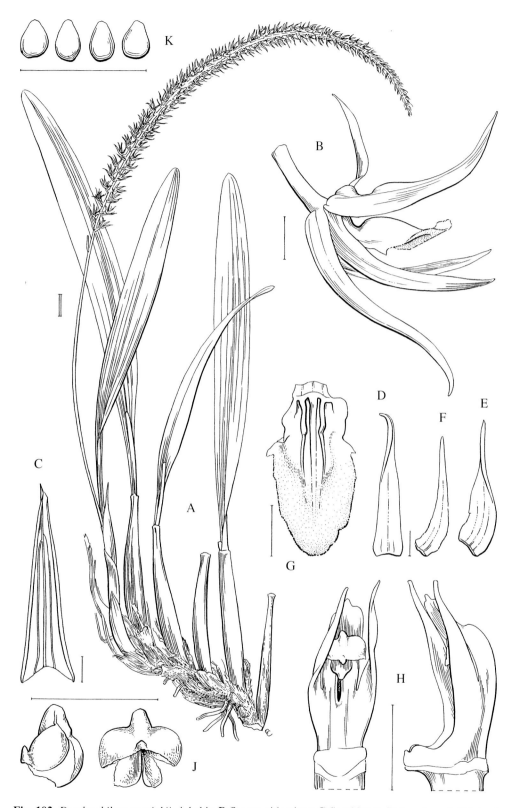

Fig. 102. *Dendrochilum geesinkii.* **A** habit; **B** flower, side view; **C** floral bract; **D** dorsal sepal; **E** lateral sepal; **F** petal; **G** labellum, front view; **H** column, front and side views; **J** anther-cap, back and side views; **K** pollinia. Scale: single bar = 1 mm; double bar = 1 cm. **A–K** from *Geesink* 9180 (holotype). Drawn by Eleanor Catherine.

portion of stigmatic cavity, narrowly triangular, obtuse to acute, slightly exceeding apical hood, 2.1–2.2 mm long; rostellum ovate, obtuse; stigmatic cavity narrowly oblong; anther-cap cucullate, dorsally obtusely carinate.

DISTRIBUTION. Borneo.

SARAWAK. Limbang District, route to Batu Lawi, Bario, 3 Aug. 1985, *Awa & Lee* S.50586 (AAU, K!, KEP!, L!, MEL, SING).

KALIMANTAN TIMUR. Apo Kayan, between Long Bawan and Panado, 9 July 1981, *Geesink* 8989 (BO, KYO, L!) & 22 July 1981, *Geesink* 9180 (holotype L!, isotypes BO, K!, KYO).

HABITAT. Hill forest on sandstone; mixed hill dipterocarp forest on ridge top. 1290–1400 m.

This species is related to *D. crassifolium* Ames but may be distinguished by the much longer pseudobulbs, slightly longer floral bracts, and slightly smaller flowers with narrower petals, obscure, auriculate labellum side lobes and stelidia exceeding the column apical hood.

The specific epithet honours the late Rob Geesink of the Rijksherbarium, Leiden, The Netherlands, who collected the type during a joint Indonesian-Japanese expedition sponsored by LIPI, Jakarta and the Ministry of Education, Tokyo.

63. DENDROCHILUM MURUDENSE

Dendrochilum murudense (J.J. Wood) J.J. Wood, **comb. et stat. nov.** Type: Malaysia, Sarawak, Marudi District, Kelabit Highlands, Mount Murud, 2400 m, 7 April 1970, *Nooteboom & Chai* 01995 (holotype K!, isotype SAR!, material labelled 01995 at L is *D. crassifolium*). **Fig. 103.**

D. crassifolium Ames var. *murudense* J.J. Wood in Wood & Cribb, *Checklist orch. Borneo*: 166, 168, fig. 22 A & B (1994).

DESCRIPTION. Epiphyte. *Rhizome* abbreviated, to 4 cm long. *Roots* branching, flexuose, minutely papillose, 0.5–1.5 mm in diameter. *Cataphylls* 2–3, ovate-elliptic, acute, speckled, 1–3 cm long. *Pseudobulbs* crowded or up to 1 cm apart, oblong-ovate or conical, 0.6–2 × 0.8 cm, greenish-yellow. *Leaf*: petiole deeply sulcate, 0.8–1.5 cm long; blade elliptic or oblong-elliptic, obtuse and mucronate, often constricted 0.5–1 cm below apex, thick, tough and coriaceous, 4.4–8 × 1–2 cm, main nerves 5–7. *Inflorescences* erect, produced with the nearly to fully expanded leaf, subdensely many-flowered, flowers borne 4.5–6 mm apart; peduncle 3.5–6.5 cm long; non-floriferous bracts 1–3, ovate-elliptic, acute or obtuse and mucronate, 4–7 mm long; rachis quadrangular, 16–20 cm long; floral bracts oblong to ovate, mucronate, 4–7 mm long. *Flowers* greenish. *Pedicel with ovary* narrowly clavate, 2.5–3 mm long. *Sepals* and *petals* 3–nerved. *Dorsal sepal* oblong-elliptic, acuminate, 9–10 × 2.1–2.8 mm. *Lateral sepals* ovate-elliptic, acute to acuminate, 9–9.2 × 2.2–3 mm. *Petals* narrowly elliptic, acuminate, minutely erose, 8–8.1 × 0.9–2.1 mm. *Labellum* stipitate to column-foot by a tiny claw, 3–lobed, 3–nerved, 6.6 mm long, 2–2.1 mm wide across side lobes; side lobes triangular, acute to acuminate, dorsal margin erose to lacinulate, 3 × c. 0.8 mm; mid-lobe elliptic, acuminate, minutely erose, particularly distally, 4 × 2.8–3 mm; disc with 2 fleshy, raised, smooth keels united by a basal ridge, with a shallow triangular flange each side at base, terminating on distal portion of mid-lobe, median nerve elevate for a short distance. *Column* slender, incurved, 4–4.5 mm long; foot 0.5–1 mm long; apical hood oblong to elliptic, rounded, entire, 0.8 mm

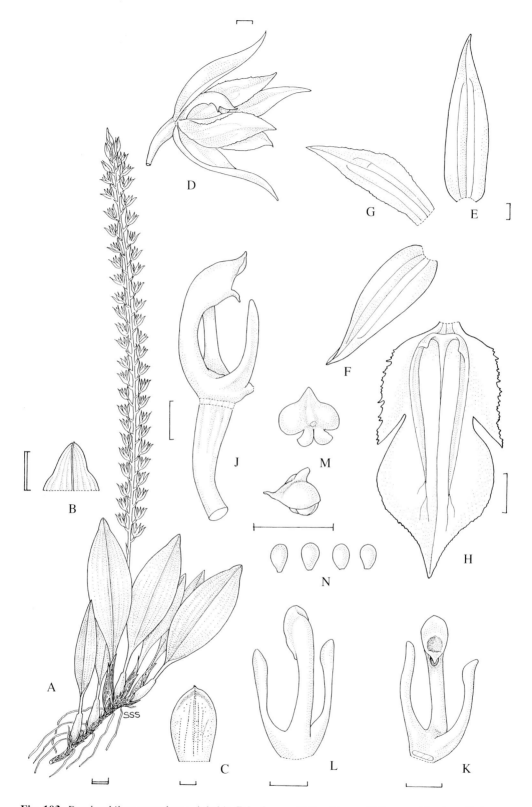

Fig. 103. *Dendrochilum murudense*. **A** habit; **B** leaf apex; **C** floral bract; **D** flower, side view; **E** dorsal sepal; **F** lateral sepal; **G** petal; **H** labellum, front view; **J** pedicel with ovary and column, anther-cap removed, side view; **K** column, anther-cap removed, front view; **L** column, anther-cap removed, back view; **M** anther-cap, back and side views; **N** pollinia. **A–N** from *Nooteboom & Chai* 01995 (holotype). Scale: single bar = 1 mm; double bar = 1 cm. Drawn by Susanna Stuart-Smith.

wide; stelidia basal, ligulate, obtuse, fleshy, reaching rostellum, 2–2.5 mm long; rostellum broadly triangular, acute; stigmatic cavity small, circular to elliptic; anther-cap cucullate, with dorsal wart.

DISTRIBUTION. Borneo.

SARAWAK. Marudi District, summit ridge of Mount Murud, 14 April 1995, *Beaman* 11459 (K!). Marudi District, Kelabit Highlands, Mount Murud, 7 April 1970, *Nooteboom & Chai* 01995 (holotype K!, isotype SAR!). Marudi District, Mount Murud, second sandstone summit, 12 Sept. 1982, *Yii* S.44430 (L!, mixed collection with *D. galbanum*).

HABITAT. Upper montane mossy forest on sandstone. 2250–2400 m.

After some consideration I have decided to elevate this plant to species status. Although closely related to *D. crassifolium*, several characters point to its separation. The column is much more slender and has obtuse fleshy stelidia which arise from the base. The leaves are proportionally much smaller, recalling those of *D. lewisii*, and are often somewhat constricted below the apex. The flowers are larger and of a greenish hue and are probably visited by a different pollinator.

The specific epithet refers to the type locality, Mount Murud (2422 m), the highest mountain in Sarawak.

64. DENDROCHILUM KAMBORANGENSE

Dendrochilum kamborangense Ames, *Orchidaceae* 6: 57–58, pl. 84, bottom (1920). Type: Malaysia, Sabah, Mount Kinabalu, Kemburongoh (Kamborangah), 17 Nov. 1915, *J. Clemens* 205 (holotype AMES!, isotypes BM!, BO!, K!, SING!). **Fig. 104, plate 9D & E.**

D. sp.: Sato, *Flowers and Plants of Mount Kinabalu*: 37, bottom (1991).

DESCRIPTION. Epiphyte. *Rhizome* abbreviated, up to 5 cm long, branching. *Roots* branching, flexuose, forming a dense mass, minutely papillose, 0.3–1 mm in diameter. *Cataphylls* 4, ovate-elliptic, obtuse or acute, speckled, becoming fibrous, persistent, 1–9 cm long. *Pseudobulbs* crowded, oblong-ovoid, or subcylindrical, finely longitudinally wrinkled, approximate, 1–4 × 0.6 –1.5 cm, yellowish. *Leaf*: petiole sulcate, 1.5–7.5 cm long; blade linear to narrowly elliptic, cuneate at base, obtuse or subacute, sometimes slightly constricted below apex, thin-textured, (9–)16–22(–28) × 0.9–1.6(–2) cm, main nerves 5–7. *Inflorescences* gently curving, produced with the nearly expanded leaf, subdensely many-flowered, *c.* 1.5 cm in diameter, flowers borne 2.5–4 mm apart; peduncle 13–26 cm long; non-floriferous bracts 1–2, ovate, acute, 3–5 mm long; rachis quadrangular, 12–18(–23) cm long; floral bracts ovate, subacute, 3 mm long, brown. *Flowers* fragrant or unscented, sepals and petals lemon-yellow, saffron-yellow, yellowish green or creamy green, labellum yellowish green at base, remainder brown, often with paler brown, yellow or green margins and a pale yellow median line in lower half, or less often entirely yellow with orange to ochre side lobes, column greenish yellow. *Pedicel with ovary* clavate, 3 mm long. *Sepals* and *petals* 3–nerved. *Dorsal sepal* oblong-elliptic, acute, carinate distally, 6–6.1(–7) × (2–)2.4–2.5 mm. *Lateral sepals* ovate-elliptic, acute, carinate distally, 5.5–6(–7) × (2–)2.5 mm. *Petals* oblong-elliptic to elliptic, minutely denticulate, acute, 5.3–6 × 2.2–2.5 mm. *Labellum* stipitate to column-foot by a very short claw, 3–lobed, 3–nerved, 4–5 mm long, 2–2.1 mm wide across side lobes, fleshy; side lobes erect, semi-elliptic, auriculate, irregularly serrate, 0.8–0.9 mm wide; mid-lobe broadly trullate to rotundate, acute, margin irregular to minutely erose, 2.9–3 × 2–3 mm; disc with 2

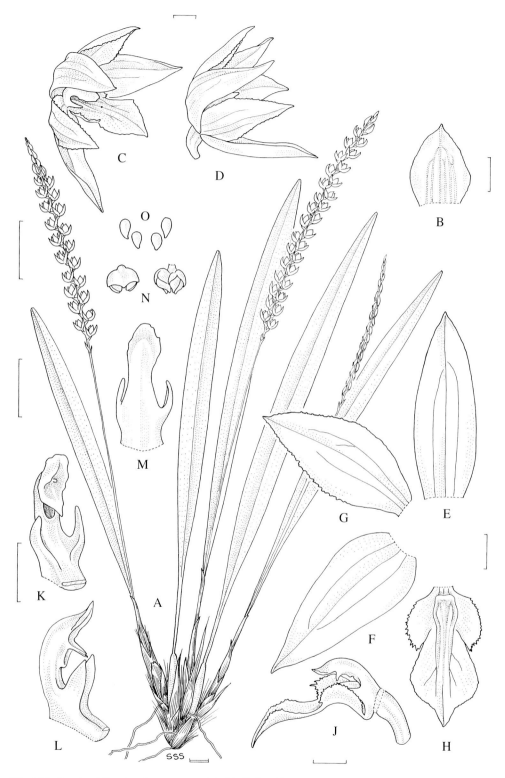

Fig. 104. *Dendrochilum kamborangense*. **A** habit; **B** floral bract; **C** flower, viewed from above; **D** flower, side view; **E** dorsal sepal; **F** lateral sepal; **G** petal; **H** labellum, front view; **J** pedicel with ovary, labellum and column, side view; **K** column, anther-cap removed, oblique view; **L** column, anther-cap removed, side view; **M** column, back view; **N** anther-cap, back and interior views; **O** pollinia. **A** from *Collenette* 21534, **B–O** from *Bailes & Cribb* 695. Scale: single bar = 1 mm; double bar = 1 cm. Drawn by Susanna Stuart-Smith.

fleshy keels united at base by a transverse ridge, terminating at or below middle of mid-lobe, median nerve elevate, slightly shorter. *Column* gently incurved, 2–3 mm long; foot prominent; apical hood oblong to subquadrate, tridentate, erose or denticulate; stelidia borne near middle of column, narrowly triangular, obtuse, divergent, reaching rostellum, 0.7–0.8 mm long; rostellum ovate-triangular, acute; stigmatic cavity oblong, lower margin elevate; anther-cap cucullate, smooth.

DISTRIBUTION. Borneo.

SABAH. Mount Kinabalu: Mesilau Cave, 1983, *Bailes & Cribb* 695 (K!); Layang Layang, *Barkman* 64 (Sabah Parks Herbarium, Kinabalu Park); Kemburongoh, 5 Aug. 1933, *Carr* 3622, SFN 27908 (K!, SING!), *Carr* 3751 (SING!) & *Carr* SFN 36567 (SING!); 2400 m, 29 April 1933, *Carr* 3742 (SING!); Eastern Shoulder, 9 July 1961, *Chew et al.* RSNB 182 (AMES!, K!, L!, SING!); Kemburongoh, 17 Nov. 1915, *J. Clemens* 205 (holotype AMES!, isotypes BM!, BO!, K!, SING!); Marai Parai Spur, 2 Dec. 1915, *J. Clemens* 385 (AMES!, BM!, E!, K!, SING); Kemburongoh, 13 Nov. 1931, *J. & M.S. Clemens* 27143 (BM!, BO!); Paka-paka Cave, 12 Nov. 1931, *J. & M.S. Clemens* 27146 (BM!, E!, K!); Tenompok, 13 Jan. 1932, *J. & M.S. Clemens* 27860 (BM!, E!, K!); Silau Basin, Mesilau, 9 April 1932, *J. & M.S. Clemens* 29288 (AMES!, BO!); Tenompok Orchid Garden, 9 Nov. 1933, *J. & M.S. Clemens* 50235 (BM, mixed collection with *D. crassifolium* and *D. exasperatum*!, G!); Gurulau Spur, 1 Dec. 1933, *J. & M.S. Clemens* 50654 (BM!, E!, K!, L!); Janet's Halt/Sheila's Plateau, 5 Sept. 1963, *Collenette* 21534 (AMES!, BO!, K!, L!); Mesilau Cave/Janet's Halt, 29 Aug. 1963, *Fuchs & Collenette* 21404 (K!, L!); Kemburongoh, 16 Nov. 1931, *Holttum* SFN 36567 (AMES!); Summit Trail, 15 Oct. 1958, *Jacobs* 5731 (B, BH, CANB, G!, K!, L!, S, SAN, US); Mount Kinabalu, locality unknown, 16 Jan. 1983, *Jumaat* 3397 (UKMS!); Mount Kinabalu, 2160 m, 28 Aug. 1913, *native collector* 90 (E!); Summit Trail, Sept. 1981, *Sato* 2100 (UKMS!) & *Sato* 2158 (UKMS!); Summit Trail, Oct. 1981, *Sato et al.* 1436 (UKMS!); Marai Parai Spur, 9 Oct. 1985, *SNP* 2369 (Sabah Parks Herbarium, Kinabalu Park); Eastern Ridge, 20 Sept. 1986, *SNP* 2397 (Sabah Parks Herbarium, Kinabalu Park). Mount Tembuyuken, 10 May 1990, *SNP* 4709 (Sabah Parks Herbarium, Kinabalu Park). Mount Kinabalu, Kemburongoh, 25 Sept. 1966, *Togashi* s.n. (TI!).

HABITAT. Upper montane mossy forest, often in exposed places on ultramafic substrate. 1500–2900 m.

This frequently collected species, described by Ames as "a very beautiful and showy orchid", is one of the largest flowered at the elevation it inhabits. It has so far only been recorded from Mount Kinabalu where it is found commonly, particularly on ridge-tops, on both ultramafic and non-ultramafic substrates. It has the general appearance and flower colour of *D. trusmadiense* (section *Eurybrachium*), but the column and labellum are quite different.

The specific epithet refers to Kemburongoh (Kamborangah), the type locality on Mount Kinabalu.

65. DENDROCHILUM JIEWHOEI

Dendrochilum jiewhoei J.J. Wood **sp. nov.** sectionis *Platyclinidis D. kamborangensi* Ames foederata, sed pseudobulbis foliisque brevioribus, floribus inflorescentiae infimis postremo aperientibus, floribus minoribus petalis integris, lobis lateralibus labelli triangularibus usque anguste triangularibus, lobo medio minore oblongo-triangulari usque triangulari-ovato, columna breviora pede minus prominenti atque stelidiis basalibus cucullum apicalem

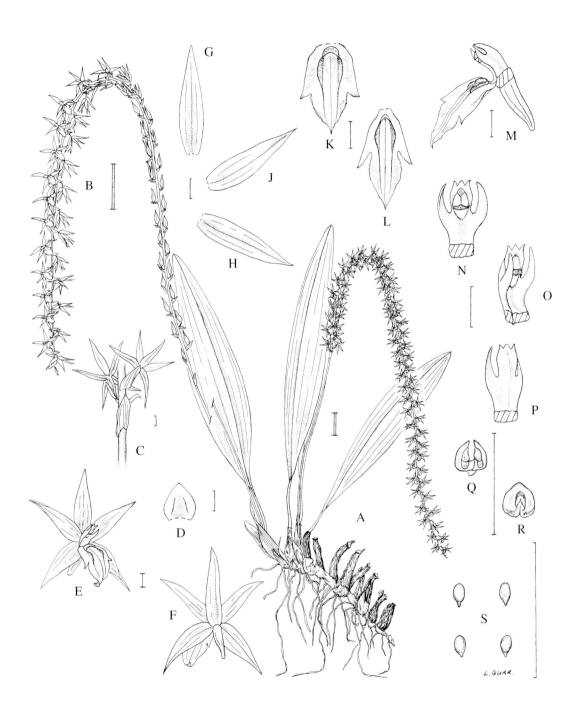

Fig. 105. *Dendrochilum jiewhoei*. **A** habit; **B** inflorescence; **C** junction of rachis with peduncle showing one non-floriferous bract and two flowers; **D** floral bract; **E** flower, oblique view; **F** flower, back view; **G** dorsal sepal; **H** lateral sepal, **J** petal; **K** & **L** labella, front view; **M** pedicel with ovary, labellum and column, side view; **N** column, front view; **O** column, oblique view; **P** column, back view; **Q** anther-cap, front view; **R** anther-cap, back view; **S** pollinia. **A–S** from *Mjöberg* s.n. (holotype). Scale: single bar = 1 mm; double bar = 1 cm. Drawn by Linda Gurr.

aequantibus vel eo brevioribus instructa distinguenda. Typus: Malaysia, Sarawak, Kuching District, Mount Penrissen, 1920's, *Mjöberg* s.n. (holotypus AMES, isotypus K). **Fig. 105.**

DESCRIPTION. Epiphyte. *Rhizome* up to *c.* 7 cm long, 2 mm in diameter. *Roots* elongate, sparsely to much-branched, smooth, 0.3–1 mm in diameter. *Cataphylls* 2–3, ovate-elliptic, obtuse to acute, finely nerved, pale brown, with obscure darker brown speckling, becoming fibrous. *Pseudobulbs* crowded on rhizome, ovoid to cylindrical, finely wrinkled in dried state, (0.6–)1–*c.* 1.8 × *c.* 0.4–0.8 cm, yellowish-olive. *Leaf*: petiole slender, sulcate, 1.4–2 cm long; blade narrowly elliptic to oblong-elliptic, constricted *c.* 0.8–1.4 cm below apex, obtuse and mucronate, thin-textured, 6.5–11.5 × 0.9–1.8 cm, main nerves 5–7. *Inflorescences* gently curving, densely many-flowered, 1–1.4 cm in diameter, flowers borne 2–3 mm apart, lowermost opening last; peduncle filiform, (5–)7–8.5 cm long; non-floriferous bracts 1–4, ovate, obtuse to acute, 1.8–4 mm long; rachis quadrangular, (3.5–)13–20 cm long; floral bracts ovate-oblong, obtuse and mucronate, 2–3 × 2.1 mm long, pale brown. *Flowers* wide-opening, stellate, colour not recorded. *Pedicel with ovary* slender, *c.* 2–2.4 mm long. *Sepals* and *petals* 3-nerved. *Dorsal sepal* narrowly elliptic, acute to acuminate, 4.5–4.6(–5.1) × 1.1–1.5 mm. *Lateral sepals* narrowly elliptic to somewhat ovate-elliptic, acute to acuminate, 4.1–4.2(–5) × 1.3–1.5 mm. *Petals* narrowly elliptic, acute to acuminate, margins entire, 3.9–4.1 × 1.2 mm. *Labellum* very shortly clawed, shortly 3-lobed, 3-nerved, very minutely papillose, *c.* 3.2–4 mm long, *c.* 1.8–2.1 mm wide across side lobes; side lobes triangular to narrowly triangular, minutely erose, free apical portion 0.4–0.8 mm long; mid-lobe oblong-triangular to triangular-ovate, acute to acuminate, margin slightly irregular, *c.* 1.4–1.5(–2) × 1.2–1.3 mm; disc with 2 keels joined at and forming a cavity at the base, extending *c.* 0.9–1.1 mm along blade, median nerve rather prominent for a short distance. *Column* 1.4–1.5 mm long; foot 0.1–0.2 mm long; apical hood rounded, slightly irregular to obscurely tridentate; stelidia basal, falcate, acute, equalling or shorter than apical hood, 1–1.1 mm long; rostellum small, ovate, obtuse; stigmatic cavity without a raised margin; anther-cap cucullate, smooth.

DISTRIBUTION. Borneo.

SARAWAK. Kuching District, Mount Penrissen, 1920's, *Mjöberg* s.n. (holotype AMES!, isotype K!).

HABITAT. Lower montane forest. 1320 m.

As so often happens, new taxa have a habit of turning up after the completion or publication of a manuscript. So it was with *D. jiewhoei* which was among a batch of previously unstudied specimens, collected in Sarawak in the 1920's by the Swedish zoologist Eric Mjöberg, on loan to Kew from AMES for determination. It is surprising that an undescribed species should have originated from Mount Penrissen, which is relatively well collected compared to many areas of Sarawak. *D. jiewhoei* is known only from the type and it is curious that no subsequent collections have been seen. Although possibly a rare endemic to Mount Penrissen, it is more likely to have simply been overlooked elsewhere.

D. jiewhoei is possibly most closely allied to the Kinabalu endemic *D. kamborangense*. It may be distinguished from *D. kamborangense* by its shorter pseudobulbs and leaves, the lowermost buds on the inflorescence opening last, and the smaller flowers which have entire petals. The labellum has triangular to narrowly triangular, minutely erose side lobes and a smaller, oblong-triangular to triangular-ovate mid-lobe. The column is shorter, with a less prominent foot, and basal stelidia which are equal to or shorter than the apical hood.

The specific epithet honours Mr Tan Jiew Hoe of Singapore, who has generously provided financial support to help cover the printing costs of the colour plates for this volume.

66. DENDROCHILUM LACINILOBUM

Dendrochilum lacinilobum J.J. Wood & A. Lamb in Wood & Cribb *Checklist orch. Borneo*: 181, 183, fig. 19 (1994). Type: Malaysia, Sabah, Crocker Range, Keningau & Tambunan District border, Ulu Apin Apin, 1500 m, 5 Nov. 1991, *Lamb & Surat in Lamb* AL 1390/91 (holotype K!). **Fig. 106, plate 10D.**

DESCRIPTION. Epiphyte. *Rhizome* branching, up to *c* 8 cm long. *Roots* branching, flexuose, forming a dense mass, minutely papillose, 0.2–1 mm in diameter. *Cataphylls* 3, ovate-elliptic, acute, distinctly mottled and speckled, soon disintegrating to fibres, 1–4 cm long. *Pseudobulbs* crowded and forming large clumps, narrowly fusiform, becoming finely wrinkled, 1.5–4 × 0.3–0.5 cm. *Leaf*: petiole sulcate, (2–)4–5 cm long; blade narrowly elliptic, or ligulate-elliptic or oblong-elliptic, obtuse and mucronate, thin-textured, 6–11(–17) × 1–1.8 cm, main nerves 5–7. *Inflorescences* gently curving, produced with the nearly expanded leaf, densely many-flowered, flowers often over 50 in number, borne 2–2.5 mm apart; peduncle (9–)15–16 cm long, pale green; non-floriferous bracts solitary, or 2, ovate, obtuse to acute, 2.5–4 mm long; rachis quadrangular, 10–13 cm long, yellowish; floral bracts ovate-oblong, obtuse, 2–3 × 2 mm; brown. *Flowers* sweetly scented, *c*. 9 mm across, creamy white, labellum sometimes very pale green to lemon, keels yellowish cinnamon or ochre, column pink. *Pedicel with ovary* clavate, 1 mm long. *Sepals* and *petals* 3–nerved, petals often obscurely so. *Dorsal sepal* oblong, subacute or acute, 4.5 × 1.5 mm. *Lateral sepals* obliquely oblong to oblong-elliptic, acute, 4.5 × 1.5–1.6 mm. Petals oblong, acute, minutely erose-papillose, 4 × 1.1–1.2 mm. *Labellum* stipitate to column-foot by a tiny claw, 3–lobed, 3 mm long when flattened; side lobes spreading, oblong, irregularly laciniate, 0.8–1 mm long; mid-lobe strongly decurved, ovate-elliptic, rounded or subacute, minutely toothed at base, surface minutely papillose, 2 mm long, 1 mm wide at base, 1.5 mm wide across middle; disc with 2 flange-like basal keels, each *c*. 0.4 mm wide, which curve toward the middle and meet, tapering off and terminating at base of mid-lobe, and with 2 low, raised keel-like, minutely papillose ridges extending from base of mid-lobe and terminating near its apex, with a furrow between. *Column* slightly incurved, 2 mm long; foot prominent; apical hood ovate, rounded, entire; stelidia borne opposite stigmatic cavity, oblong, subacute, reaching rostellum; rostellum not prominent; stigmatic cavity narrowly ovate-oblong; anther-cap ovate, cucullate, 0.2 × 0.2 mm.

DISTRIBUTION. Borneo.

SABAH. Tenom District, Mount Anginon near Sapong, April 1965, *H.F. Comber* 4032 (K!). Crocker Range, Keningau & Tambunan District border, Ulu Apin Apin, 5 Nov. 1991, *Lamb & Surat in Lamb* AL 1390/91 (holotype K!). Along Keningau to Kimanis road, watershed of Crocker Range, Dec. 1986, *Vermeulen & Duistermaat* 663 (K!, L!). Tenom District, above Kallang Waterfall, *c*. 10 km north of Tenom, 11 Oct. 1986, *de Vogel* 8172 (L!) & *de Vogel* 8173 (L!).

HABITAT. In thick moss on trunks and branches of trees in hill and lower montane forest on sandstone and shale ridges. 900–1600 m.

D. lacinilobum is similar in habit to *D. gramineum* (Ridl.) Holttum from Peninsular Malaysia and *D. kamborangense* Ames from Borneo. It differs from both in having creamy white sepals and petals, distinct irregularly laciniate labellum side lobes, two flange-like basal keels each about 0.4 mm wide and stelidia borne opposite the stigmatic cavity. It is further distinguished from *D. gramineum* by its broader petals and slightly broader labellum and from *D. kamborangense* by its smaller flowers with narrower petals and labellum.

Fig. 106. *Dendrochilum lacinilobum*. **A** habit; **B** flower, front view; **C** dorsal sepal; **D** lateral sepal; **E** petal; **F** labellum, front view; **G** pedicel with ovary, labellum and column, side view; **H** column, front view; **J** anther-cap, back view; **K** pollinia. **A–K** from *Lamb & Surat* in *Lamb* AL 1390/91 (holotype). Scale: single bar = 1 mm; double bar = 1 cm. Drawn by Eleanor Catherine.

The specific epithet is derived from the Latin *laciniatus*, slashed into narrow divisions with tapering pointed incisions and *lobus*, lobe, referring to the distinctive side lobes of the labellum.

67. DENDROCHILUM MUCRONATUM

Dendrochilum mucronatum J.J. Sm. in *Brittonia* 1: 106–107 (1931). Type: Malaysia, Sarawak, Lundu District, Mount Pueh (Poi), Sept. 1929, *J. & M.S. Clemens* 22587 (holotype L!, isotypes AMES!, BO!, K!, SING). **Figs. 107 & 111 P–W.**

DESCRIPTION. Epiphyte. *Rhizome* branching, up to 10 cm long. *Roots* with a few branches, flexuose, smooth to very minutely papillose, elongate, 0.2–0.5 mm in diameter. *Cataphylls* 3–4, ovate-elliptic, acute, soon disintegrating to fibres, 0.4–5 cm long. *Pseudobulbs* borne at an acute angle to rhizome, 3–7 mm apart, oblong–ovoid, wrinkled when dry, approximate, 1–1.6 × 0.6–0.8 cm. *Leaf:* petiole narrow, canaliculate, 2.3–3 cm long; blade linear, ligulate, acute and mucronate, rigid, thinly coriaceous, 7–18 × 0.25–4 cm, main nerves 3–5. *Inflorescences* erect to gently curving, densely many-flowered, flowers borne 2–2.5(–3.5) mm apart; peduncle 5.5–8 cm long; non-floriferous bracts absent or solitary, acuminate, 4–6 mm long; rachis quadrangular, 9–16.5 cm long; floral bracts triangular-ovate, acute, often setaceous-mucronate, 2.5–3.4 × 1.5 mm. *Flowers* cream, flushed green. *Pedicel with ovary* narrowly clavate, 1.5–2.4 mm long. *Sepals* and *petals* spreading, 3–nerved. *Dorsal sepal* narrowly elliptic, acuminate, concave below, apex recurved, 5.4–5.5 × 1.2–1.3 mm. *Lateral sepals* obliquely obovate-elliptic, acuminate or acute, 5.3–5.4 × 1.5 mm. *Petals* narrowly linear-elliptic, acute, narrowed at base, minutely erose, 5 × 1 mm. *Labellum* stipitate to column-foot by a short claw, 3–lobed, 3–nerved, recurved, 4 mm long; side lobes narrowly triangular, tooth-like, slightly minutely serrate or erose, 1.5 × 1.4 mm; mid-lobe obovate, apex triangular, acute, margin minutely irregular to almost erose, 2.5 mm long, 1 mm wide at base, 1.4 mm wide above; disc with 2 prominent parallel fleshy keels united at base and forming an M with the shortly elevate median nerve, papillose distally and terminating just below middle of mid-lobe, rarely with an extrorse basal tooth either side. *Column* gently incurved, 2.5 mm long; foot short; apical hood subquadrate, truncate, irregularly toothed; stelidia median, borne just below stigmatic cavity, obliquely narrowly elliptic, subacute, slightly longer than apical hood, rostellum ovate, subacute; stigmatic cavity narrowly oblong; anther-cap ovate-triangular, cucullate, apex incurved. *Capsule* ellipsoid-globose, column persistent, 5–7 × 4–5 mm.

DISTRIBUTION. Borneo

SARAWAK. Lundu District, Mount Rumput, Aug. 1912, *J.W. Anderson* 196 (SING!). Lundu District, Mount Pueh (Poi) 19 Sept. 1929, *J. & M.S. Clemens* 20307 (AMES, mixed collection with *D. gibbsiae*!, K, mixed collection with *D. gibbsiae*!, L!, SAR!); Sept. 1929, *J. & M.S. Clemens* 22587 (holotype L!, isotypes AMES!, BO!, K!, SING!). Lundu District, Mount Pueh, *Mjöberg* s.n. (AMES!). Lundu District, Mount Pueh, 24 Sept. 1955, *Purseglove & Shah* P.4728 (K!, SAR!, SING!).

KALIMANTAN TIMUR. Apo Kayan, Mount Sungai Pendan, 13 Oct. 1991, *de Vogel & Cribb* 9157 (K!).

HABITAT. Upper montane mossy forest. 1200–1800 m.

237

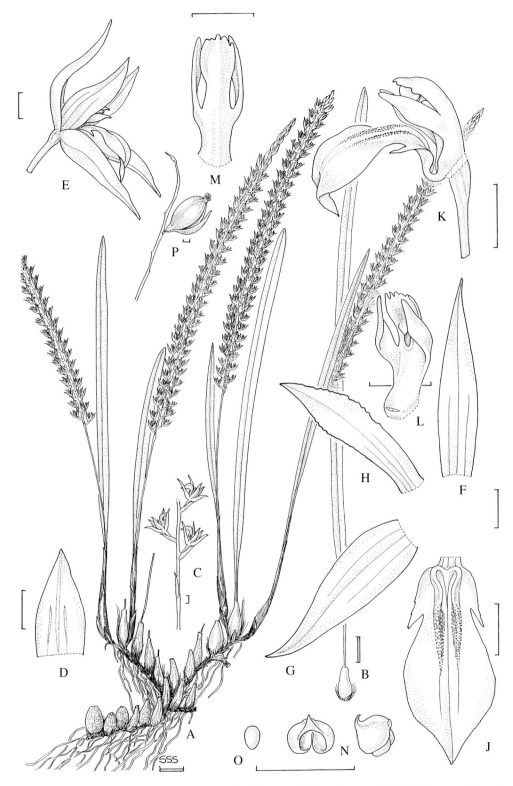

Fig. 107. *Dendrochilum mucronatum*. **A** habit; **B** pseudobulb and mature leaf; **C** lower portion of inflorescence; **D** floral bract; **E** flower, side view; **F** dorsal sepal; **G** lateral sepal; **H** petal; **J** labellum, front view; **K** pedicel with ovary, labellum and column, side view; **L** column, anther-cap removed, oblique view; **M** column, back view; **N** anther-cap, front and side views; **O** pollinium; **P** fruit capsule. **A & C** from *Purseglove & Shah* P. 4728, **B, D–O** from *J. & M.S. Clemens* 22587 (holotype), **P** from *J. & M.S. Clemens* 20307. Scale: single bar = 1 mm; double bar = 1 cm. Drawn by Susanna Stuart-Smith.

De Vogel & Cribb 9157 (see Fig 110 P–W, pl. 17C) differs in having more distinct, serrate labellum side lobes and sometimes minute extrorse teeth developed at the base of the keels in the manner of *D. crassifolium*.

The Latin specific epithet *mucronatum* means to possess a short, straight point and refers to the leaves, sepals, petals and labellum of this species.

68. DENDROCHILUM TENUITEPALUM

Dendrochilum tenuitepalum J.J. Wood in Wood & Cribb, *Checklist orch. Borneo*: 195–196, fig. 24 N & O (1994). Type: Malaysia, Sabah, Sipitang District, ridge between Maga River headwaters and Malabid River headwaters, 1600 m, Dec. 1986, *Vermeulen & Duistermaat* 1008 (holotype L!, isotype K!). **Fig. 108.**

DESCRIPTION. Epiphyte. *Rhizome* branching, up to *c.* 4 cm long. *Roots* with a few branches, flexuose, wiry, smooth to very minutely papillose, elongate, 0.2–0.8 mm in diameter. *Cataphylls* 4, ovate-elliptic, acute to acuminate, 0.5–7 cm long. *Pseudobulbs* borne at an acute angle to rhizome, 5 mm apart, ovate-elliptic or oblong-elliptic, straight to gently curved, wrinkled when dry, 1–1.8 × 0.4 –0.6 cm. *Leaf*: petiole narrow, canaliculate, 2.5–5 cm; blade linear, ligulate, obtuse and mucronate or apiculate, cuneate, conduplicate below, thinly coriaceous, 10.5–21.5 × 0.4 –0.6 cm. *Inflorescences* gently curving, subdensely many-flowered, flowers borne 3–4 mm apart; peduncle 6–9 cm long; non-floriferous bracts absent; rachis quadrangular, 14–18 cm long; floral bracts ovate-triangular, acute to acuminate, 2–5 mm long. *Flowers* unscented, very pale greenish. *Pedicel with ovary* narrowly clavate, 2.5–2.6 mm long. *Sepals* and *petals* spreading, 3–nerved. *Dorsal sepal* narrowly triangular-lanceolate, acuminate, 6–6.5 × 0.8–0.9 mm. *Lateral sepals* narrowly triangular-lanceolate, somewhat falcate, acuminate, 7–7.1 × 1–1.1 mm. *Petals* linear, ligulate, acuminate, 7–7.1 × 1 mm. *Labellum* stipitate to column-foot by a short foot, entire, 3–nerved, narrowly ovate-elliptic, acute, minutely papillose, margin erect proximally, irregular in places, 2.8 × 1 mm; disc with 2 low papillose, parallel keels, not united at base, terminating at or just beyond middle of labellum. *Column* gently incurved, 2 mm long; foot small; apical hood ovate-oblong, obtuse, minutely irregularly toothed; stelidia borne opposite stigmatic cavity, oblong, ligulate, obtuse to subacute, often minutely serrulate distally, slightly longer than apical hood, *c.* 1.8 mm long; rostellum prominent, acute; stigmatic cavity large, oblong, with a bilobed fleshy flange on lower margin; anther-cap ovate, cucullate, acute.

DISTRIBUTION. Borneo.

SABAH. Sipitang District, ridge between Maga River headwaters & Malabid River headwaters, Dec. 1986, *Vermeulen & Duistermaat* 1008 (holotype L!, isotype K!).

HABITAT. Open, mossy low forest. 1600 m.

D. tenuitepalum, which is known only from the type, is distinguished from the closely related *D. mucronatum* J.J. Sm. from Sarawak by its longer and slightly narrower sepals and petals, non-erose petals, entire, narrowly ovate-elliptic, minutely papillose labellum and bilobed flange on the lower margin of the stigmatic cavity.

The specific epithet is derived from the Latin *tenui*, slender, thin, and *tepalum*, a division of the perianth, either sepal or petal, referring to the long, narrow sepals and petals.

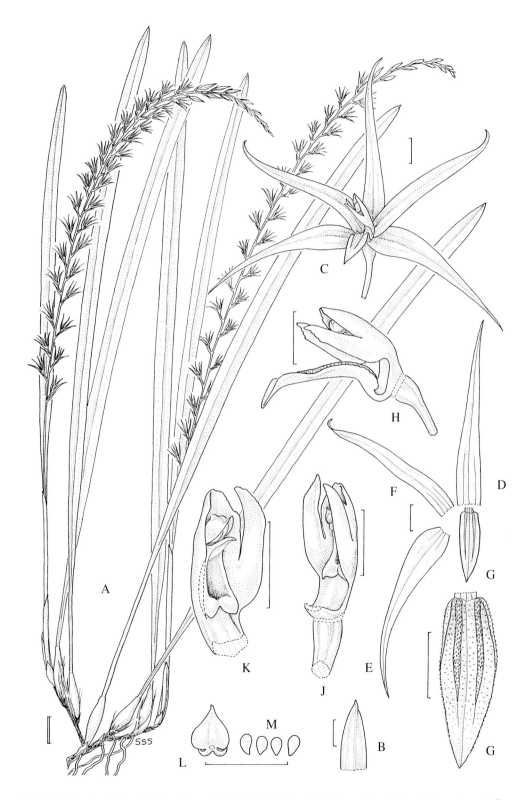

Fig. 108. *Dendrochilum tenuitepalum*. **A** habit; **B** floral bract; **C** flower, front view; **D** dorsal sepal; **E** lateral sepal; **F** petal; **G** labellum, front view; **H** pedicel with ovary, labellum and column, side view; **J** ovary and column, oblique view; **K** column, stelidium removed, oblique view; **L** anther-cap, back view; **M** Pollinia. **A–M** from *Vermeulen & Duistermaat* 1008 (holotype). Scale: single bar = 1 mm; double bar = 1 cm. Drawn by Susanna Stuart-Smith.

69. DENDROCHILUM TENOMPOKENSE

Dendrochilum tenompokense Carr in *Gdns. Bull. Straits Settl.* 8: 225 (1935). Type: Malaysia, Sabah, Mount Kinabalu, Tenompok, 1600 m, July 1933, *Carr 3623*, SFN 27891 (holotype SING!, isotypes AMES, K!).

DESCRIPTION. Epiphyte. *Rhizome* branching, up to 10 cm long. *Roots* branching, flexuose, wiry, smooth or very minutely papillose, 0.2–1.5 mm in diameter. *Cataphylls* 4–5, ovate-elliptic, acute to acuminate, 0.5–3 cm long, often reddish brown. *Pseudobulbs* often forming huge clumps, crowded or up to 1.3 cm apart, ovoid or oblong-elliptic, smooth, becoming finely wrinkled, approximate, 0.5–1.8 × 0.4–1.3 cm. *Leaf*: petiole slender, sulcate, 0.2–1.6(–2) cm long; blade narrowly linear-lanceolate to ligulate, or oblong-elliptic, obtuse and mucronate, somewhat conduplicate, thinly coriaceous to rather rigid, 1.5–7(–10.5) × 0.3–1 cm, main nerves 5. *Inflorescences* erect to gently curving, laxly to subdensely *c.* 12– to many-flowered, upper flowers sometimes opening first, borne 2–5 mm apart; peduncle 1.7–5.5 cm long; non-floriferous bracts absent or solitary, acuminate, 4–5 mm long; rachis quadrangular, 5–10(–25) cm long; floral bracts narrowly ovate to ovate-elliptic, 2–4(–5) × 1.5–1.7 mm. *Flowers* pale lemon-yellow, or pale greenish cream with pale green labellum and column, or translucent creamy white, flushed green, or white, or pale green with brighter green labellum, or yellowish green. *Pedicel with ovary* very slender, clavate, 2–3.5 mm long. *Sepals* and *petals* spreading, 3–nerved, sometimes with a few brown ramentaceous trichomes at base. *Dorsal sepal* oblong-elliptic, acute, concave, slightly carinate, 5–7 × 1.5–2 mm. *Lateral sepals* oblong-elliptic or ovate-elliptic, slightly oblique, acute to acuminate, slightly carinate, 5–7 × 1.5–2.1 mm. *Petals* narrowly oblong to oblong-elliptic, acute, sometimes slightly concave, margin sometimes irregular to erose, 4.3–6.9 × 1.6–2.2 mm. *Labellum* stipitate to column-foot by a short claw, shallowly 3–lobed, 3–nerved, 3.1–5.1 mm long, 1.5–2 mm wide across side lobes, 1.1–2.9 mm wide across mid-lobe, decurved; side lobes triangular or shallowly rounded, obtuse to acute, posterior margin minutely erose to irregularly toothed, upper surface usually papillose to almost papillose-hairy except along margins, 1.2–1.5 mm long; mid-lobe broadly elliptic or broadly ovate from a cuneate base, obtuse to acute, minutely papillose, 2.4–3 mm long; disc with 2 prominent papillose-hairy or minutely papillose, fleshy parallel keels which are incurved and joined at base by a smooth thickening, terminating on lower half of mid-lobe, median nerve sometimes elevate. *Column* strongly incurved, often very slender below, 2–3.5 mm long; foot 0.6–0.9 mm long; apical hood oblong, truncate, irregularly toothed or tridentate; stelidia borne either side of stigmatic cavity, narrowly elliptic or obliquely oblong, falcate, obtuse to acute, equal to slightly longer than apical hood, occasionally slightly shorter, 1–2 × 0.5–0.6 mm; rostellum narrowly ovate, obtuse to acute; stigmatic cavity narrowly oblong to almost rectangular, lower margin usually swollen and elevate, bilobed, or developed into a minutely papillose bilobed flange; anther-cap ovate, cucullate, acute, 0.6–0.9 × 0.6–0.9 mm.

KEY TO THE VARIETIES OF *D. TENOMPOKENSE*

Leaves narrowly linear-lanceolate to ligulate, 0.3–0.8 cm wide. Lower margin of stigmatic cavity usually swollen and elevate, bilobed. Labellum minutely papillose proximally, median nerve between keels elevate to varying degrees. Sepals and petals glabrous ....
.... var. *tenompokense*

241

Leaves oblong-elliptic, 0.6–1 cm wide. Lower margin of stigmatic cavity developed into a prominent bilobed flange. Labellum often densely papillose to almost papillose-hairy proximally, median nerve between keels not elevate. Sepals and petals with a few brown ramentaceous trichomes or papillae at the base ... var. *papillilabium* (J.J. Wood) J.J. Wood

a. var. **tenompokense**. Figs. 109 & 110 A–O, plate 17A & B.

DISTRIBUTION. Borneo.
SARAWAK. Marudi District, Gunung Mulu National Park, west ridge above Camp 3, 1 Oct. 1976, *Jermy* 13200 (K!). Marudi District, Gunung Mulu National Park, path from sub-camp 3 to sub-camp 4, near sub-camp 4, 19 Oct. 1977, *Lewis* 346 (K!, SAR!).
SABAH. Penampang District, Tunggul Togudon K 48 Jalan Tambunan/Penampang, 26 Aug. 1989, *Asik* SAN 127822 (K!, SAN). Mount Kinabalu, along road to Power Station, *Barkman* 262 (Sabah Parks Herbarium, Kinabalu Park). Mount Kinabalu, Mempening Trail, Nov. 1997, *Barrett* (colour slide!). Crocker Range, Penampang District, Km. 51.8 on Kota Kinabalu to Tambunan road, 7 Sept. 1983, *Beaman* 6950 (K!). Mount Kinabalu: Tenompok, July 1933, *Carr* 3623, SFN 27891 (holotype SING!, isotypes AMES, K!); Kemburongoh, 1933, *Carr* 3760, SFN 27140 (SING!); 2400 m, 15 Nov. 1931, *J. & M.S. Clemens* 27140 (BM!, SING!); Tenompok Orchid Garden, 9 Nov. 1933, *J. & M.S. Clemens* 50236 (AMES!, BM!, K!, L!) & 17 Nov. 1933, *J. & M.S. Clemens* 50360 (BM!, K!); Tenompok, 1931–32, *J. & M.S. Clemens* s.n. (AMES!). Crocker Range, Sinsuron road, Mount Alab, 14 Nov. 1989, *Cribb* 89/24 (K!) & 21 Sept. 1985, *Hani* SH 0045 (UKMS!). Mount Kinabalu, Kemburongoh, 6 Nov. 1959, *Meijer* SAN 20367 (AMES!, K!, SAN). Mount Alab, 21 Sept. 1985, *Nair* AN 064 (UKMS!). Mount Alab, south ridge leading towards summit, 31 Oct. 1986, *de Vogel* 8662 (L!), *de Vogel* 8663 (L!) & *de Vogel* 8664 (L!).
HABITAT. Lower montane mossy forest; oak-laurel forest; *Leptospermum/Dacrydium* forest about 8 metres high over withered sandstone and shale on steep east facing slopes; preferring to grow high in the canopy. 1500–2000 m.

The specific epithet refers to the type locality.

b. var. **papillilabium** (J.J. Wood) J.J. Wood **comb. & stat. nov. Fig. 111.**

Dendrochilum papillilabium J.J. Wood in Wood & Cribb, *Checklist orch. Borneo*: 191, fig. 24 L&M (1994). Type: Malaysia, Sabah, Crocker Range, Keningau to Kimanis road, 1500–1600 m, Dec. 1986, *Vermeulen & Duistermaat* 666 (holotype L!).
DISTRIBUTION. Borneo.
SABAH. Beluran District, Bukit Monkobo, 15 March 1982, *Aban* SAN 95230 (K!, L!, SAN), specimen in packet attached to *Aban* SAN 95241 (L, mixed collection with *D. simplex*!, SAN) & *Aban* SAN 95243 (SAN!). Crocker Range, Keningau to Kimanis road, Mile 16, 21 June 1978, *Dewol & Abas* SAN 89075 (SAN!). Crocker Range, Keningau to Kimanis road, Dec. 1986, *Vermeulen & Duistermaat* 666 (holotype L!). Crocker Range, Kimanis road, 10 Oct. 1985, *Wood* 573 (K!) & *Wood* 583 (K!).
HABITAT. Lower montane forest; mossy forest on sandstone ridges, with rattans, *Loxocarpus,* etc.; roadside banks on shale, with *Gahnia* spp., *Lycopodium* spp., *Melastoma* spp., *Nepenthes fusca* Danser, ferns, etc. 1200–1600 m.

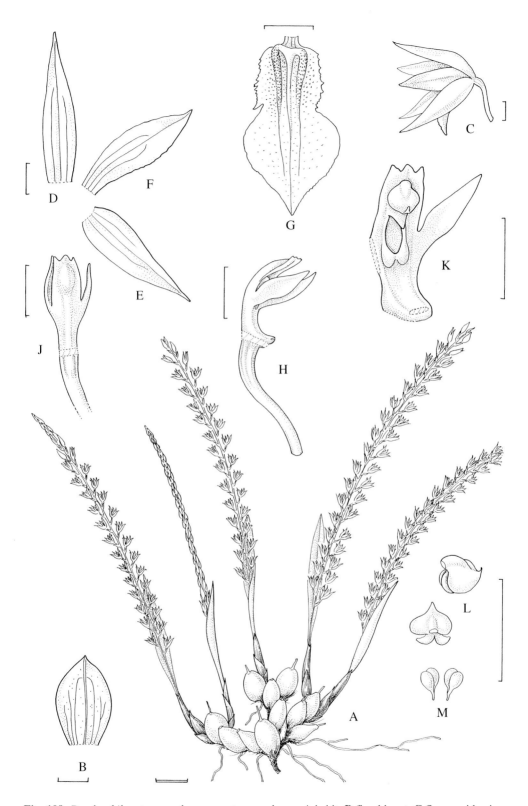

Fig. 109. *Dendrochilum tenompokense* var. *tenompokense*. **A** habit; **B** floral bract; **C** flower, side view; **D** dorsal sepal; **E** lateral sepal; **F** petal; **G** labellum, front view; **H** pedicel with ovary and column, side view; **J** ovary and column, back view; **K** column, stelidium removed, oblique view; **L** anther-cap, back and side views; **M** pollinia. **A–M** from *Carr* C. 3623, SFN 27891 (holotype). Scale: single bar = 1 mm; double bar = 1 cm. Drawn by Susanna Stuart-Smith.

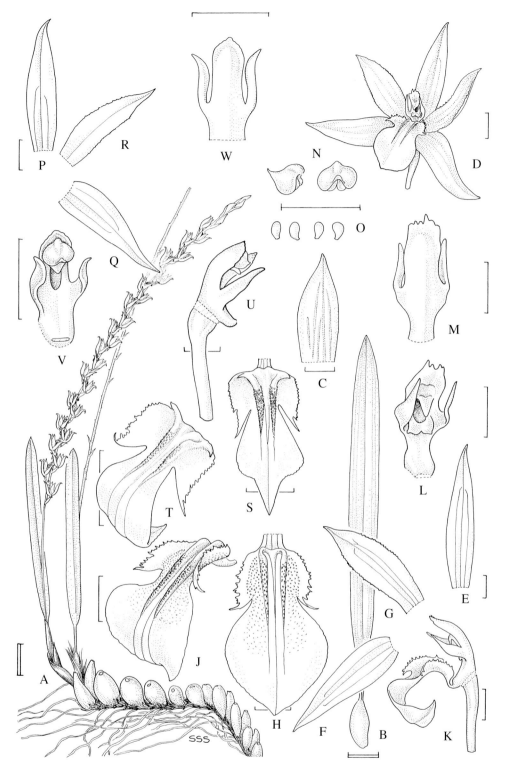

Fig. 110. *Dendrochilum tenompokense* var. *tenompokense*. **A** habit; **B** pseudobulb and mature leaf; **C** floral bract; **D** flower, oblique view; **E** dorsal sepal; **F** lateral sepal; **G** petal; **H** labellum, front view; **J** labellum, oblique view; **K** pedicel with ovary, labellum and column, side view; **L** column, anther-cap removed, oblique view; **M** column, back view; **N** anther-cap, side and front views; **O** pollinia. *D. mucronatum* (variant). **P** dorsal sepal; **Q** lateral sepal; **R** petal; **S** labellum, front view; **T** labellum, oblique view; **U** pedicel with ovary and column, side view; **V** column, oblique view; **W** column, back view. **A** from *de Vogel* 8662, **B** from *Lewis* 346, **C** from *de Vogel* 8663, **D–O** from *Jermy* 13200, **P–W** from *de Vogel & Cribb* 9157. Scale: single bar = 1 mm; double bar = 1 cm. Drawn by Susanna Stuart-Smith.

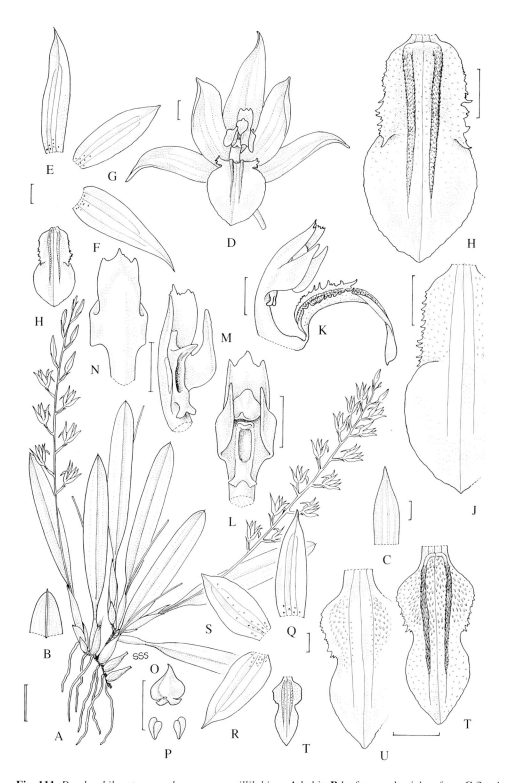

Fig. 111. *Dendrochilum tenompokense* var. *papillilabium*. **A** habit; **B** leaf apex, abaxial surface; **C** floral bract; **D** flower, front view; **E** dorsal sepal; **F** lateral sepal; **G** petal; **H** labellum, front view; **J** labellum, back view; **K** labellum and column, side view; **L** column, front view; **M** column, anther-cap removed, oblique view; **N** column, back view; **O** anther-cap, back view; **P** pollinia; **Q** dorsal sepal; **R** lateral sepal; **S** petal; **T** labellum, front view; **U** labellum, back view. **A–P** from *Vermeulen & Duistermaat* 666 (holotype of *D. papillilabium*), **Q–U** from *Aban* SAN 95230. Scale: single bar = 1 mm; double bar = 1 cm. Drawn by Susanna Stuart-Smith.

The specific epithet is derived from the Latin *papillatus*, having papillae and *labium*, a lip, referring to the papillose labellum.

D. tenompokense is allied to the Peninsular Malaysian *D. linearifolium* Hook. f. and records of *D. linearifolium* from Borneo probably refer to *D. tenompokense*, or possibly *D. galbanum*. *D. linearifolium* is similar in habit, but has fewer flowers per inflorescence, brown keels on the labellum and basal stelidia. *D. tenompokense* is a variable species, particularly with regard to stature, flower size, degree of labellum side lobe development, labellum papillosity and details of column structure. The type collection is a small slender plant, while, e.g., *Meijer* SAN 20367 is a more robust example with inflorescences up to 25 cm long. A distinguishing character of all is the swollen bilobed lower margin of the stigmatic cavity similar to that found in *D. tenuitepalum*.

The extent of variation is difficult to assess given the limited number of collections available for study. After critical study, however, the demarcation between *D. papillilabium* and *D. tenompokense* became increasingly ill-defined. Although intermediates may exist, I have decided, for the time being, to recognise *D. papillilabium* as a variety in order to define populations having broader leaves, somewhat larger flowers with a usually densely papillose labellum and a prominent flange-like ventral extension of the stigmatic cavity wall.

Barkman (pers. comm.) comments that *D. tenompokense* var. *tenompokense* is very common around Park Headquarters on Mount Kinabalu.

70. DENDROCHILUM LUMAKUENSE

Dendrochilum lumakuense J.J. Wood in Wood & Cribb, *Checklist orch. Borneo*: 184–185, fig. 24 G&H (1994). Type: Malaysia, Sabah, Sipitang District, Mount Lumaku, 1800 m, Dec. 1963, *J.B. Comber* 108 (holotype K!). **Fig. 112.**

DESCRIPTION. Epiphyte. *Rhizome* to *c.* 5 cm long. *Roots* branching, flexuose, smooth, 0.2–1 mm in diameter. *Cataphylls* 3, ovate-elliptic, acuminate, 0.5–2.5 cm long, greyish fawn, finely speckled black. *Pseudobulbs* crowded, up to 3 mm apart, ovoid or oblong, wrinkled in dried material, approximate, 0.8–1.5 × 0.6–1 cm. *Leaf*: petiole sulcate, rigid, 0.5–1 cm long; blade oblong-elliptic, ligulate, conduplicate at base, subacute, minutely apiculate, thinly coriaceous, 3–5 × 0.7–0.8 cm, main nerves 5–7. *Inflorescences* erect to gently curving, laxly or subdensely many-flowered, flowers borne 2–2.5 mm apart; peduncle 3.5–4 cm long; non-floriferous bracts absent; rachis quadrangular, 8–18 cm long; floral bracts oblong-ovate, acute, apiculate, 3 mm long. *Flowers* unscented, green. *Pedicel with ovary* clavate, 3 mm long. *Sepals* and *petals* 3–nerved. *Dorsal sepal* linear-oblong, acute, 7–8 × 1–1.2 mm. *Lateral sepals* linear-lanceolate, subfalcate, acute, 6.5–7 × 1–1.2 mm. *Petals* narrowly oblong-elliptic, spathulate, acute or subacute, 6–6.8 × 1.8–2 mm. *Labellum* entire, 3–nerved, oblong-subpandurate, obtuse, curved in lower portion, 3–3.1 × 1.1–1.2 mm; disc with 2 parallel papillose keels, not united at base, terminating about $^3/4$ along labellum, papillose between. *Column* gently incurved, 2 mm long; foot short; apical hood irregularly toothed; stelidia borne either side of stigmatic cavity, narrowly elliptic, obliquely oblong, falcate, acute, equalling apical hood; rostellum short, truncate; stigmatic cavity broadly oblong, subquadrate, lower margin elevate; anther-cap ovate, cucullate, with an apical wart.

DISTRIBUTION. Borneo.

SABAH. Mount Lumaku, 1800 m, Dec. 1963, *J.B. Comber* 108 (holotype K!).

HABITAT. Lower montane forest, growing in the canopy of a 15 metre high tree in 80% sun. 1800 m.

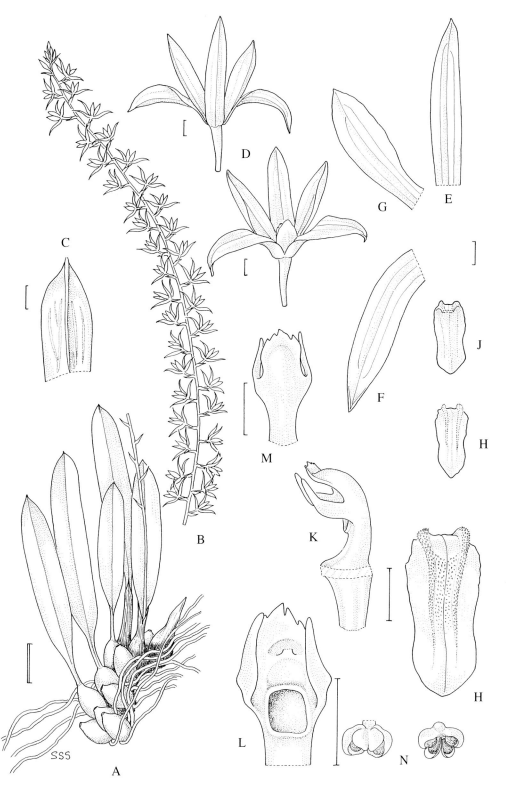

Fig. 112. *Dendrochilum lumakuense*. **A** habit; **B** inflorescence; **C** floral bract; **D** flower, front and back views; **E** dorsal sepal; **F** lateral sepal; **G** petal; **H** labellum, front view; **J** labellum, back view; **K** ovary and column, wide view; **L** column apex, anther-cap removed, front view; **M** column, back view; **N** anther-cap, front and interior views. **A–N** from *J.B. Comber* 108 (holotype). Scale: single bar = 1 mm; double bar = 1 cm. Drawn by Susanna Stuart-Smith.

D. lumakuense is clearly related to *D. tenompokense*, but easily distinguished by its entire labellum, recalling those of subgenus *Dendrochilum*, and non-bilobed lower margin of the stigmatic cavity. It is known only from the type.

The specific epithet refers to the type locality, Mount Lumaku (1966 m) in Sipitang District, Southwest Sabah.

71. DENDROCHILUM GRACILE

Dendrochilum gracile (Hook. f.) J.J. Sm. in *Recl. Trav. bot. néerl.* 1: 69–70 (1904). Type: Peninsular Malaysia, Perak, Larut, *King's collector* 3280 (holotype K!).

DESCRIPTION. Epiphyte. *Rhizome* to *c.* 6 cm long. *Roots* branching, flexuose, minutely papillose, 0.2–1 mm in diameter. *Cataphylls* 3–4, ovate-elliptic, obtuse to acute, 0.4–5 cm long, finely speckled brown. *Pseudobulbs* crowded, or up to 5 cm apart, cylindrical or narrowly conical, sometimes ovoid-pyriform, wrinkled in dried material, approximate, 1.3–5(–6) × 0.4–0.7(–1.6) cm, olive-green. *Leaf*: petiole sulcate to canaliculate, 0.3–4.5 cm long; blade narrowly elliptic to elliptic, acute or obtuse, thin-textured, papyraceous, (3.8–)9–15(–18) × (0.7–)1.1–3(–4.3) cm, main nerves 5–9. *Inflorescences* pendulous, laxly or subdensely few- to many-flowered, flowers borne 3–5.5 mm apart, uppermost sometimes opening first; peduncle filiform, (5–)8–12.5 cm long; non-floriferous bracts solitary or up to 5 or 6, ovate-elliptic, obtuse to subacute, 2–6 mm long; rachis quadrangular, (4–)8–12.5(–30) cm long; floral bracts ovate or oblong-ovate or triangular-ovate, obtuse or shortly acute, 3–5 mm long. *Flowers* scented, pale green, yellowish green to greenish brown, mid-lobe of labellum with two longitudinal brown, reddish brown or blackish brown streaks, sometimes converging to form a patch, column greenish white. *Pedicel with ovary* clavate, (1.7–)2–3.3 mm long. *Sepals* and *petals* 3–nerved, sometimes sparsely papillose at base. *Dorsal sepal* narrowly elliptic or ovate-elliptic, often strongly concave, acute, 5.8–8.7 × 1.6–2 mm. *Lateral sepals* narrowly obliquely ovate-elliptic, slightly concave, acute, 5.1–8.5 × 1.8–2.4 mm. *Petals* narrowly obliquely elliptic, often minutely erose, acute, 5–8.2 × 1.1–2 mm. *Labellum* stipitate to column-foot by a short narrow claw, 3–lobed, 3–nerved, recurved just above base, 3.5–7.7 mm long, 1.5–4.7 mm wide across side lobes, 2.3–3.5 mm wide across mid-lobe; side lobes narrowly oblong-triangular, often subtruncate at base, irregularly serrate to laciniate, free apical portion narrowly linear to narrowly triangular, sometimes subfalcate, acuminate, free apical portion 0.7–1 mm long; mid-lobe ovate, suborbicular to subrhombic, or subspathulate, cuneate to narrowly unguiculate at base, usually shortly apiculate or subacute, margin irregular to erose, 3.3–4.3 mm long; disc with 2 fleshy keels united at base with elevate median nerve to form an M-shape, terminating near base or at about middle of mid-lobe, keels often swollen at base or each bearing a thorn-like, acute basal projection which is occasionally shortly bifid, median nerve elevate proximally, often becoming elevate again on mid-lobe. *Column* incurved, 2.9–4 mm long; foot prominent, up to 0.6 mm long; apical hood ovate, ± entire, or irregularly bidentate, or tridentate to 5–toothed; stelidia borne centrally, arising just below stigmatic cavity, linear-subulate, falcate, acute, slightly twisted, slightly shorter than apical hood, *c.* 1.7–2 mm long; stigmatic cavity narrowly oblong or oblong-triangular, lower margin slightly elevate; rostellum narrowly triangular-ovate, acute; anther-cap cucullate, cordate, with a conical apical wart, 1 × 0.7 mm.

KEY TO THE VARIETIES OF *D. GRACILE*

Keels on labellum without thorn-like basal projections, although sometimes swollen ....
.... var. *gracile*
Keels on labellum each with a thorn-like, acute basal projection ...... var. *bicornutum* J.J. Wood

a. var. **gracile. Figs. 113–116, plate 5B.**

Platyclinis gracilis Hook. f., *Hooker's Icon.* pl. 21: pl. 2016 (1880).
Acoridium gracile (Hook. f.) Rolfe in *Orchid Rev.* 12: 220 (1904).
Dendrochilum heterotum Pfitzer in Engler, *Pflanzenreich* 4, 50, 2 B, 7: 103, fig. 35 N–O
    (1907). Type: Indonesia, Java, *Zollinger* 1619 (holotype K–LINDL!).
Platyclinis gracilis Hook. f. var. *angustifolia* Ridl., *Mat. fl. Malay. Penins.* 1: 27 (1907). Type:
    Peninsular Malaysia, Perak, Bujong Malacca, *Ridley* 9816 (holotype SING).
Dendrochilum tardum J.J. Sm. in *Bull. Dép. Agric. Indes Néerl.* 15: 6–7 (1908). Type:
    Indonesia, Kalimantan, cult. Bogor, ex. *Hallier* (holotype BO!), **syn. nov.**
Dendrochilum lyriforme J.J. Sm. in *Brittonia* 1: 107–108 (1931). Type: Malaysia, Sarawak,
    Mount Pueh (Poi), *J. & M.S. Clemens* 20313 (holotype L!, isotypes K!, SAR!, SING!).
D. gracile (Hook. f.) J.J. Sm. var. *angustifolium* (Ridl.) Holttum, *Rev. fl. Malaya* 1: 233
    (1953).
[*D. simile* sensu Yong, Orchid Portraits. *Wild orchids of Malaysia & Southeast Asia*: 2 figs.
    s.n. on p. 59, excl. text (1990).]

DISTRIBUTION. Peninsular Malaysia, Sumatra, Java, Borneo, Lesser Sunda Islands.
SARAWAK. Lundu District, Mount Pueh (Poi)/Mt Rumput, 20 Aug. 1912, *J.W. Anderson*
173 (K!, SING!). Kuching District, Mount Penrissen, West Side, 4 Dec. 1994, *Beaman* 11106
(K!). Lundu District, Mount Pueh (Poi), Sept. 1929, *J. & M.S. Clemens* 20313 (holotype of *D.
lyriforme* L!, isotypes K!, SAR!, SING!). Belaga District, Usun Apau Plateau, upper Kelayan
River, 10 Oct. 1994, *Lai et al.* S.69869 (AAU, K!, KEP, SAR). Marudi District, trail from
Bario to Pa Berang, March 1997, *Leiden* cult. (*Roelfsema et al.*) 970344 (K!, L!). Lundu
District, Mount Pueh, 1920's, *Mjöberg* s.n. (AMES!, K!). Kapit District, Batu Tiban Hill,
1920's, *Mjöberg* s.n. (AMES!). Kuching District, Mount Penrissen, Nov. 1909, *Museum
collectors* s.n. (K!, SAR). Lundu District, Mount Pueh, 24 Sept. 1955, *Purseglove & Shah*
P.4747 (K!, L!, SING). Kuching District, 1988, Leiden cult. (*de Vogel*) 27371 (L!).
SABAH. Tawau District, Tawau River Forest Reserve, along Tawau River, 11 Nov. 1968,
Kokawa & Hotta 890 (KYO, L!).
KALIMANTAN TIMUR. Kutai, Mount Beratus, near Balikpapan, Terrace Berikan bulu, 11
July 1952, *Meijer* 696 (BO, L!), 12 July 1952, *Meijer* 715 (BO, mixed collection with *D.
dolichobrachium*!) & 18 July 1952, *Meijer* 842 (BO, K!, L!).
KALIMANTAN. Province and locality unknown, cult. Bogor, ex *Hallier* (holotype of *D.
tardum* BO!).
HABITAT. Hill and lower montane mossy forest, on sandstone; kerangas forest. 300–1800
m.

Reported by Comber (1990) as fairly common all over Java, this species appears, from
extant collections, to be a much rarer plant in Borneo. Leaf shape, non-floriferous bract
number and floral details are variable throughout its range. Plants from Sarawak, named as *D.*

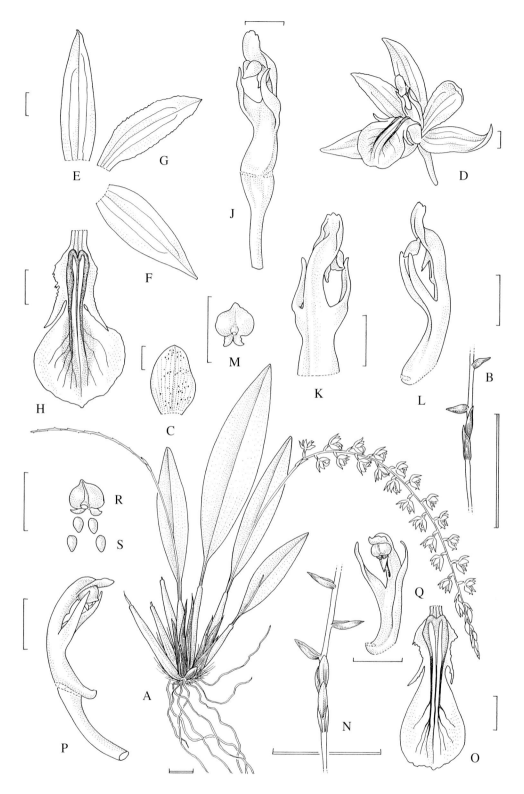

Fig. 113. *Dendrochilum gracile* var. *gracile*. **A** habit; **B** junction of rachis with peduncle showing two non-floriferous bracts and two floral bracts; **C** floral bract; **D** flower, oblique view; **E** dorsal sepal; **F** lateral sepal; **G** petal; **H** labellum, front view; **J** pedicel with ovary and column, oblique view; **K** column, back view; **L** column, side view; **M** anther-cap, back view; **N** junction of rachis with peduncle showing four non-floriferous bracts and three floral bracts; **O** labellum, front view; **P** pedicel with ovary and column, side view; **Q** column, oblique view; **R** anther-cap, back view; **S** pollinia. **A–M** from *Meijer* 696, **N–S** from *Meijer* 715. Scale: single bar = 1 mm; double bar = 1 cm. Drawn by Susanna Stuart-Smith.

Fig. 114. *Dendrochilum gracile* var. *gracile*. **A** habit; **B** junctions of rachis with peduncle showing one and two non-floriferous bracts, two floral bracts and two buds; **C** floral bract; **D** pedicel with ovary, labellum and column, side view, drawn from a bud; **E** anther-cap, back view; **F** floral bract; **G** flower, side view; **H** dorsal sepal, **J** lateral sepal; **K** petal; **L** labellum, front view; **M** pedicel with ovary, labellum and column, side view; **N** pedicel with ovary and column, oblique view; **O** column, anther-cap removed, oblique view; **P** column, back view; **Q** anther-cap, back view; **R** pollinia. **A–E** from *Kokawa & Hotta* 890, **F–R** from *Lewis* 158. Scale: single bar = 1 mm; double bar = 1 cm. Drawn by Susanna Stuart-Smith.

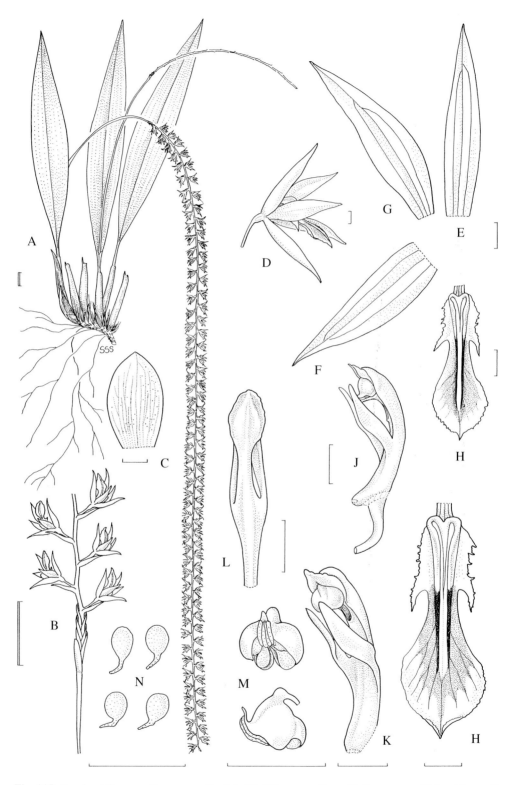

Fig. 115. *Dendrochilum gracile* var. *gracile*. **A** habit; **B** lower portion of inflorescence; **C** floral bract; **D** flower, side view; **E** dorsal sepal; **F** lateral sepal; **G** petal; **H** labellum, front view; **J** pedicel with ovary and column, side view; **K** column, oblique view; **L** column, back view; **M** anther-cap, interior view showing pollinia, and side view; **N** pollinia. **A–N** from *J.W. Anderson* 173 (form referable to *D. lyriforme*). Scale: single bar = 1 mm; double bar = 1 cm. Drawn by Susanna Stuart-Smith.

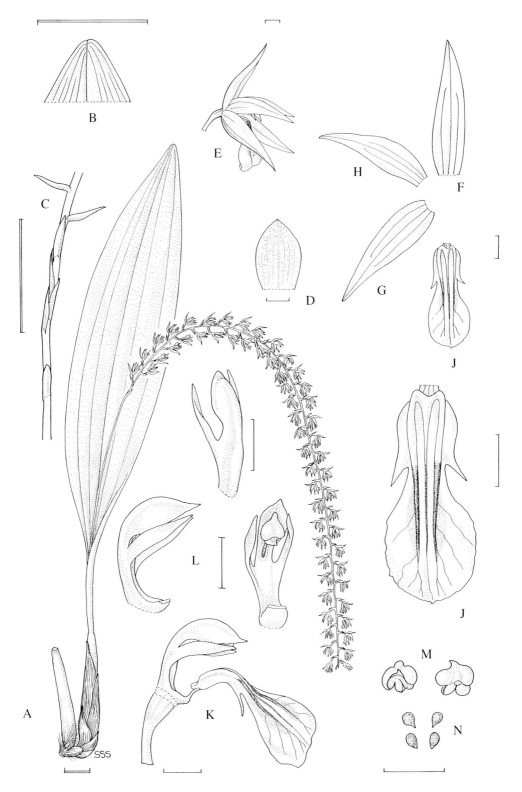

Fig. 116. *Dendrochilum gracile* var. *gracile*. **A** habit; **B** leaf apex; **C** junction of rachis with peduncle showing five non-floriferous bracts and two floral bracts; **D** floral bract; **E** flower, side view; **F** dorsal sepal; **G** lateral sepal; **H** petal; **J** labellum, front view; **K** pedicel with ovary, labellum and column, side view; **L** column, oblique, side and back views; **M** anther-cap, interior and back views; **N** pollinia. **A–N** from cult. Bogor, ex *Hallier* (holotype of *D. tardum*). Scale: single bar = 1 mm; double bar = 1 cm. Drawn by Susanna Stuart-Smith.

lyriforme by J.J. Smith, differ in having a laxer inflorescence bearing a group of up to six imbricate non-floriferous bracts and rather larger flowers. Others from Peninsular Malaysia, including the type from Perak, as well as *Kokawa & Hotta* 890 from Sabah, have a solitary non-floriferous bract. *Meijer* 696, 715 and 842 from Kalimantan Timur have from two to four non-floriferous bracts and rather short, obtuse leaves. *Kokawa & Hotta* 890 has longer, acute leaves. I cannot see any appreciable differences between *D. lyriforme* and the type of *D. gracile* apart from the number of non-floriferous bracts and flower size. These seem too variable to justify separation from *D. gracile*.

D. tardum, known only from the type, is said to differ from *D. gracile* by its broad, very obtuse leaves with one or two crenate apical teeth, the oval mid-lobe and acuminate, five-toothed apical hood. Examination of the type reveals no crenate apical teeth on the leaves. J.J. Smith may have been referring to the subapical constriction of the leaf margin present in several species of *Dendrochilum*. Similarly, I can find no other appreciable differences separating *D. tardum* from *D. gracile*.

The Latin specific epithet *gracile*, meaning thin or slender, refers to the general habit.

b. var. **bicornutum** *J.J. Wood* **var. nov.** a varietate typica projectura acuta acanthoïdea basi utrinsque in labello carinarum duarum distinguitur. Typus: Brunei, Temburong District, Amo subdistrict, depression on west flank of ridge Northeast of Mount Retak, 1300 m, 11 March 1991, *Sands et al.* 5327 (holotypus BRUN!, isotypus K!). **Fig. 117, plate 5C.**

Leaves coriaceous; petiole 3–4.5 cm long; blade 11–18 × 1.1–1.5 cm. *Inflorescence* flowering from top down, flowers borne 3–4 mm apart; peduncle 7.5–12.5 cm long; non-floriferous bracts solitary; rachis up to 31 cm long. *Flowers* yellowish green or pale green, labellum with two longitudinal reddish or blackish brown stripes, often converging on the mid-lobe, column greenish cream, anther-cap cream. *Sepals* and *petals* with a few minute basal papillae. *Dorsal sepal* 5.8–6 × 1.8–2 mm. *Lateral sepals* 5.1–6.8 × 1.8–2.1 mm. *Petals* 5–5.9 × 1.1–1.8 mm. *Labellum* 3.5–5 mm long, 2.4–2.9 mm wide across mid-lobe, 1.1–1.6 mm wide at base; side lobes narrowly triangular, acuminate, irregularly erose to serrate, free apical portion 0.8–1 mm long; mid-lobe suborbicular to subrhombic, cuneate at base, apiculate, margin irregular to erose; keels 2, united at base, low, fleshy, terminating on proximal half of mid-lobe, each with a thorn-like, acute basal projection which is occasionally shortly bifid, median nerve prominent. *Column* 2.9–3 mm long.

DISTRIBUTION. Borneo.

BRUNEI. Mount Pagon, Temburong River, April 1958, *Ashton* A. 250 (K!). Temburong District, Amo Subdistrict, depression on west flank of ridge Northeast of Mount Retak, 11 March 1991, *Sands et al.* 5327 (holotype BRUN!, isotype K!). Temburong District, Mount Pagon ridge from helicopter pad to the top, 8–11 Jan. 1992, *de Vogel* 1825 (L!) & *de Vogel* 1826 (L!).

SABAH. Mount Trus Madi, cult. Tenom Orchid Centre, photo taken 20 Dec. 1992 by *C.L. Chan* (!).

HABITAT. Upper montane swamp forest on sandstone; montane ridge forest 5–20 m high on sandstone. 1300–1500 m.

Differs from var. *gracile* in having distinct thorn-like, acute basal projections on the keels of the labellum.

The varietal epithet is derived from the Latin, *bi*, two, and *cornutus*, horned, and refers to the basal projections on the labellum keels.

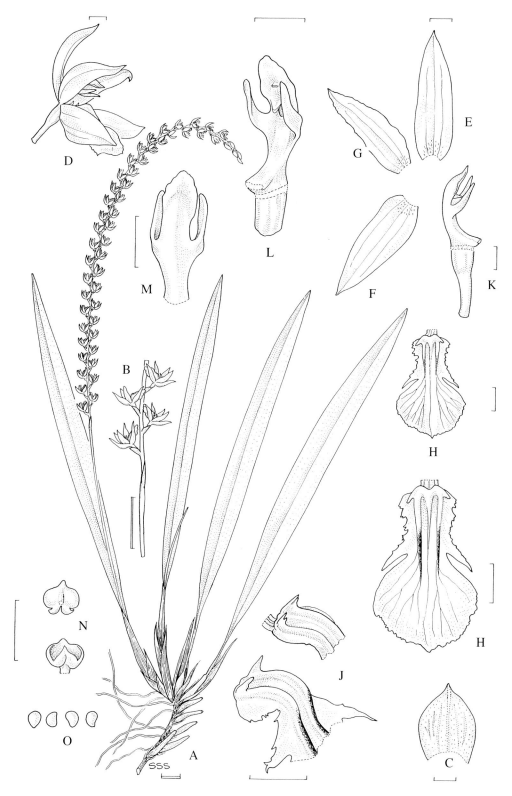

Fig. 117. *Dendrochilum gracile* var. *bicornutum*. **A** habit; **B** lower portion of inflorescence; **C** floral bract; **D** flower, side view; **E** dorsal sepal; **F** lateral sepal; **G** petal; **H** labellum, front view; **J** close-up of base of labellum showing basal projections on keels; **K** pedicel with ovary and column, side view; **L** column, oblique view; **M** column, back view; **N** anther-cap, back and interior views; **O** pollinia. **A–O** from *Sands et al.* 5327 (holotype). Scale: single bar = 1 mm; double bar = 1 cm. Drawn by Susanna Stuart-Smith.

72. DENDROCHILUM FLOS-SUSANNAE

Dendrochilum flos-susannae J.J. Wood in *Orchid Rev.* 105 (1215): 142, 144, fig. s.n. on p. 143 (1997). Type: Malaysia, Sarawak, Sarikei Division, Julau District, Lanjak Entimau Protected Forest, upper Ensirieng River, primary mixed dipterocarp forest, 200 m, January 1993, *Leiden* cult. (*Schuiteman, Mulder & Vogel*) 932780 (holotype L, isotype K). **Fig. 118, plate 4D.**

DESCRIPTION. Epiphyte. *Habit* described as similar to *D. gracile* (Hook. f.) J.J. Sm. *Inflorescences* pendulous, laxly 16– to 32–flowered, flowers borne 5–12 mm apart; peduncle up to 21.5 cm long; non-floriferous bracts solitary, ovate-elliptic, acute, 1–1.5 cm long; rachis quadrangular, 13–28 cm long; floral bracts ovate-elliptic, acute, apiculate, 8–13 × 5–6 mm. *Flowers* slightly fragrant, sepals and petals light brownish olive, labellum light orange at very base, side lobes dark chocolate-brown, mid-lobe light brownish olive with dark chocolate-brown keels separated by a thin orange line, column very pale salmon-pink, anther-cap cream, pollinia pale yellow. *Pedicel with ovary* clavate, minutely papillose, particularly on ovary, 4–6 mm long. *Sepals* 3–nerved. *Dorsal sepal* lanceolate, acuminate, concave proximally, reflexed distally, 14–17 × (3–)3.8–3.9 mm. *Lateral sepals* slightly obliquely lanceolate, acuminate, slightly concave proximally, somewhat carinate and reflexed distally, 13–17 × 3–4 mm. *Petals* lanceolate, acuminate, margin slightly irregular, sparsely papillose-hairy at base, 12–15 × 3–3.1 mm. *Labellum* stipitate to column-foot by a 1 mm long claw, rather concave, 3–nerved, 11–13.8 mm long, 3.8–4 mm wide across side lobes; side lobes erect to spreading, obliquely triangular-acuminate, exterior margins erose to serrulate, 5 mm long, free apical portion 2–2.1 mm long; mid-lobe rhomboid from a narrow, linear-oblong claw and cuneate base, acute, margin minutely erose, whole blade shallowly v-shaped in cross-section, 7.5–10 mm long, (2–)2.5–2.6 mm wide across claw, blade (5–)5.5–5.6 mm wide at middle; disc with 2 keels not united by a transverse thickening, extending from base of side lobes, terminating and ± meeting midway along mid-lobe, separated by a deep groove which is narrow proximally, but becoming much wider distally, creating a low raised keel on underside of labellum. *Column* gently curved distally, 6–7 mm long; foot *c.* 0.9 mm long; apical hood oblong, margin irregular, 1.4–1.5 mm wide; stelidia borne just below stigmatic cavity, narrowly falcate, acute, reaching to lower portion of apical hood, 2–2.9 mm long; stigmatic cavity narrowly oblong, lower margin elevate; rostellum triangular-ovate, acute; anther-cap ovate-cordate, cucullate, very minutely papillose, *c.* 1 × 1–1.1 mm.

DISTRIBUTION. Borneo.

SARAWAK. Sarikei Division, Julau District, Lanjak Entimau Protected Forest, upper Ensirieng River, Jan. 1993, *Leiden* cult. (*Schuiteman, Mulder & Vogel*) 932780 (holotype L!, isotype K!).

HABITAT. Primary mixed dipterocarp forest. 200 m.

D. flos-susannae, which is allied to *D. megalanthum* Schltr. from Sumatra, is the largest flowered species of *Dendrochilum* in Borneo. It is distinguished from *D. megalanthum* by its even larger flowers with a labellum having a longer-clawed, rhomboid mid-lobe. It also differs from the variable *D. gracile* (Hook. f.) J.J. Sm., and from *D. oxylobum* Schltr., by its much larger flowers. The labellum keels are not united by a transverse thickening with an elevate median nerve to form an M-shaped structure as in *D. gracile*.

A search of the living collections at the Hortus Botanicus in Leiden, The Netherlands, failed to locate extant living material from which a description of the pseudobulbs and leaves could be made. Schuiteman (pers. comm.) confirms their habit and appearance to be very similar to *D. gracile*.

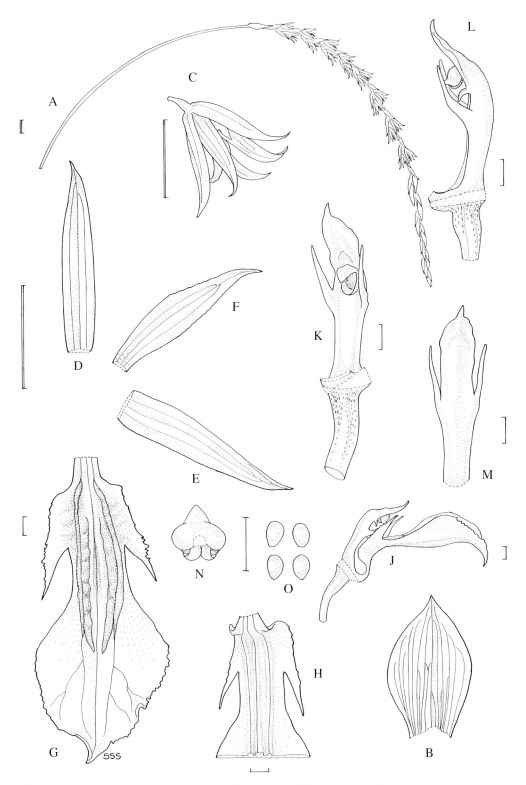

Fig. 118. *Dendrochilum flos-susannae*. **A** inflorescence; **B** floral bract; **C** flower, side view; **D** dorsal sepal; **E** lateral sepal; **F** petal: **G** labellum, front view; **H** lower portion of labellum, back view; **J** pedicel with ovary, labellum and column, side view; **K** ovary and column, anther-cap removed, oblique view; **L** ovary and column with anther-cap, side view; **M** column, back view; **N** anther-cap, back view; **O** pollinia. **A–O** from *Leiden* cult. (*Schuiteman, Mulder & Vogel*) 932780 (holotype). Scale: single bar = 1 mm; double bar = 1 cm. Drawn by Susanna Stuart-Smith.

The first part of the specific epithet is derived from the Latin *flos*, a flower. The second part is for free-lance artist Susanna Stuart-Smith who has patiently illustrated most of the species figured in this revision.

73. DENDROCHILUM IMBRICATUM

Dendrochilum imbricatum Ames, *Orchidaceae* 6: 54–55, pl. 82, fig. I, 1 (1920). Type: Malaysia, Sabah, Mount Kinabalu, Kiau, 900 m, 8 Nov. 1915, *J. Clemens* 179 (holotype AMES!, isotypes BM!, BO!, K!, NY!, SING!). **Fig. 119, plate 8A–C.**

DESCRIPTION. Epiphyte. *Rhizome* stout, up to 6 cm long. *Roots* branching, flexuose, smooth, elongate, 1–2 mm in diameter. *Cataphylls* 4, oblong-elliptic, obtuse to acute, 1.5–18 cm long, brownish with darker brown mottling. *Pseudobulbs* crowded, narrowly pyriform or cylindrical, smooth or finely striated, 4.5–8 × 0.5–1.5 cm. *Leaf*: petiole tough, slender, sulcate to canaliculate, 7–18 cm long, up to 4 mm wide; blade oblong-elliptic to elliptic, acute to acuminate, coriaceous, often constricted to varying degrees distally, 23–52 × 3.8–6 cm, main nerves 7–9, median nerve prominent, raised on abaxial surface. *Inflorescences* pendulous, laxly to subdensely many-flowered, flowers borne 4–5 mm apart; peduncle porrect to arching, 26–56 cm long; non-floriferous bracts 3–4, ovate-elliptic, obtuse, 0.8–1 cm long; rachis pendulous, quadrangular, 38–50 cm long, 1.5–1.8 mm wide; floral bracts broadly ovate-elliptic, rounded, imbricate when inflorescence is immature, equal to or exceeding the flowers, 0.8–1.9 × 0.8–0.9 cm, brown, sometimes flushed pink. *Flowers* strongly sweet-scented, sepals and petals pale green or yellowish green, side lobes of labellum pale green or yellowish, margin chocolate-brown, mid-lobe cream with a chocolate-brown basal patch, disc yellowish, keels chocolate-brown, column white or cream, orange at base, anther-cap cream. *Pedicel with ovary* 2 mm long. *Sepals* and *petals* 3–nerved. *Dorsal sepal* oblong-elliptic, acute, slightly carinate distally, 7–9 × 2.9–3 mm. *Lateral sepals* oblong-elliptic, acute, slightly carinate distally, (7–)8–9 × 3 mm. *Petals* oblong or elliptic, acute, (7–)8–8.5 × 2.8 mm. *Labellum* not or barely stipitate to column-foot, 3–lobed, 3–nerved, reflexed, 4–4.1 mm long, 2.4–2.5 mm wide across side lobes; side lobes triangular, acute, subsetaceous distally, free apical portion 0.9–1 mm long; mid-lobe rounded, suborbicular to spathulate, cuneate at base, often obscurely apiculate, 2.7–2.8 × 3–3.2 mm, with several branching secondary nerves; disc with 2 low keels extending from just above base and terminating on claw of mid-lobe, median nerve only slightly elevate. *Column* somewhat papillose below, 3–3.2 mm long; foot short; apical hood rounded, obscurely retuse or irregularly obtusely 3–lobed; stelidia borne just below stigmatic cavity, subulate, just equalling rostellum, 1 mm long; stigmatic cavity oblong; rostellum ovate, obtuse to subacute; anther-cap ovate, cucullate.

DISTRIBUTION. Borneo.

SARAWAK. Limbang District, Pa Mario River, Ulu Limbang, route to Batu Lawi, Bario, 8 Aug. 1985, *Awa & Lee* S. 50704 (K!, L!, SAR, SING). Marudi District, Mount Murud, Oct. 1922, *Mjöberg* 64 (AMES!).

SABAH. Mount Kinabalu: Liwagu River, *Barkman* 259 (Sabah Parks Herbarium, Kinabalu Park); Tenompok, 22 Aug. 1933, *Carr* 3653, SFN 27998 (SING!); Kiau, 8 Nov. 1915, *J. Clemens* 179 (holotype AMES!, isotypes BM!, BO!, K!, NY!, SING!) & *J. Clemens* 318 (AMES); Penibukan, 30 Dec. 1932, *J. & M.S. Clemens* 30578 (BM!); Tenompok Orchid Garden, Nov. 1933, *J. & M.S. Clemens* 50162 (BM!, K!, L!); Mamut River, 7 Oct. 1966, *Collenette* 1042 (K!); Liwagu River, 17 Oct. 1983, *Phillipps & Lamb* s.n. (drawing by *C.L.*

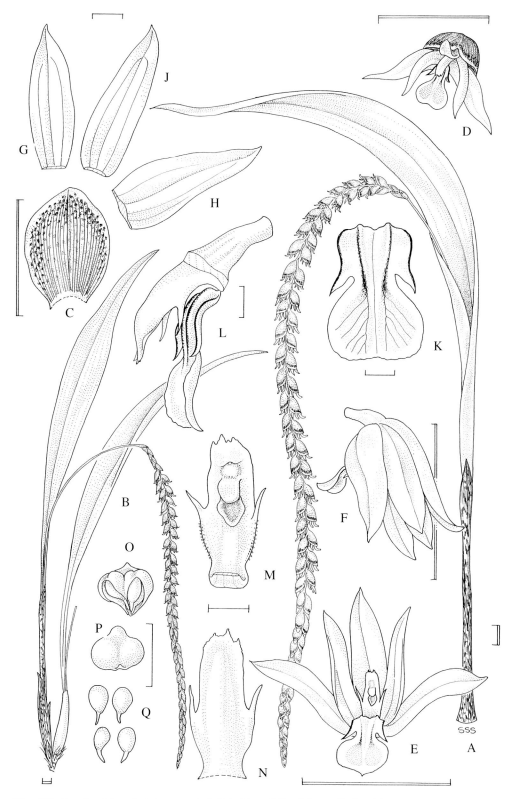

Fig. 119. *Dendrochilum imbricatum.* **A** & **B** habits; **C** floral bract; **D** floral bract and flower, oblique view; **E** flower, front view; **F** flower, side view; **G** dorsal sepal; **H** lateral sepal; **J** petal; **K** labellum, front view; **L** pedicel with ovary, labellum and column, side view; **M** column with anther-cap removed, front view; **N** column, back view; **O** anther-cap with pollinia, interior view; **P** anther-cap, back view; **Q** pollinia. **A** & **C–Q** from *Wood* 618, **B** from *Awa & Lee* S.50704. Scale: single bar = 1 mm; double = 1 cm. Drawn by Susanna Stuart-Smith.

Chan no. 63 K!); Mountain Garden, *SNP* 1587 (Sabah Parks Herbarium, Kinabalu Park). Sipitang District, Long Pa Sia to Long Samado trail, *c*. 4 km Southwest of Long Pa Sia, 22 Oct. 1986, *de Vogel* 8452 (L!). Mount Kinabalu, Park Headquarters, 20 Oct. 1985, *Wood* 618 (K!).

KALIMANTAN TIMUR. Apo Kayan, between Long Bawan and Panado, 16 July 1981, *Geesink* 9074 (BO, KYO, L!) & 23 July 1981, *Geesink* 9190 (BO, KYO, L!).

HABITAT. Hill forest; lower montane forest; kerangas forest; dry, rather low and open forest on steep ridges, with small open patches of grasses and *Gleichenia* spp., on soil probably derived from sandstone and shales. 900–1530 m, recorded between 1900 and 2400 m on Mount Murud by *Mjöberg*.

D. imbricatum is a distinctive plant easily distinguished from other Bornean species by the large floral bracts which partially conceal the flowers.

The specific epithet is derived from the Latin *imbricatus,* to overlap, and refers to the floral bracts.

74. DENDROCHILUM LONGIFOLIUM

Dendrochilum longifolium Rchb. f. in *Bonplandia* 4: 329 (1856). Type: Origin and collector unknown, imported by Consul *Schiller*, cult. *Stange* s.n. (holotype Herb. Rchb. f. 49081 W!). **Fig. 120.**

D. fuscum Teijsm. & Binnend. in *Natuurk. Tijdsschr. Ned.–Indië* 24: 305–306 (1862). Type: "Indiam Orientalem", *Lobb*, cult. Hort. Bot. Bog., (presumed type material, Hort. Bot. Bog. 57, SING).

D. bracteosum Rchb. f. in Walp., *Ann. bot. syst.* 6: 241 (1863). Type: Peninsular Malaysia, *Finlayson* s.n. (holotype Herb. Lindl., K!).

Platyclinis longifolia (Rchb. f.) Hemsl. in *Gdnrs. Chron.* 2, 16: 656 (1881).

Acoridium bracteosum (Rchb. f.) Rolfe in *Orchid Rev.* 12: 220 (1904).

A. longifolium (Rchb. f.) Rolfe in *Orchid Rev.* 12: 220 (1904).

Dendrochilum clemensiae Ames, *Orchidaceae* 2: 109–110, fig. s.n. on p. 110 (1908). Type: Philippines, Mindanao, Lake Lanao, Camp Keithly, *M.S. Clemens* 642 (holotype PNH, destroyed, photograph at AMES!).

Platyclinis bartonii Ridl. in *J. Straits Brch. R. Asiat. Soc.* 50: 128 (1908). Type: Papua New Guinea, *Barton* 5 (holotype K!).

Dendrochilum bartonii (Ridl.) Schltr., *Reprium. Spec. nov. Regni veg. Beih.* 1: 106 (1911).

D. longifolium Rchb. f. var. *papuanum* J.J. Sm. in Beaufort *et al., Nova Guinea* 8: 527 (1911). Type: Indonesia, Irian Jaya, "An dem Noord-Fluss", *Djibida*, cult. Hort. Bot. Bog. 112 (holotype BO, ? destroyed).

Platyclinis longifolia (Rchb. f.) Hemsl. f. *papuana* Ridl. in *Trans. Linn. Soc. Lond.* 2, 9: 202 (1916). Type: Indonesia, Irian Jaya, *Wollaston*, cult. *Rothschild* s.n., June 1914 (holotype not located).

Dendrochilum amboinense J.J. Sm. in *Bull. Jard. Bot. Buitenz.* 3, 1: 95 (1919), *nom. nud.* Type: Indonesia, Maluku, Ambon, collector unknown, cult. Hort. Bot. Bog. II M a 35 (holotype BO, probably destroyed, isotypes K!, L, SING).

D. longifolium Rchb. f. var. *fuscum* J.J. Sm. in *Bull. Jard. Bot. Buitenz.* 3, 1: 96 (1919), *nom. nud.* Type: Origin and collector unknown, cult. Hort. Bot. Bog. II M b 20 (holotype BO, probably destroyed).

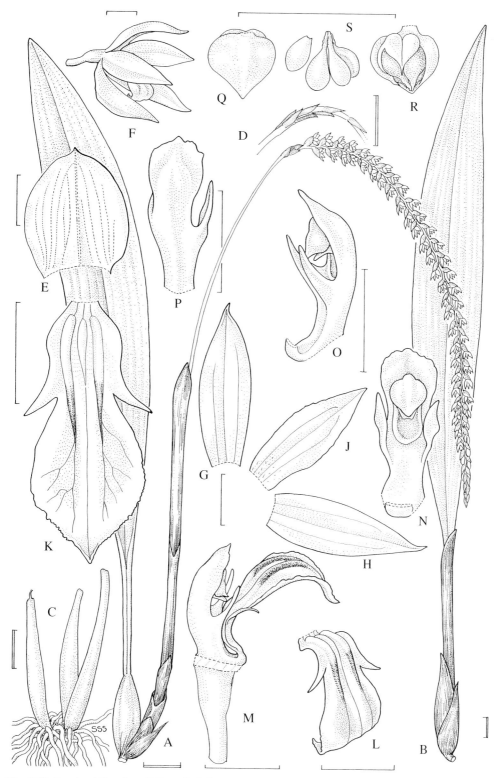

Fig. 120. *Dendrochilum longifolium.* **A & B** habits; **C** pseudobulbs; **D** junction of rachis with peduncle showing seven non-floriferous bracts; **E** floral bract; **F** flower, side view; **G** dorsal sepal; **H** lateral sepal; **J** petal; **K** labellum, front view; **L** labellum, oblique view; **M** pedicel with ovary, labellum and column, side view; **N** column, front view; **O** column, side view; **P** column, back view; **Q** anther-cap, back view; **R** anther-cap, interior view showing pollinia; **S** pollinia. **A & C** from *Bacon* 93, **B** from *de Vogel* 8402, **D–S** from *Cribb* 89/66. Scale: single bar = 2 mm; double bar = 2 cm. Drawn by Susanna Stuart-Smith.

D. longifolium Rchb. f. var. *buruense* J.J. Sm. in *Bull. Jard. Bot. Buitenz.* 3, 9: 451 (1928). Type: Indonesia, Maluku, Buru, Mount Togha, 1900 m, *Stresemann* 377 (holotype L).

DESCRIPTION. Epiphyte, occasionally lithophytic or terrestrial. *Rhizome* short, up to *c.* 4 cm long. *Roots* branching, robust, smooth, 1–3 mm in diameter. *Cataphylls* 4–6, oblong-elliptic, obtuse to acute, 0.5–20 cm long, reddish brown, often speckled brown. *Pseudobulbs* crowded, or up to 3.5 cm apart, cylindrical to narrowly pyriform or narrowly ovate-elliptic, smooth to finely wrinkled, 3–8 × 1–1.5 cm. *Leaf:* petiole rigid, sulcate, 3–15 cm long; blade narrowly elliptic to oblong-elliptic, subacute to acute, thick and coriaceous, 21–50 × 2.3–6.5 cm, main nerves 10–15, median nerve prominent, raised on abaxial surface. *Inflorescences* pendulous, laxly to subdensely many-flowered, flowers borne 3–5 mm apart; peduncle porrect to arching, 15–56 cm long; non-floriferous bracts 1–3, ovate-elliptic, subacute to acute, 7–9 mm long; rachis pendulous, quadrangular, (16–)22–35 cm long, yellowish green; floral bracts oblong-ovate to ovate-elliptic, obtuse to apiculate, 6–8 mm long, pale green, flushed pink, or yellowish green, turning brown. *Flowers* scented, sepals and petals pale green or yellowish, often ochre distally, labellum greenish with two brown parallel streaks, side lobes sometimes brown, column pale green. *Pedicel with ovary* clavate, 3.5–3.8 mm long. *Sepals* and *petals* 3–nerved. *Dorsal sepal* oblong-elliptic, acute, 5–6.5 × 2–2.1 mm. *Lateral sepals* narrowly ovate-elliptic, acute, (5–)6–7 × 2–2.1 mm. *Petals* narrowly elliptic, acute, slightly erose, sparsely hairy at base, (4–)5–6 × 2 mm. *Labellum* stipitate to column-foot by a short claw, 3–lobed, 3–nerved, reflexed, 4.5 mm long, 1.5–1.6 mm wide near base; side lobes narrowly triangular, acute, free apical portion 0.8–0.9 mm long, somewhat decurved; mid-lobe elliptic, acute, somewhat erose, 3 × 2.2 mm, with a few branching secondary nerves; disc with 2 low fleshy keels along outer nerves, extending from just above base and terminating on lower portion of mid-lobe, median nerve elevate and thickened for a short distance. *Column* 2.8–3 mm long; foot distinct; apical hood oblong, curved, dorsally somewhat carinate, irregular to erose; stelidia borne just below stigmatic cavity, narrowly triangular, acute, slightly twisted, equalling or extending beyond rostellum, 0.9–1 mm long; stigmatic cavity oblong, lower margin somewhat elevate; rostellum triangular-ovate, acute; anther-cap cordate, cucullate.

DISTRIBUTION. Myanmar (Burma), Peninsular Malaysia, Singapore, Sumatra, Java, Borneo, Sulawesi, Maluku, Philippines, New Guinea, east as far as New Britain.

SARAWAK. Marudi District, Mount Api, Ulu Melinau, Tutoh, 29 Sept. 1971, *J.A.R. Anderson* S. 30808 (K!, L!, SAR). Kuching District, Braang, 18 Nov. 1888, *Haviland* 46 (BM!,SING!).

SABAH. Sipitang District, Long Pa Sia, Oct. 1967, *Bacon* 78, cult. Royal Botanic Gardens, Edinburgh, C. 6530 (E!) & *Bacon* 93, C. 6531 (E!). Nabawan, 68 km east of Keningau, 21 Nov. 1989, *Cribb* 89/66 (K!). Crocker Range, Kimanis road, 9 Nov. 1979, *Lamb* SAN 91572 (drawing only at K!). Lohan River, 1987, *Leiden* cult. (*Vermeulen*) 26684 (K!, L!). Crocker Range, Keningau to Kimanis road, Dec. 1986, *Vermeulen & Duistermaat* 687 (K!, L!). Sipitang District, Maga River, confluence with Pa Sia River, 19 Oct. 1986, *de Vogel* 8402 (L!).

KALIMANTAN TIMUR. Northern part, 1981, *Leiden* cult. (*Geesink*) 20313B (L!). Apo Kayan, Maru, foot of Mount Malem, Long Bawan, 18 Aug. 1981, *Ueda & Darnaedy* B 11544 (L!).

KALIMANTAN. Province and locality unknown, *Amdjah* s.n., cult. hort. Bogor no. 108 (L!).

Fig. 121 (previous pages). Gunung Mulu National Park, Mount Api, Sarawak, limestone pinnacles. *D. longifolium* occurs in the forest in the vicinity of these bizarre formations. (Photo: Hans P. Hazebroek)

HABITAT. Primary riverine forest up to 30 metres high with undergrowth of small trees, on flat alluvial terrain, sandy soil derived from sandstone; podsol forest; limestone boulders; roadside cuttings on sandstone and shale outcrops, partially covered with grass and small bushes, fully exposed to sun. 400–1500 m.

This is the most widespread of the species native to Borneo and has been known in cultivation for a long time.

The specific epithet refers to the long leaves.

75. DENDROCHILUM OXYLOBUM

Dendrochilum oxylobum Schltr. in *Reprium. Spec. nov. Regni veg.* 9: 431 (1911). Type: Malaysia, Sarawak, Kuching District, Kuching, Nov. 1865, *Beccari* 1125 (holotype FI!, isotype drawings at K!, L). **Figs. 122 & 123, plate 12A & B.**

D. viridifuscum J.J. Sm. in *Bull. Jard. Bot. Buitenz.* 2, 25: 11–12 (1917). Type: Indonesia, Kalimantan Tengah, Kotawaringin, *van Nouhuys*, cult. Hort. Bot. Bog. 16 (holotype BO!, drawing at K!).

DESCRIPTION. Epiphyte. *Rhizome* abbreviated, up to 4 cm long. *Roots* branching, flexuose, smooth, 0.2–1.5 mm in diameter. *Cataphylls* 4–5, oblong-elliptic, acute, 0.5–15 cm long, green to chestnut-brown, sometimes speckled brown. *Pseudobulbs* crowded, oblong-ovoid or cylindrical, approximate, 3–7 × 0.4–1.6 cm, shiny pale green. *Leaf*: petiole sulcate, (1.5–)4.5–15 cm long; blade narrowly elliptic, ensiform, acute, sometimes constricted below apex, coriaceous, (12.7–)19–36 × 1.6–4.6 cm, main nerves 5–7, median nerve elevate and prominent on abaxial surface, suffused brown when immature. *Inflorescences* laxly to subdensely many-flowered, flowers borne 3.5–6 mm apart; peduncle 15–30 cm long; non-floriferous bracts 1–2, ovate-elliptic or oblong, acute or obtuse and mucronate, 5–11 mm long; rachis quadrangular, decurved, 8.5–28 cm long; floral bracts oblong to oblong-ovate, acute or apiculate, 5–9 mm long, cinnamon brown or flesh-coloured. *Flowers* scented, sepals and petals brownish ochre, orange brown or brownish, or pale green, suffused brown, labellum dark chocolate-brown, lemon-yellow at base, margin of mid-lobe paler, column white, brown dorsally, yellow at base. *Pedicel with ovary* narrowly clavate, glabrous or pubescent, 3–5 mm long. *Sepals* and *petals* 3–nerved. *Dorsal sepal* narrowly elliptic, acute, concave, margin convex, (8–)8.2–8.3 × 3 mm. *Lateral sepals* oblong-elliptic, acute, dorsally thickened, somewhat carinate distally, 8.4 × 3 mm. *Petals* oblong-elliptic or ovate-oblong, acute, sometimes minutely erose, sparsely hairy at base, 7.5 × 2.5 mm. *Labellum* stipitate to column-foot by a short claw, 3–lobed, 3–nerved, reflexed, 6.5 mm long, 2 mm wide across side lobes; side lobes linear-subulate to narrowly triangular, acute, exterior margin sometimes erose, free apical portion 1–1.5 mm long; mid-lobe rhomboid, cuneate at base, triangular-apiculate, erose, 4–4.5 × 2.5–3.3 mm; disc with 2 parallel keels extending from just above claw and terminating on lower portion of mid-lobe, with a third less prominent shorter keel in between which is itself sulcate or sometimes divided into 2 keels. *Column* curved distally, convex, 4–4.5 mm long; foot distinct; apical hood oblong, tridentate or irregularly toothed, recurved; stelidia borne just below stigmatic cavity, subulate, falcate, acute, exceeding rostellum, *c.* 2.7 mm long; stigmatic cavity quadrangular, lower margin elevate, recurved; rostellum ovate-triangular, acute, convex; anther-cap ovate, cucullate, apex recurved, acute.

DISTRIBUTION. Borneo.

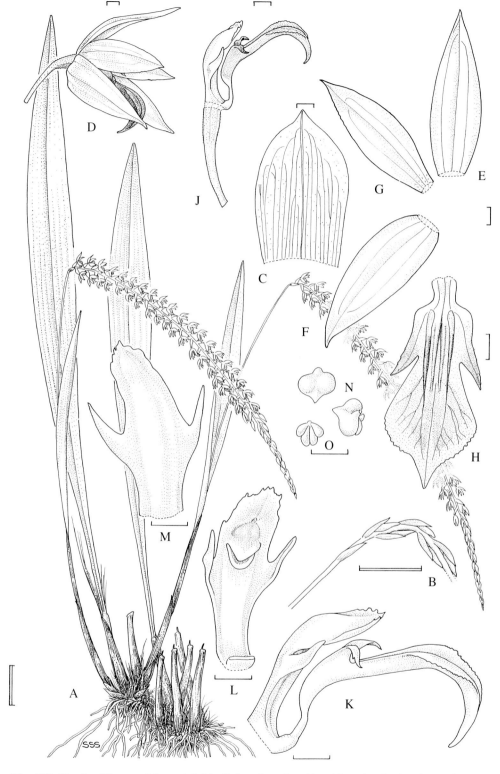

Fig. 122. *Dendrochilum oxylobum*. **A** habit; **B** junction of rachis with peduncle showing solitary non-floriferous bract, five floral bracts and five buds; **C** floral bract; **D** flower, side view; **E** dorsal sepal; **F** lateral sepal; **G** petal; **H** labellum, front view; **J** pedicel with ovary, labellum and column, side view; **K** labellum and column, side view; **L** column, anther-cap removed, oblique view; **M** column, back view; **N** anther-cap, back and side views; **O** pollinia. **A** from *Chan & Lohok* s.n. and *Lamb* AL 506/85, **B–O** from *Lamb* AL 506/85. Scale: single bar = 1 mm; double bar = 1 cm. Drawn by Susanna Stuart-Smith.

SARAWAK. Kuching District, Kuching, Nov. 1865, *Beccari* 1125 (holotype FI!, drawing at K!, L). Kuching District, Kuching, Nov. 1906, *Hewitt* 46 (SAR!) & May 1907, *Hewitt* s.n. (K!).

SABAH. Pensiangan District, 6 km from Nabawan, Nov. 1988, cult. Tenom Orchid Centre, *Chan & Lohok* s.n. (K!) & Jan. 1985, *Lamb* AL 425/85 (K!). Crocker Range, Kimanis road, Jan. 1985, *Lamb* AL 506/85 (K!). Crocker Range, Keningau to Kimanis road, Oct. 1986, cult. *Tenom Orchid Centre*, TOC 1117 (L!). Road from Keningau to Sepulot, 6 km past Nabawan, near old airstrip, 9 Oct. 1986, *de Vogel* 8150 (L!).

KALIMANTAN BARAT. Locality unknown, cult. Bogor Botanic Garden, Java, Dec. 1997, *Cribb* (photo !).

KALIMANTAN SELATAN. Tanah Laut District, 37 km Northeast of Hutan Kintap Base Camp, 20 km N of Kintap, 15 April 1985, *Leeuwenberg & Rudjiman* 13405 (L!).

KALIMANTAN TENGAH. Kotawaringin, *van Nouhuys*, cult. Hort. Bot. Bog. 16 (holotype of *D. viridifuscum* BO!). Cult. Hort. Bot. Bog. (BO!, K!, L, 6 sheets!).

KALIMANTAN TIMUR. Neighbourhood of Balikpapan, 1981, *Leiden* cult. (*Franken & Roos*) 21035 (K!, L!), 21085 (K!, L!) & 21093 (L!).

HABITAT. Dipterocarp/*Dacrydium* forest on podsolic white sandy soil, with pools of stagnant brown water; dipterocarp forest on laterite, recorded as epiphytic on branches of *Shorea* spp.; lower montane mossy forest on sandstone. Sea level–900 m.

D. oxylobum is an attractive species most closely allied to the widespread *D. longifolium* but distinguished by its larger, brownish-ochre flowers with differently aligned keels on the labellum.

The specific epithet is derived from the Greek *oxy*, sharp, and the Latin *lobus*, a lobe, in reference to the shape of the labellum.

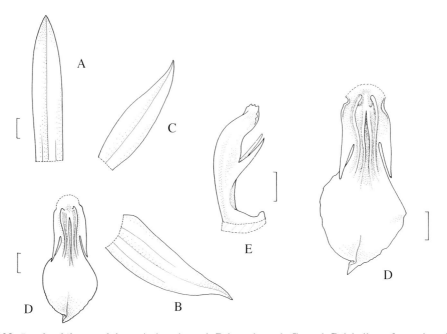

Fig. 123. *Dendrochilum oxylobum*. **A** dorsal sepal; **B** lateral sepal; **C** petal; **D** labellum, front view; **E** column, side view. **A–E** from *Beccari* 1125 (isotype). Scale: single bar = 1 mm Drawn by Susanna Stuart-Smith (after Vermeulen).

267

76. DENDROCHILUM PAPILLITEPALUM

Dendrochilum papillitepalum J.J. Wood in *Orchid Rev.* 103 (1201): 8, fig. 4 (1995). Type: Malaysia, Sarawak, Limbang District, upper Sipayan River, Mount Pagon, 530 m, 5 Aug. 1984, *Awa & Lee* S. 47676 (holotype K!, isotypes KEP!, SAR). **Fig. 124.**

DESCRIPTION. Epiphyte. *Rhizome* 6 cm long. *Roots* branching, wiry, very minutely papillose, 0.6–0.8 mm in diameter. *Cataphylls* ? 3, up to 6 cm long, brown, speckled darker brown. *Pseudobulbs* crowded, approximate, fusiform, 2–3 × 0.6–0.7 cm. *Leaf*: petiole narrow, sulcate, 3–5.5 cm long; blade narrowly elliptic-ligulate, acute, 18–25 × (1–)1.3–1.6 cm, main nerves 4–5. *Inflorescences* pendulous, subdensely many-flowered, flowers borne 2–2.5 mm apart; peduncle filiform, 18–21 cm long; non-floriferous bracts 4–7, oblong, obtuse, 2 mm long; rachis quadrangular 34–42 cm long; floral bracts oblong-ovate, obtuse, 3 × 1.6 mm. *Flowers* pale green, flushed reddish. *Pedicel with ovary* narrowly clavate, 2–3 mm long. *Sepals* and *petals* papillate on adaxial surface. *Dorsal sepal* oblong-elliptic, acute, 3–nerved, 4–4.1 × 1.1 mm. *Lateral sepals* narrowly elliptic, acute, 3–nerved, 4–4.1 × 1.1 mm. *Petals* linear-elliptic, acute, 1-nerved, 3.3 × 0.8–0.9 mm. *Labellum* stipitate to column-foot by a narrow claw, 3–lobed, 3–nerved, 2.5–2.6 mm long, 1.5 mm wide across side lobes; side lobes rounded, auriculate, free apical portion narrowly triangular-acuminate, outer margin serrate to lacerate; mid-lobe oblong-ovate to ovate-elliptic, acute to shortly acuminate, minutely erose, 1.1–1.3 mm wide; disc with 2 small swollen flanges between which is a narrow, linear depression extending to base of mid-lobe. *Column* straight, 1.6 mm long, *c.* 0.3–0.4 mm wide at apex; foot obscure; apical hood entire, obtuse, somewhat truncate; stelidia arising just above base, spathulate, rounded to truncate, auriculate, reaching to base of stigmatic cavity; stigmatic cavity narrowly oblong; rostellum narrowly ovate, acute; anther-cap cucullate, acute, surface uneven.

DISTRIBUTION. Borneo.

SARAWAK. Limbang District, upper Sipayan River, Mount Pagon, 5 Aug. 1984, *Awa & Lee* S. 47676 (holotype K!, isotypes KEP!, SAR).

HABITAT. Riparian forest, growing one metre above ground level. 530 m.

A curious plant known only from the type and distinguished from other Bornean species by the papillate sepals and petals and auriculate stelidia.

The specific epithet is derived from the Latin *papillatus*, having papillae, i.e. nipple-like protuberances, and *tepalum*, a division of the perianth, i.e. either sepal or petal.

77. DENDROCHILUM GLOBIGERUM

Dendrochilum globigerum (Ridl.) J.J. Sm. in *Recl. Trav. bot. Néerl.* 1: 64 (1904). Type: Malaysia, Sarawak, Kuching District, summit of Mount Serapi, *Haviland* 169 (lectotype K!, syntype SAR!). **Fig. 126.**

Platyclinis globigera Ridl. in *J. Linn. Soc., Botany* 31: 266 (1896).
Acoridium globigerum (Ridl.) Rolfe in *Orchid Rev.* 12: 220 (1904).
Platyclinis minor Ridl. in *J. Straits Brch. R. Asiat. Soc.* 49: 30 (1907). Type: Malaysia, Sarawak, Kuching District, Mount Santubong, *Hewitt* s.n. (holotype SING, isotypes BM!, K!).

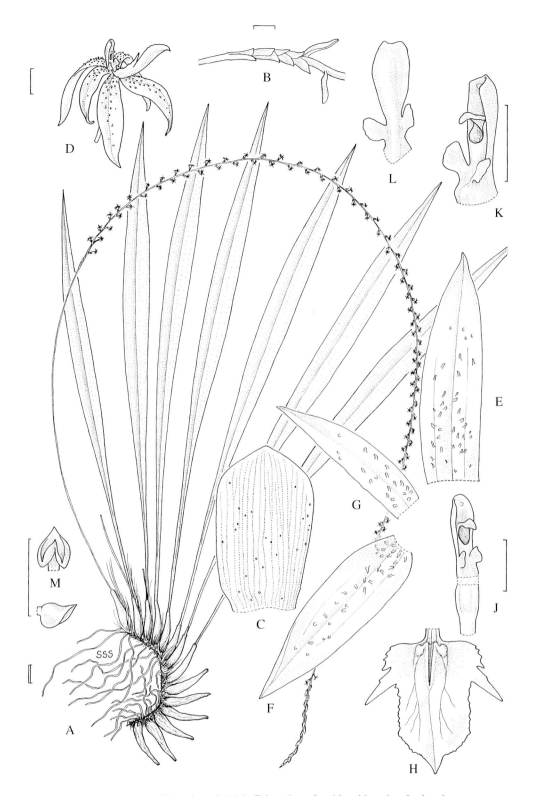

Fig. 124. *Dendrochilum papillitepalum*. **A** habit; **B** junction of rachis with peduncle showing seven non-floriferous bracts and two floral bracts; **C** floral bract; **D** flower, side view; **E** dorsal sepal; **F** lateral sepal; **G** petal; **H** labellum, front view; **J** pedicel with ovary and column, anther-cap removed, oblique view; **K** column, anther-cap removed, oblique view; **L** column, back view; **M** anther-cap, interior and side views. **A–M** from *Awa & Lee* S. 47676 (holotype). Scale: single bar = 1 mm; double bar = 1 cm. Drawn by Susanna Stuart-Smith.

Fig. 125. Kubah National Park, Sarawak, from 780 metres on Mount Serapi. The type collection of *D. globigerum*, a species restricted to the mountains of western Sarawak, was collected by Haviland on Mount Serapi. (Photo: Hans P. Hazebroek)

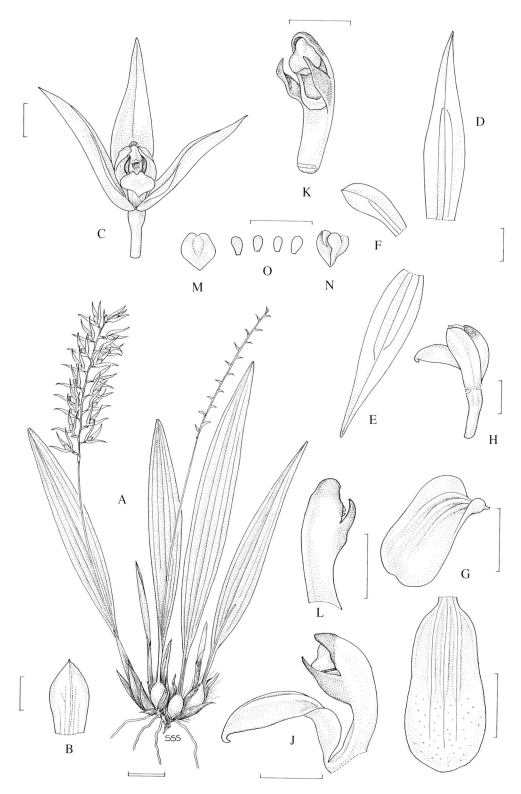

Fig. 126. *Dendrochilum globigerum.* **A** habit; **B** floral bract; **C** flower, front view; **D** dorsal sepal; **E** lateral sepal; **F** petal; **G** labellum, front and oblique views; **H** flower with sepals removed, side view; **J** labellum and column, side view; **K** column, oblique view; **L** column, back view; **M** anther-cap, back view; **N** anther-cap, side view; **O** pollinia. **A–O** from *I. Sinclair & Argent* 126. Scale: single bar = 1 mm; double bar = 1 cm. Drawn by Susanna Stuart-Smith.

P. minima Ridl. in *J. Straits Brch. R. Asiat. Soc.* 49: 30–31 (1907). Type: Malaysia, Sarawak,
Sri Aman District, Tiang Laju, Undup, *Hewitt* 14 (holotype SING!, isotype K!).
Dendrochilum minimum (Ridl.) Ames in *J. Straits Brch. R. Asiat. Soc. Special no.*: 148 (1921).
D. minus (Ridl.) Ames in *J. Straits Brch. R. Asiat. Soc. Special no.*: 148 (1921).

DESCRIPTION. Epiphyte or lithophyte. *Rhizome* 7–10 cm long, branching. *Roots* branching,
wiry, smooth, forming a dense mass, up to 1 mm in diameter. *Cataphylls* 3–4, 0.5–3 cm long,
ovate, acute or apiculate, soon becoming fibrous, speckled. *Pseudobulbs* crowded,
approximate, 0.5–1.5 × 0.6–0.8 cm, globose, ovoid or oblong-ovoid, obtuse. *Leaf*: petiole
0.7–2.5 cm long, narrow, sulcate; blade 3–9 × 0.6–1.5 cm, narrowly elliptic, sometimes
constricted below apex, obtuse, thin-textured, 5–nerved. *Inflorescences* erect to gently
curving, laxly *c.* 8– to 20–flowered, flowers borne 2–3 mm apart; peduncle filiform, 3.5–10
cm long; non-floriferous bracts solitary, oblong-ovate, acute, 2–3 mm long; rachis
quadrangular, 2–5 cm long, pale orange; floral bracts oblong-ovate, acute, apiculate or obtuse,
2–3 mm long. *Flowers* sweetly scented, pedicel yellowish, sepals and petals creamy white,
labellum brown with whitish base and margins, column apex and anther-cap brown. *Pedicel
with ovary* narrowly clavate, 1.9–2 mm long. *Sepals* 3–nerved. *Dorsal sepal* narrowly elliptic,
acuminate, carinate, 5–6 × 1–1.5 mm. *Lateral sepals* narrowly elliptic, acuminate, carinate, 6
× 1.1–1.5 mm. *Petals* oblong-elliptic or subspathulate, obtuse, 1–nerved, rarely with a
secondary nerve, 2.4–2.5 × 1 mm. *Labellum* entire, narrowly clawed, decurved, oblong to
subpandurate, obtuse, lower margins erect, rather fleshy textured, 2.7–2.8(–3) × 1–1.5 mm;
disc with 3 low rib-like keels extending from base and terminating ³/4 along blade. *Column*
gently curving, 1.8–2 mm long, foot distinct; apical hood ovate, obtuse or denticulate; stelidia
arising opposite or just below stigmatic cavity, triangular-acuminate, falcate, slightly shorter
than apical hood, *c.* 1mm long; stigmatic cavity oblong, margins elevate; rostellum ovate,
acute; anther-cap *c.* 2 × 2 mm, ovate, cucullate, very minutely papillose.

DISTRIBUTION. Borneo.
SARAWAK. Kuching District, Mount Santubong, summit, 4 June 1995, *Beaman* 11574
(K!). Kuching District, Mount Serapi, ridge running Northwest from radar installation, 6 May
1996, *Beaman* 12068 (K!). Kuching District, Mount Matang, May 1866, *Beccari* s.n. (FI!). Sri
Aman District, Tiang Laju, Undup, April 1867, *Beccari* 3342 (FI!, G!, K!, W!). Kuching
District, Mount Penrissen, May 1910, *Brooks* 107 (BM!). Kuching District, Mount Santubong,
19 March 1967, *Chew* 1410 (HBG!, K!, L!, SAR!, SING!). Kuching District, Mount Matang,
15 May 1961, *Collenette* 695 (K!). Bau District, Mount Raya, 7 April 1965, *Elsener* H114
(L!). Kuching District, summit of Mount Serapi, June 1890, *Haviland* 169 (lectotype of
Platyclinis globigera K!, syntype SAR!). Kuching District, Mount Santubong, 1907, *Hewitt*
s.n. (holotype of *Platyclinis minor* SING!, isotypes BM!, K!). Sri Aman District, Tiang Laju,
Undup, *Hewitt* 14 (holotype of *Platyclinis minima* SING!, isotype K!). Kuching District,
Mount Matang, April 1908, *Hewitt* s.n. (SAR!, SING!). Kuching District, Mount Matang, 22
April 1987, *Lee* S. 54013 (AAU, K!, L!, SAR, SING). Locality unknown, coll. 1958, cult. &
comm. 13 April 1962, *Mason* 986 (K!) & comm. 5 May 1970, *Mason* 92 (K!). Kuching
District, Mount Santubong, 13 March 1914, *native collector* D 162 (E!, SAR!). Locality
unknown, collected through Sarawak Museum for Bureau of Science, Manila, Philippines,
Sarawak Museum native collector 1580 (K!) & 1584 (BM!). Kuching District, Mount
Santubong, March 1982, *I. Sinclair & Argent* 126 (E!). Kuching District, Mount Santubong,
23 Feb. 1949, *J. Sinclair* 5604, SFN 38351 (BO!, E!, K!, SING!).

273

HABITAT. Hill and lower montane forest, often on sandstone substrate; low-stature cloud forest; on mossy rocks, at the base and on the branches of small shrubs, tree trunks, often on ridges. 700–900 m.

This distinctive little species appears, from extant collections, to be restricted to the mountains of western Sarawak.

The specific epithet is derived from the Latin *globiger*, globe-carrying, i.e. with a spherical organ, and refers to the pseudobulbs.

78. DENDROCHILUM SIMPLEX

Dendrochilum simplex J.J. Sm. in *Bull. Dép. Agric. Indes Néerl.* 22: 13–14 (1909). Type: Indonesia, Kalimantan Tengah, Liangangang (Liang Gagang), *Hallier* 2646 (holotype BO!, isotypes K!, L!). **Figs. 127 & 128, plate 15D.**

D. remotum J.J. Sm. in *Bull. Jard. Bot. Buitenz.* 2, 3: 57–58 (1912). Type: Malaysia, Sarawak, Limbang District, Mount Batu Lawi, Ulu Limbang, *Moulton* 10 (holotype SAR!).

DESCRIPTION. Terrestrial, epiphytic or lithophytic. *Rhizome* elongate, up to *c.* 40 cm long, 1–2.5 mm in diameter, branching, covered when young with persistent chestnut-brown sheaths. *Roots* elongate, primary roots producing a tangle of branching secondary roots distally, wiry, primary roots usually minutely papillose, secondary roots smooth, 0.1–0.2 mm in diameter. *Cataphylls* 5–6, 0.3–3.6 cm long, ovate-elliptic, acute to acuminate, persistent, reddish brown or chestnut-brown. *Pseudobulbs* remote, borne 1–5.5 cm apart, 0.8–2.75 × 0.3–0.5 cm, terete, narrowly-oblong to narrowly fusiform, smooth, spreading. *Leaf*: petiole 0.15–1.5 cm long, sulcate, canaliculate; blade 2.5–16 × 0.5–1.8 cm, ovate-elliptic to narrowly elliptic, often constricted below apex, acute to shortly acuminate, sometimes narrowly obtuse, thin-textured, main nerves 5–7. *Inflorescences* gently curving, laxly to subdensely many-flowered, flowers borne 1.5–2.2(–2.5) mm apart; peduncle filiform, 2–7.5 cm long; non-floriferous bracts 1–6, ovate, apiculate to acuminate, 2–3 mm long; rachis quadrangular, somewhat compressed, 4–14 cm long; floral bracts ovate-orbicular or broadly ovate, apiculate, acute, sometimes obtuse, somewhat erose, concave, 2–2.9 mm long. *Flowers* fragrant, creamy white to white, pale green, yellowish green, creamy gold or yellowish cream, labellum often greenish at base. *Pedicel with ovary* 0.75–2 mm long, slender, geniculate or straight. *Sepals* and *petals* 3–nerved. *Dorsal sepal* narrowly oblong to linear-lanceolate, apex often recurved, acute, somewhat concave, 3.7–4 × 0.7–1.1 mm. *Lateral sepals* obliquely narrowly lanceolate, falcate, acute, concave, dorsally somewhat carinate, 3–3.9 × 0.7–1 mm. *Petals* narrowly lanceolate, falcate, acute, minutely erose-crenulate, concave, 2.9–3.8 × 0.6–1 mm. *Labellum* entire or shallowly 3–lobed, shortly clawed, decurved, oblong, shortly acute, 3–nerved, 1.1–1.6 × 0.6–1 mm; side lobes (when present) erect, small, shallowly rounded, somewhat crenulate; mid-lobe (when present) semi-orbicular to ovate, *c.* 0.4 mm long; disc with two low keels united below to form a U-shape. *Column* gently curving, 1.2–1.75 mm long; foot 0.2–0.4 mm long; apical hood oblong-quadrangular, obtuse to acute, or variously toothed; stelidia arising either side of stigmatic cavity, lanceolate, subulate, subfalcate, acute to acuminate, equal to or slightly shorter than apical hood, *c.* 0.3–1 mm long; stigmatic cavity oblong-elliptic, lower margin elevate; rostellum recurved, shortly triangular, acute, convex; anther-cap minute, cucullate. *Capsule c.* 2 × 1.5–2 mm, globose, orange, remains of sepals and petals persistent.

Fig. 127. *Dendrochilum simplex.* **A** habit; **B** floral bract; **C** flower, side view; **D** labellum, front view; **E** labellum and column, side view; **F** column, front view; **G** pollinia; **H** floral bract; **J** flower, side view; **K** dorsal sepal; **L** lateral sepal; **M** petal; **N** labellum, front view; **O** pedicel with ovary, labellum and column with part of lateral sepal and petal, side view; **P** column, oblique view; **Q** column, side view; **R** column, foot and anther-cap removed, oblique view; **S** column, back view; **T** anther-cap, front and back views; **U** pollinia. **A–G** from *Hallier* 2646 (holotype), **H–U** from *Vermeulen & Lamb* 324. Scale: single bar = 1 mm; double bar = 1 cm. Drawn by Susanna Stuart-Smith.

Fig. 128. *Dendrochilum simplex*. **A** habit; **B** distal portion of inflorescence; **C** floral bract; **D** flower, side view; **E** dorsal sepal; **F** lateral sepal; **G** petal; **H** labellum, front view; **J** pedicel with ovary, labellum and column, side view; **K** column, side view; **L** column, foot and stelidium removed, side view; **M** column, foot removed, back view; **N** pollinia; **O & P** labelli, front views. **A–N** from *Moulton* 10 (holotype of *D. remotum*), **O** from *Leiden* cult. (*de Vogel*) 913965; **P** from *Leiden* cult. (*de Vogel*) 913963. Scale: single bar = 1 mm; double bar = 1 cm. Drawn by Susanna Stuart-Smith.

DISTRIBUTION. Borneo.

SARAWAK. Limbang District, upper Limbang River, route to Mount Batu Lawi, Bario, 12 Aug. 1985, *Awa & Lee* S. 50792 (AAU, K!, KEP!, L!, MEL, SING). Mengiong/Balleh Rivers, Upper Entulu, 26 July 1987, *Lee* S. 54798 (K!, SAR). Kuching District, Mount Penrissen near Padawan, 1993, *Leiden* cult. (*Schuiteman et al.*) 933140 (L!). Kapit District, Hose Mountains, northern part, base of ridge leading to Bukit Batu, 1 Dec. 1991, *Leiden* cult. (*de Vogel*) 913963 (K!, L!), 913964 (K!, L!) & 913965 (L!). Marudi District, Mount Murud, Oct. 1922, *Mjöberg* 49 (AMES!). Limbang District, Mount Batu Lawi, Ulu Limbang, 29 May 1911, *Moulton* 10 (holotype of *D. remotum* SAR!).

SABAH. Mt Monkobo, Beluran District, 15 March 1982, *Aban* SAN 95241 (K!, L!, SAN; material in packet on sheet at L is *D. tenompokense* var. *papillilabium*, probably SAN 95230). Pig Hill, 20 Sept. 1996, *Barkman* s.n. (K!). Mount Kinabalu: Marai Parai Spur, 22 Nov. 1915, *J. Clemens* 278 (AMES, BM!); Penataran Basin, 31 Aug. 1933, *J. & M.S. Clemens* 40196 (BM!, E!, L!); Penibukan, 3 Oct. 1933, *J. & M.S. Clemens* 40548 (BM!), & 2 Nov. 1933, *J. & M.S. Clemens* 50105 (AMES!, BM!, K!); Tenompok Orchid Garden, Nov. 1933, *J. & M.S. Clemens* 50347 (AMES!, BM!, E!, K!) Crocker Range, Sinsuron road, Sept. 1979, *Collenette* 48/79 (E!). Penampang District, Tunggul Togudon, km 48 Jalan Tambunan/Penampang, 26 Aug. 1989, *Fidilis* SAN 127819 (K!, SAN!). Mount Kinabalu, locality unknown, *Leiden* cult. (*Roelfsema*) 950222 (L!). Mount Alab, 1987, *Leiden* cult. (*Vermeulen*) 26488 (L!). Keningau to Sapulut road, 6 km past Nabawan, near old airstrip, June 1986, *Vermeulen & Lamb* 324 (K!, L!).

KALIMANTAN TENGAH. Liangangang (Liang Gagang), *Hallier* 2646 (holotype of *D. simplex* BO!, isotypes K!, L!).

HABITAT. Lower montane forest, on sandstone and ultramafic substrates; primary submontane forest; podsol forest with *Dacrydium* spp. and dipterocarps on very wet sandy soil, often associated with *Bromheadia finlaysoniana* (Lindl.) Miq., *Bulbophyllum nabawanense* J.J. Wood & A. Lamb, *Nepenthes ampullaria* Jack and aroids; rocky places; riparian forest. 400–2000 m.

A graceful orchid unlikely to be confused with other Bornean species, *D. simplex* varies in habit according to habitat. Robust terrestrial populations growing in shady podsol forest in Sabah have leaves up to 16 cm long. Others from Sarawak, referred to *D. remotum* by J.J. Smith, have smaller leaves and flowers. Shallow lobing of the labellum, reported in *D. remotum*, also occurs in larger flowered populations.

The Latin specific epithet refers to the simple, undivided labellum found in many populations.

79. DENDROCHILUM HAMATUM

Dendrochilum hamatum Schltr. in *Reprium. Spec. nov. Regni veg.* 9: 340 (1911). Type: Malaysia, Sarawak, Sri Aman District, near Undup, Sept. 1865, *Beccari* 476 (holotype B, destroyed, isotype FI!). **Figs. 129 & 130, plate 6D & E.**

DESCRIPTION. Epiphyte. *Rhizome* elongate, to 10 cm long. *Roots* filiform, branching, flexuous, glabrous. *Pseudobulbs* cylindrical, somewhat dilated towards base, borne 1–1.5 cm apart, 1–1.5 × 0.4 cm. *Leaf*: petiole 0.5–1.5 cm long; blade 3.5–9.5 × 0.3–0.4 cm, erect, linear-ligulate, acute. *Inflorescences* erect, 10– to 16–flowered, rather lax, flowers borne about

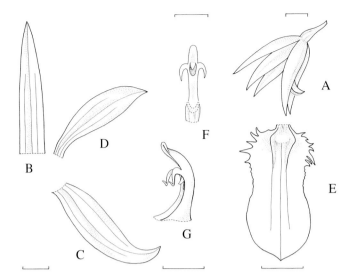

Fig. 129. *Dendrochilum hamatum*. **A** flower, side view; **B** dorsal sepal; **C** lateral sepal; **D** petal; **E** labellum, front view; **F** column and upper portion of ovary, front view; **G** column, side view. **A–G** from *Beccari* 476 (isotype). Scale: single bar = 1 mm. Drawn by Susanna Stuart-Smith (after Schlechter and Vermeulen ined.).

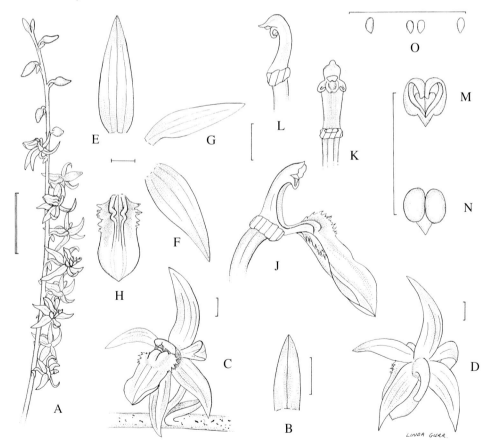

Fig. 130. *Dendrochilum hamatum*. **A** inflorescence; **B** floral bract: **C** portion of rachis, floral bract and flower, oblique view; **D** flower, back view; **E** dorsal sepal; **F** lateral sepal; **G** petal; **H** labellum, front view; **J** pedicel with ovary, labellum and column, anther-cap removed; **K** column, anther-cap removed, front view; **L** column, anther-cap removed, side view; **M** anther-cap, interior view; **N** anther-cap, back view; **O** pollinia. **M–O** from *Bogor* cult. 980.I.669. Scale: single bar = 1 mm; double bar = 1 cm. Drawn by Linda Gurr.

3–4(–5) mm apart; peduncle and rhachis sparsely ramentaceous; peduncle *c*. 1.7–4 cm long; non-floriferous bracts absent; rachis *c*. 3.5–5.5 cm long; floral bracts narrowly triangular-ovate to elliptic, acute, 1.8–2 mm long, very sparsely ramentaceous. *Flowers* glabrous, wide-opening, sepals and petals pale lemon-yellow, whitish at base, labellum darker lemon-yellow, deeper mustard-yellow proximally, disc whitish, keels palest lemon, column lemon-yellow, stelidia and margin of apical hood white. *Pedicel with ovary* clavate, 2.5–3.5 mm long. *Sepals* and *petals* 3–nerved. *Dorsal sepal* narrowly elliptic or oblong-elliptic, acute, 5 × 1–1.5 mm. *Lateral sepals* slightly obliquely narrowly elliptic, acute, 4.2–5 × *c*. 1.2–1.8 mm. *Petals* obliquely narrowly elliptic or ligulate, slightly falcate, acute, margin very minutely serrulate, 4–4.2 × 0.9–1.2 mm. *Labellum* oblong-ligulate to subpandurate, narrowest at the middle, subacute, proximal margins irregularly toothed, *c*. 2.5–3.2 × *c*. 1.8–2 mm, very shortly clawed, claw *c*. 2 mm long, blade sharply deflexed at claw, 3–nerved; disc with 2 short basal keels which are not united at the base. *Column* slender, curved, 2.1–2.5 mm long; foot prominent; apical hood elongate, rounded, entire or very shortly 5–toothed; stelidia borne either side of stigmatic cavity at base of hood, deflexed, subulate, abruptly upcurved distally and hamate; stigmatic cavity narrowly oblong; anther-cap *c*. 0.3–0.4 × 0.3–0.4 mm, cordate, cucullate.

DISTRIBUTION. Borneo.

SARAWAK. Sri Aman District, near Undup, Sept. 1865, *Beccari* 476 (holotype B, destroyed, isotype FI!).

KALIMANTAN BARAT. Locality unknown, comm. Jan. 1999, *Bogor* cult. 980.I.669 (BO, K!), 995.IX.30 (BO, K!) & 995.IX.301 (BO, K!).

HABITAT. Not recorded.

D. hamatum was, until recently, known only from the type collected over 130 years ago by Odoardo Beccari in southern Sarawak close to the border with Kalimantan. Cribb, however, has recently photographed a plant in cultivation at Bogor Botanic Garden originating from West Kalimantan which is clearly referable to this species. Fresh inflorescences in alcohol have become available for study, and examination of the isotype located in Florence herbarium (FI), Schlechter's sketches and description together with colour slides taken in Bogor by Cribb, confirm its affinity with *D. pubescens*. Both species share a similar habit, labellum morphology and flower colour. *D. hamatum*, however, lacks the distinctive leaf and floral pubescence of *D. pubescens* and has smaller flowers with more distinctly hamate stelidia on the column.

The specific epithet is derived from the Latin *hamatus*, barbed, hooked at the tip, and refers to the shape of the stelidia.

80. DENDROCHILUM PUBESCENS

Dendrochilum pubescens L.O. Williams in *Bot. Mus. Leafl. Harv. Univ.* 6: 58–59 (1938). Type: Malaysia, Sarawak, Marudi District, Tama Abu Range (Mount Temabok), Upper Baram (sphalm. Barami) Valley, 900 m, 8 Nov. 1920, *Moulton* SFN 6763 (holotype AMES!, isotype SING!). **Fig. 131, plate 14C & D.**

DESCRIPTION. Epiphyte. *Rhizome* elongate, up to 45 cm long and 0.4 cm in diameter, simple or with an occasional short branch, covered when young with persistent, pale brown, speckled darker brown sheaths. *Roots* wiry, flexuous, with a few branches, minutely papillose,

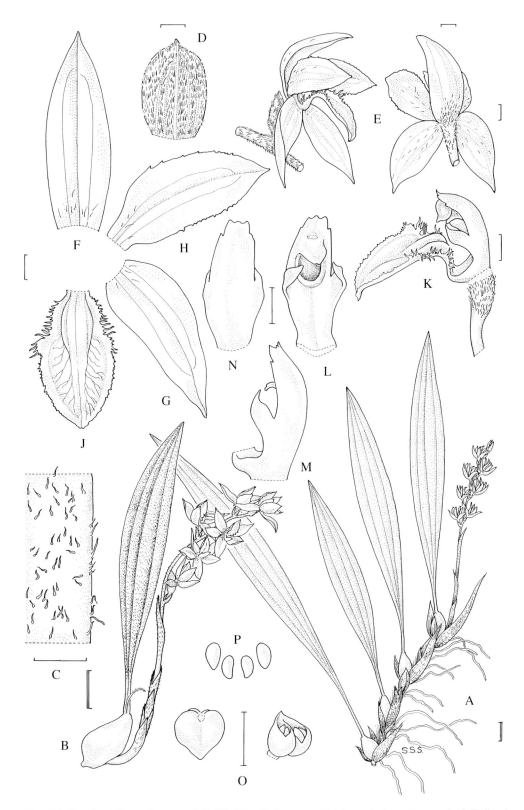

Fig. 131. *Dendrochilum pubescens*. **A & B** habits; **C** close-up of trichome indumentum on leaf; **D** floral bract; **E** flowers, side and back views; **F** dorsal sepal; **G** lateral sepal; **H** petal; **J** labellum, front view; **K** pedicel with ovary, labellum and column, side view; **L** column, anther-cap removed, front view; **M** column, anther-cap removed, side view; **N** column, back view; **O** anther-cap, back and interior views; **P** pollinia. **A** from *Johns* 7341, **B–P** from *Thomas* 197. Scale: single bar = 1 mm; double bar = 1 cm. Drawn by Susanna Stuart-Smith.

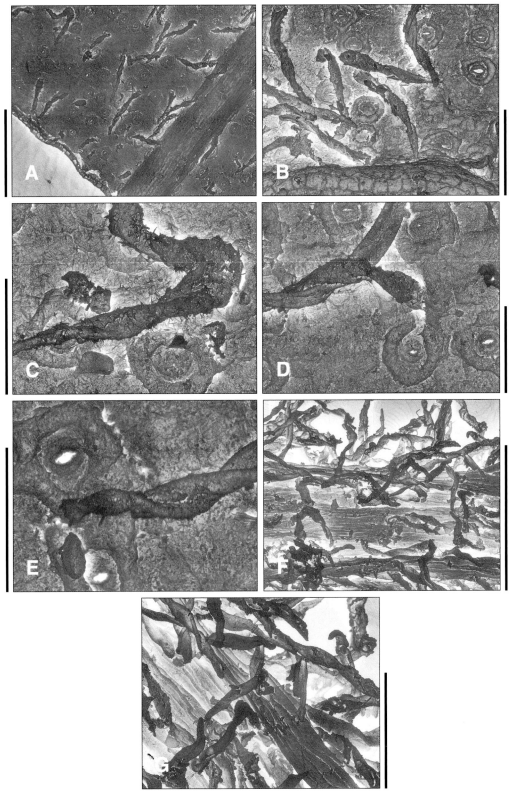

Fig. 132. *Dendrochilum pubescens* (*Thomas* 197, K): **A** abaxial surface of leaf showing trichomes; **B** margin of leaf, abaxial surface showing stomata and trichomes; **C** two-branched v-shaped trichome; **D** two unbranched trichomes; **E** unbranched trichome; **F** portion of rachis showing dense covering of trichomes; **G** close-up of rachis trichomes. **A & F**: scale bar equals 500 μm; **B & G**: scale bar equals 200 μm; **C–E**: scale bar equals 100 μm.

hirsute in places, *c.* 0.3–2 mm in diameter. *Cataphylls* 4–5, *c.* 1.8–6 cm long, ovate-elliptic, acute, shortly black or brown furfuraceous-pubescent. *Pseudobulbs* borne 0.5–5 cm apart, ovoid, somewhat compressed, smooth, 0.5–2.5 × 1 cm. *Leaf:* petiole canaliculate, pubescent, 0.6–4 cm long; blade oblanceolate or narrowly elliptic, acute to acuminate, constricted below apex, thin-textured but stiff, 4–22 × 0.5–3.5 cm, 7–nerved, 3 of which are prominent, shortly black or brown furfuraceous pubescent on abaxial surface, slightly less so on adaxial surface. *Inflorescence* erect at first, becoming gently curved or pendulous, laxly 3– to 10–flowered, flowers borne 2–6 mm apart; peduncle, rachis and floral bracts shortly black or brown pubescent; peduncle 5–14 cm long; non-floriferous bracts absent; rachis quadrangular, 2–5 cm long; floral bracts ovate, apiculate, 3–5 mm long. *Flowers* green to brownish ochre or translucent yellow, labellum green to orange-brown, column pale green or white. *Pedicel with ovary* clavate, curved, 3 mm long, black or brown pubescent on ovary. *Sepals* and *petals* 3–nerved, somewhat fleshy, with a few sparse brown hairs mostly towards the base on both surfaces. *Dorsal sepal* oblong-elliptic, acute, concave and curving forward, 7–8 × 3–3.5 mm. *Lateral sepals* slightly obliquely ovate-elliptic, acute, 6.9–7 × 3 mm. *Petals* oblong-elliptic, acute, shortly clawed, erose, 6–6.8 × 2.5–3 mm. *Labellum* 3–nerved, lateral nerves with numerous secondary nerves, very shortly clawed, oblong-ovate, acute, entire, margin lacerate proximally, erose-dentate distally, fleshy, thin-textured along margin, shallowly concave, minutely papillose, 5–5.2 × 2.5–3 mm; disc fleshy, with 2 lateral keels, not united at base. *Column* gently curving, 3–3.8 mm long; foot 1 mm long; apical hood oblong, shortly tridentate or unevenly toothed; stelidia borne just below stigmatic cavity, falcate, acute, a little incurved, shorter than apical hood, 1–1.1 mm long; stigmatic cavity oblong, lower margin thickened and rather swollen; rostellum large, ovate-triangular, obtuse to acute; anther-cap cordate-cucullate, *c.* 0.9 × 1 mm. *Capsule* glabrous, ellipsoid, 2 × 0.8–1 cm, dull greenish brown.

DISTRIBUTION. Borneo.

SARAWAK. Marudi District, Tama Abu Range (Mount Temabok), Upper Baram Valley, 8 Nov. 1920, *Moulton* SFN 6763 (holotype AMES!, isotype SING!). Kapit District, Hose Mountains, northern part, base of ridge leading to Bukit Batu, 11 Dec. 1991, *Leiden* cult. (*de Vogel*) 914389 (L!).

BRUNEI. Temburong River Valley, along ridges above helicopter pad, LP286, to unused ridge top pad and above this on ridge towards Mount Retak, 26 April 1992, *Johns* 7341 (BRUN, K!). Locality unknown, *Leiden* cult. (*de Vogel*) 27686 (L!). Belait Melilas, Ingei River, Operation Raleigh path from camp at Ingei River to hot springs, 7 Dec. 1992, *Thomas* 197 (BRUN, K!). Batu Melintang to hot spring and first waterfall, 4 Jan. 1989, *de Vogel* 8888 (L!).

HABITAT. Mixed dipterocarp ridge forest to 40 metres high on loamy substrate; mixed dipterocarp forest turning to kerangas; lower montane dipterocarp forest; on tree trunks near waterfalls. Sea level–900 m.

D. pubescens is at once distinguished from other mainland Bornean species by the dense brown to black pubescence on the leaves, sheaths, inflorescence and sepals.

81. DENDROCHILUM VESTITUM

Dendrochilum vestitum J.J. Sm. in *Reprium. Spec. nov. Regni veg.* 30: 331–332 (1932). Type: Indonesia, Bunguran Island, southern summit of Mount Ranai, 800–1000 m, *van Steenis* 1435 (holotype L!, isotype BO!). **Fig. 133.**

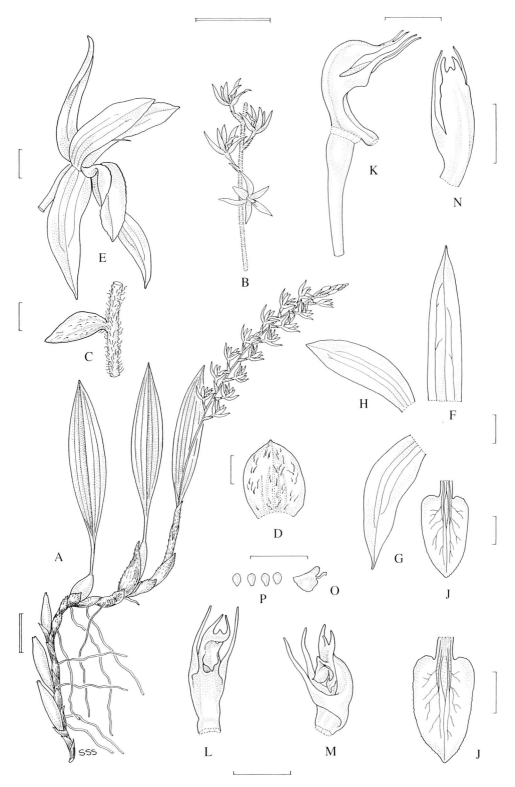

Fig. 133. *Dendrochilum vestitum*. **A** habit; **B** lower portion of inflorescence; **C** portion of rachis with floral bract; **D** floral bract; **E** flower, side view; **F** dorsal sepal; **G** lateral sepal; **H** petal; **J** labellum, front view; **K** pedicel with ovary and column, side view; **L** column, anther-cap removed, oblique view; **M** column, oblique view; **N** column, back view; **O** anther-cap, side view; **P** pollinia. **A–P** from *van Steenis* 1435 (holotype). Scale: single bar = 1 mm; double bar = 1 cm. Drawn by Susanna Stuart-Smith.

DESCRIPTION. Epiphyte. *Rhizome* elongate, *c.* 10–15 cm long, 0.2–0.3 cm in diameter, branching, covered when young with persistent greyish brown sheaths. *Roots* wiry, flexuous, branching, smooth or minutely papillose, *c.* 0.5–1.8 mm in diameter. *Cataphylls* 4–5, 0.4–3 cm long, ovate-elliptic, obtuse to shortly apiculate, black pubescent. *Pseudobulbs* borne 1–1.5 cm apart, conical to fusiform, obtuse, 1.3–1.5 × 0.9 cm. *Leaf*: petiole canaliculate, black-pilose when young, 0.4–1.6 cm long; blade narrowly elliptic, obtuse to subacute, narrowly cuneate below, coriaceous, 3–5.5 × 0.8–1 cm, main nerves 5, clothed in a short, sparse black pubescence when young, less so when mature. *Inflorescence* erect or slightly curved, laxly 1- to 20–flowered, flowers borne 2.5– 3 mm apart; peduncle and rachis densely clothed in an adpressed black-pilose pubescence; peduncle 3.5–4 cm long; non-floriferous bracts absent; rachis quadrangular, 3.7–4 cm long; floral bracts sparsely brown furfuraceous-pubescent on exterior, ovate, shortly apiculate, 2.2–2.3 × 2.4 mm. *Flowers* with a white labellum flushed light green at base. *Pedicel with ovary* narrowly clavate, almost straight, glabrous, 2 mm long. *Sepals* and *petals* 3–nerved, glabrous. *Dorsal sepal* oblong-lanceolate, acute, concave, incurved, 5.3 × 1.7–1.8 mm. *Lateral sepals* obliquely oblong-lanceolate, subfalcate, acute, 5 × 1.3 mm. *Petals* porrect, narrowly obliquely oblong, subfalcate, minutely apiculate, sometimes obscurely erose, concave, 4.5 × 1.5 mm. *Labellum* 3–nerved, lateral nerves branching, entire, narrowly cordate-ovate, unguiculate, shortly acute, rounded at base, serrulate to obscurely erose, slightly concave, recurved above base, glabrous, 3.4–3.5 × 2.4 mm, claw 0.3 × 0.5 mm; disc 3–keeled, keels borne on claw just above base, terminating on lower half of blade, median keel broader than laterals, free at base. *Column* curved, 2.5 mm long; foot distinct; apical hood subquadrangular, tridentate, median tooth much shorter than lateral teeth; stelidia borne just below stigmatic cavity, linear-subulate to filiform, acute, slightly longer than apical hood; stigmatic cavity oval, margin elevate; rostellum ovate-triangular, recurved; anther-cap cucullate, obtuse, 0.5 × 0.5 mm.

DISTRIBUTION. Borneo.

BUNGURAN ISLAND. Southern summit of Mount Ranai, 15 April 1928, *van Steenis* 1435 (holotype L!, isotype BO).

HABITAT. Mossy forest; on cliffs. 800–1000 m.

D. vestitum is distinguished from the related *D. pubescens* by its smaller leaves and more numerous flowers which have a distinctly clawed labellum and column with a deeply toothed apical hood and longer stelidia.

The specific epithet is derived from the Latin *vestitus*, clothed, and refers to the pubescence, particularly on the inflorescence.

Hybrids

Morphological and distributional evidence suggests that many hybrid taxa exist on Mount Kinabalu. Hybridisation may be common at the higher elevations on the mountain and could partly explain much of the current diversity found there today (Barkman, PhD dissertation, 1998). See also under *D. acuiferum*.

A specimen of intermediate morphology from Mount Kinabalu appears to be of hybrid origin. Barkman (pers. comm.) notes *D. exasperatum* growing on the ridge above it and *D. gibbsiae* growing below.

Dendrochilum gibbsiae *Rolfe* × **D. exasperatum** *Ames*

SABAH. Mount Kinabalu, Power Station, 12 Jan. 1996, *Barkman* 223 (K!, Sabah Parks Herbarium, Kinabalu Park!). **Fig. 134, plate 17E.**
HABITAT. Montane forest. 1700–1900 m.

Another specimen from Mount Kinabalu appears intermediate between *D. pterogyne* and *D. transversum*, having the longer column of the former, but the broad labellum of the latter.

Dendrochilum pterogyne Carr × **D. transversum** Carr

SABAH. Mount Kinabalu, *R. Beaman* 94032811 (K!). **Fig. 135.**
HABITAT. Unknown.

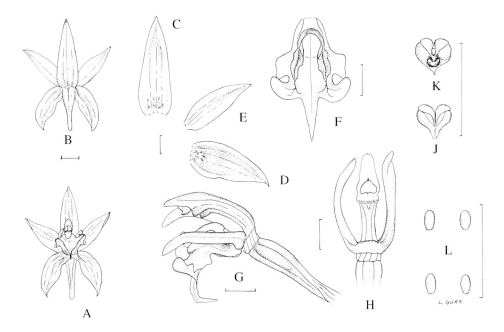

Fig. 134. *Dendrochilum gibbsiae* × *D. exasperatum*. **A** flower, front view; **B** flower, back view; **C** dorsal sepal; **D** lateral sepal; **E** petal; **F** labellum, front view; **G** pedicel with ovary, labellum and column, side view; **H** column, anther-cap removed, front view; **J** anther-cap, interior view; **K** anther-cap, back view; **L** pollinia. A–L from *Barkman* 223. Scale: single bar = 1 mm. Drawn by Linda Gurr.

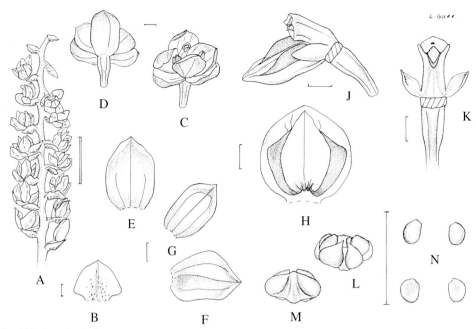

Fig. 135. *Dendrochilum pterogyne* × *D. transversum*. **A** inflorescence; **B** floral bract; **C** flower, front view; **D** flower, back view; **E** dorsal sepal; **F** lateral sepal; **G** petal; **H** labellum, front view; **J** pedicel with ovary, labellum and column, side view; **K** pedicel with ovary and column, front view; **L** anther-cap, interior view; **M** anther-cap, back view; **N** pollinia. A–N from *R. Beaman* 94032811. Scale: single bar = 1 mm; double bar = 1 cm. Drawn by Linda Gurr.

Incompletely Known Taxon

Dendrochilum trilobum (Ridl.) Ames in Merr. in *J. Straits Brch. R. Asiat. Soc. Special no.*: 149 (1921). Type: Sarawak, Kuching District, Mount Penrissen, 1150 m., *Moulton* s.n. (holotype not located).

Platyclinis triloba Ridl. in Sarawak Mus. J. 1 (2): 35 (1912).

Pseudobulbs approximate, cylindric conic, one inch long, covered with large lanceolate, acute sheaths. *Leaves* lanceolate, linear, narrowed at both ends, 10 inches long by 1/4 inch wide or smaller, narrowed gradually into the petiole which is hardly distinct. *Raceme* with the leaf very slender, about 8 inches, flowers very small, rather distant. *Bracts* 1/12 inch long, lanceolate, acute. *Flowers* 1/8 inch across. *Sepals* lanceolate, acute. *Petals* smaller. *Lip* 3–lobed, base with two thick raised ridges or wings, limb 3–lobed, broader than long, side lobes oblong, rounded, mid-lobe narrow, linear, acute. A dark coloured spot at each side of the base of the side lobes. *Column* tall, stelidia from the base, large, linear, obtuse, petaloid. Margin of clinandrium ovate, longer than the anther.

The brief and rather inadequate description provided by Ridley, reproduced above, indicates similarities with *D. gibbsiae* and the two species may be conspecific. Unfortunately, confirmation of this will have to await examination of the type, if located. Naturally, Ridley's epithet will have priority if this is so.

Excluded Taxa

Dendrochilum cornutum Blume, *Bijdr.*: 399 (1825). Type: Indonesia, Java, *Blume* s.n. (holotype L).

This species, distributed in Sumatra, Java and the Lesser Sunda Islands, was cited from Borneo by J.J. Smith (1905). Much later he excluded Borneo from its distribution in his enumeration of Sumatran orchids (1933). I have seen no examples from Borneo.

Dendrochilum latifolium Lindl. in *Edwards' Bot. Reg.* 29: 56 (1843). Type: Philippines, Luzon, sine coll., cult. *Loddiges* s.n. (holotype K–LINDL!).

The provenance of an alcohol collection (*Geesink* s.n.) of this species prepared from living material cultivated at the Hortus Botanicus, Leiden (Leiden cult. no. 20227) is given as Kalimantan Timur, N. part. Details of locality and habitat are not known. I have not seen any authentic material of *D. latifolium* from Borneo and suspect that this represents an error in provenance data. See Fig. 136 A–L.

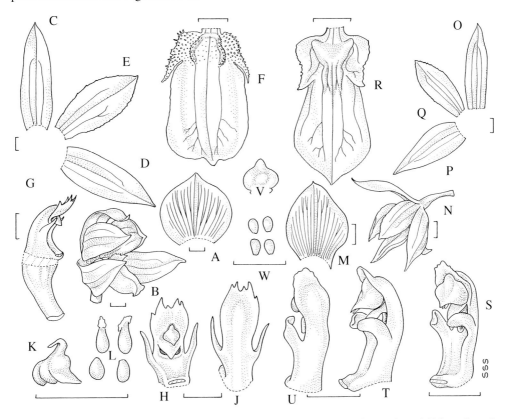

Fig. 136. *Dendrochilum latifolium.* **A** floral bract; **B** flower, oblique view; **C** dorsal sepal; **D** lateral sepal; **E** petal; **F** labellum, front view; **G** pedicel with ovary and column, side view; **H** column, oblique view; **J** column, back view; **K** anther-cap, side view; **L** pollinia. *D. uncatum.* **M** floral bract; **N** flower, side view; **O** dorsal sepal; **P** lateral sepal; **Q** petal; **R** labellum, front view; **S** column, oblique view; **T** column, side view; **U** column, back view; **V** anther-cap, back view; **W** pollinia. **A–L** from *Leiden* cult. (*Geesink*) 20227, **M–W** from *Leiden* cult. 21135. Scale: single bar = 1 mm. Drawn by Susanna Stuart-Smith.

Dendrochilum linearifolium Hook. f. in *Hooker's Icon. Pl.* 3, 9: pl. 1859 (1889). Types: Peninsular Malaysia, Perak, Batang Padang, *c.* 1500 m, *Wray* 1512 (lectotype K); Chosang, *Scortechini* 259 (syntype K); locality unknown, *Scortechini* 882 (syntype K).

This species is widespread throughout the mountains of Peninsular Malaysia where it grows on bare rocks in exposed places between 1200 and 2100 m. It is also recorded from Sumatra. There is some variability in labellum shape and the point of insertion of the stelidia. The type material includes specimens with stelidia borne at the base of the column as well as just below the stigmatic cavity. Epiphytic populations with broad leaves and larger flowers have been distinguished as *Platyclinis pulchella* by Ridley. I have seen no authentic material from Borneo and reports of *D. linearifolium* from Borneo probably refer to *D. galbanum* or forms of *D. tenompokense*.

Dendrochilum simile Blume, *Bijdr.*: 400 (1825). Types: Indonesia, Java, Buitenzorg Province, *Blume* s.n. (syntype L); Java, Tjanjor Province, *Blume* s.n. (syntype, not located).

D. simile is distributed in Peninsular Malaysia, Sumatra, Java and the Lesser Sunda Islands. I have seen no authentic material from Borneo.

There is a crude sketch of a flower in the Kew illustration collection, drawn and later presented to Kew by H.N. Ridley between 1935 and 1936, of a collection by Hewitt from Kuap (Quop), Sarawak made in 1907. Ridley provisionally determined this as *Platyclinis similis* (Blume) Benth. ex Ridl., which was later confirmed by Kew orchidologist V.S. Summerhayes. The flower colour is given as whitish becoming yellow. Ridley's sketch, however, almost certainly represents *D. havilandii* which is recorded from Kuap.

Dendrochilum uncatum Rchb. f. in *Bonplandia* 3: 222 (1855). Type: Philippines, Luzon, Zambales Province, *Cuming* 2073 (holotype Herb. Rchb. f. 49080, W, isotypes BM, K, L, P, SING, W).

D. uncatum, a familiar species in cultivation, is native to Taiwan and the Philippines. The provenance of an alcohol collection (*Franken & Roos*) of this species prepared from living material cultivated at the Hortus Botanicus, Leiden (Leiden cult. no. 21135) is given as Kalimantan Timur, neighbourhood of Balikpapan, alt. 50–200 m. De Vogel (pers. comm.) relates how Franken & Roos rescued a collection of orchids made by Academica, a private collector, from Balikpapan, who obtained his specimens from logged forests in the neighbourhood. He may also have had cultivated specimens from other non Bornean provenances. *D. uncatum* is usually found at higher elevations and, since I have never seen any authentic material from Borneo, I suspect the provenance is probably erroneous. See Fig. 135 M–W.

ACKNOWLEDGEMENTS

I am grateful for the help of many people during the preparation of this revision, in particular, Dr. Phillip Cribb, Assistant Keeper of the Kew Herbarium and Dr. Ed de Vogel (Rijksherbarium, Leiden, The Netherlands) who kindly read through the manuscript and provided useful comments and constructive criticism. Dr. Todd Barkman (formerly of University of Texas, Austin, now at Western Michigan University, Kalamazoo) very generously provided access to his pioneering work on the evolution of section *Eurybrachium*, which has helped clarify phylogenetic relationships within the genus, and whose results are summarised in this revision. I also wish to thank Dr. Henrik Pedersen (Botanical Museum, University of Copenhagen, Denmark) for fruitful collaboration, particularly over the sectional subdivision of *Dendrochilum*.

I am most thankful to Mr Tan Jiew Hoe (Singapore) for his generous financial support in making the inclusion of many more colour plates possible.

I appreciate the assistance of the directors and staff of various herbaria for the loan of specimens, particularly, AMES, BM, BO, E, HBG, L, SAN, SAR and SING. I would also like to thank Dr. Chiara Nepi and Giorgio Padovani for their assistance while working on the Beccari collection in Florence (FI) herbarium.

A special thank you must go to Susanna Stuart-Smith who patiently illustrated most of the species included in this revision. Thanks are also due to artists Eleanor Catherine, Maureen Church, Linda Gurr, and Oliver Whalley.

The colour photographs have been kindly provided by Todd J. Barkman, Kath Barrett, C.L. Chan, James B. Comber, Phillip Cribb, Gerald Cubitt, Paul Davies, Hans P. Hazebroek, Peter Jongejan, Anthony Lamb, Gwilym Lewis, Robert New, Cede Prudente, Takashi Sato, and André Schuiteman.

Teofila Beaman provided help with Sarawak locality names and their correct usage.

Maureen Bradford kindly undertook the tedious task of typing several drafts of the manuscript. Cheng Jen Wai assisted in the formatting and layout.

Last, but by no means least, I thank C.L. Chan for seeing the book through the press and for maintaining such a high quality of production throughout.

REFERENCES

Ames, O. (1908). Orchidaceae: illustrations and studies of the family Orchidaceae III. The Merrymount Press, Boston.

——————— (1920). Orchidaceae: illustrations and studies of the family Orchidaceae VI. The Merrymount Press, Boston.

——————— (1922). Orchidaceae: illustrations and studies of the family Orchidaceae VII. The Merrymount Press, Boston.

——————— (1937). Additions to the genus *Acoridium*. *Blumea* Suppl. 1: 66–77.

Arnold, R.E. (1953). *Platyclinis. Orchid Rev*. 61: 166–168.

Bailey, L.H. (1916). The standard cyclopedia of horticulture; a discussion for the amateur, and the professional and commercial grower, of the kinds, characteristics and methods of cultivation of the species of plants grown in the regions of the United States and Canada for ornament, for fancy, for fruit and for vegetables V. Macmillan & Co., New York.

Barkman, T.J. (1998). Evolution of *Dendrochilum* subgenus *Platyclinis* section *Eurybrachium* investigated in a phylogenetic context. PhD dissertation, University of Texas at Austin.

Bechtel, H., Cribb, P. & Launert, E. (1981). The manual of cultivated orchid species. Blandford Press, Poole.

Bentham, G. (1883a). Orchideae. In: Bentham, G. & Hooker, J.D., Genera plantarum III: 460–636. L. Reeve & Co., London.

——————— (1883b). Cyperaceae. In: Bentham, G. & Hooker, J.D., Genera plantarum III: 1037–1073. L. Reeve & Co., London.

Benzing, D.H. & Clements, M.A. (1991). Dispersal of the orchid *Dendrobium insigne* by the ant *Iridomyrmex cordatus* in Papua New Guinea. *Biotropica* 23: 304–607.

Bernard, N. (1909). L'evolution dans la symbiose les orchidées et leurs champignons commensaux. *Annls. Sci. nat. (Botanique)* 9, 9: 1–196 & 4 pl.

Blowers, J. W. (1957). Methods of spike production in orchids. *Orchid Rev*. 65: 180–183.

——————— (1969). Golden chain orchids – *Dendrochilums. Orchid Rev*. 77: 154–156.

Blume, C. L. (1825–1827). Bijdragen tot de flora van Nederlandsch Indië. – Batavia.

Böckeler, O. (1879). Mittheilungen über Cyperaceen. *Flora, Jena* 62: 158–160.

Burbidge, F.W. (1900). *Platyclinis uncata. Garden (London)* 58: 127–128.

Burns-Balogh, P. (1989). A reference guide to orchidology. Koeltz Scientific Books, Koenigstein.

Butzin, F. (1974). Bestimmungsschlüssel für die in Kultur genommenen Arten der Coelogyninae (Orchidaceae). *Willdenowia* 7: 245–260.

——————— (1984–1986). Subtribus: Coelogyninae. In: Schlechter, R. 1970 ff., Die Orchideen. Ihre Beschreibung, Kultur und Züchtung I. 3rd ed. by F.G. Brieger, R. Maatsch & K. Senghas: 914–958. Verlag Paul Parey, Berlin & Hamburg.

Chadefaud, M. & Emberger, L. (1960). Traité de botanique (systématique) II (1–2). Les végétaux vasculaires. Masson & Cie., Paris.

Chan, C.L. & Barkman, T.J. (1997). A new species of *Calanthe* (Orchidaceae : Epidendroideae : Arethuseae) from Mount Kinabalu, Sabah. *Sandakania* 9: 27–34.

Chase, M.W. & Palmer, J.D. (1992). Floral morphology and chromosome number in subtribe *Oncidiinae* (Orchidaceae) : evolutionary insights from a phylogenetic analysis of chloroplast DNA restriction site variation. In Soltis, P.S., Soltis, D.E. & Doyle, J.J. (eds.), Molecular Systematics of Plants. Chapman & Hall, New York.

Chase, M.W. & Palmer, J.D. (1997). Leapfrog radiation in floral and vegetative traits among twig epiphytes in the orchid subtribe *Oncidiinae*. In Givnish, T.J. & Sytsma, K.J. (eds.), Molecular Evolution and Adaptive Radiation. Cambridge Univesity Press, New York.

Comber, J.B. (1990). Orchids of Java. Bentham-Moxon Trust, Royal Botanic Gardens, Kew.

Constantin, J. (1913). Atlas en couleur des Orchidées cultivées. Librairie Générale de l'Enseignement, Paris.

Curtis, C.H. (1950). Orchids—their description and cultivation. Putnam & Company, Ltd., London.

de Dalla Torre, C.G. & Harms, H. (1900–1907). Genera siphonogamarum ad systema Englerianum conscripta. G. Engelmann, Leipzig.

Darwin, C. (1877). The various contrivances by which orchids are fertilised by insects. 2nd ed. John Murray, London.

Docters van Leeuwen, W. (1929a). Kurze Mitteilung über Ameisen-Epiphyten aus Java. *Ber. dt. bot. Ges.* 47: 90–99.

———— (1929b). Mierenepiphyten. *Trop. Natuur* 18: 57–65, 131–139.

Dressler, R.L. (1981). The orchids. Natural history and classification. Harvard University Press, Cambridge, USA.

———— (1993). Phylogeny and classification of the orchid family. Cambridge University Press, Cambridge, U.K.

Endlicher, S.L. (1843). Mantissa botanica altera. Sistens generum plantarum supplementum tertium. Frideric Beck, Vienna.

Farr, E.R., Leussink, J.A. & Stafleu, F.A. (eds.) (1979). Index nominum genericorum (plantarum) 1, III. *Regnum veg.* 100: 1–26, 1–630; 102: 1277–1896.

Frolik, F. (1913). Aus österreichischen Privatgärten. Öst. Gartenztg. 8: 43–48.

Graf, A.B. (1963). Exotica 3. Pictorial Cyclopedia of Exotic Plants. Roehrs Company, Rutherford.

Greatwood, A. (1973). Hints and tips for the amateur. *Orchid Rev.* 81: 114–117.

Greuter, W., Brummitt, R.K., Farr, E., Kilian, N., Kirk, M. & Silva, P.C. (eds.) (1993). NCU – 3. Names in current use for extant plant genera. *Regnum veg.* 129: i–xxvii, 1–1464.

Gunn, C.R., Wiersema, J.H., Ritchie, C.A. & Kirkbride, J.H. (1992). Families and genera of spermatophytes recognized by the Agricultural Research Service. *Tech. Bull. U.S. Dep. Agric.* 1796: 1–499.

Hapeman, J.R. & Inoue, K. (1997). Plant-pollinator interactions and floral radiation in *Platanthera* (Orchidaceae). In Guvnish, J.T. & Sytsma, K.J. (eds.), Molecular Evolution and Adaptive Radiation. Cambridge University Press, New York.

Hemsley, W.B. (1881). List of garden orchids. *Gdnrs'. Chron.* 2, 16: 656.

Hiroe, M. (1971). Orchid flowers I–II. Kyoto-Shoin Co., Kyoto.

Hoffmann, K.M. (1929). Cytologische Studien bei den Orchidaceen. *Ber. dt. bot. Ges.* 47: 321–326.

———— (1930). Beiträge zur Cytologie der Orchidaceen. *Planta* 10: 523–595.

Holloway, J.D. (1996). Butterflies and moths. In Wong, K.M. & Phillipps, A. (eds.), Kinabalu: Summit of Borneo. The Sabah Society, Kota Kinabalu.

Holttum, R.E. (1954). A revised flora of Malaya. An illustrated systematic account of the Malayan flora, including commonly cultivated plants II. Ferns of Malaya. Botanic Gardens, Singapore.

————— (1957). A revised flora of Malaya. An illustrated systematic account of the Malayan flora, including commonly cultivated plants I. Orchids of Malaya. 2nd ed. Botanic Gardens, Singapore.

————— (1964). A revised flora of Malaya. An illustrated systematic account of the Malayan flora, including commonly cultivated plants I. Orchids of Malaya. 3rd ed. Botanic Gardens, Singapore.

Hooker, J.D. (1886–1890). The flora of British India 5. Chenopodiaceae to Orchideae. L. Reeve & Co., Ashford.

Hsieh, A. (1955). An enumeration of the Formosan Orchidaceae. *Q. Jl. Taiwan Mus.* 8: 213–282.

Hu, S.Y. (1975). The orchidaceae of China (IX). *Q. Jl. Taiwan Mus.* 28: 125–182.

Kaiser, R. (1993). The scent of orchids. Olfactory and chemical investigations. Elsevier & Editiones Roche, Amsterdam & Basel.

Kishino, H. & Hasegawa, M. (1989). Evaluation of the maximum likelihood estimate of the evolutionary tree topologies from DNA sequence data, and the branching order in Hominoidea. *J. Mol. Evol.* 29: 170–179.

Kitayama, K. (1991). Vegetation of Mount Kinabalu Park, Sabah, Malaysia. Map of physiognomically classified vegetation. East-West Center, Honolulu, Hawaii.

————— (1997). Patterns of species diversity on an oceanic versus a continental island mountain: a hypothesis on species diversification. *J. Vegetation Sci.* 7: 879–888.

Kränzlin, F. (1908). *Dendrochilum glumaceum* Lindl. *Orchis* II: 37–38, pl. 14.

de Laubenfels, D.J. (1988). Coniferales. Flora Malesiana, series I, 10L: 337–453.

Lawler, L.J. & Slaytor, M. (1970). The distribution of alkaloids in orchids from the territory of Papua and New Guinea. *Proc. Linn. Soc. N.S.W.* 94: 237–241.

Lee, D.W. & Lowry, J.B. (1980). Plant speciation on tropical mountains: *Leptospermum* (Myrtaceae) on Mount Kinabalu, Borneo. *Bot. J. Linn. Soc.* 80: 223–242.

Lim, K.–Y. (1985). The chromosomes of orchids at Kew – 3. Miscellaneous species. *Amer. Orchid Soc. Bull.* 54: 1234–1235.

Lüning, B. (1964). Studies on Orchidaceae alkaloids I. Screening of species for alkaloids I. *Acta chem. scand.* 18: 1507–1516.

Mansfeld, R. (1937). Über das System der Orchidaceae-Monandrae. *Notizbl. bot. Gart. Mus. Berl.* 13: 666–676.

————— (1955). Über die Verteilung der Merkmale innerhalb der Orchidaceae-Monandrae. *Flora, Jena* 142: 65–80.

Meinecke, E.P. (1894). Beiträge zur Anatomie der Luftwurzeln der Orchideen. *Flora, Jena* 78: 133–203 & 2 pl.

Miethe, E. (1913). Einige kulturwürdige Platyclinisarten. *Gartenwelt, Berl.* 17: 437–439.

Möbius, M. (1887). Ueber den anatomischen Bau der Orchideenblätter und dessen Bedeutung für das System dieser Familie. *Jb. wiss. Bot.* 18: 530–607, pl. 21–24.

Nilsson, L.A. (1978). Pollination ecology of *Epipactis palustris* (Orchidaceae). *Bot. Notiser* 131: 355–368.

Northen, R.T. (1962). Home orchid growing. 2nd ed. D. van Nostrand Company, Inc., Princeton.

Patton, J.L. & Smith, M.F. (1992). mt DNA phylogeny of Andean mice : a test of diversification across ecological gradients. *Evolution* 46: 174–183.

Pedersen, H.Ae. (1995). Anthecological observations on *Dendrochilum longibracteatum* – a species pollinated by facultatively anthophilous insects. *Lindleyana* 10: 19–28.

——————— (1997). The genus Dendrochilum (Orchidaceae) in the Philippines – a taxonomic revision. *Opera Bot.* 131: 1–205.

——————— Wood, J.J. & Comber, J.B. (1997). A revised subdivision and bibliographical survey of *Dendrochilum* (Orchidaceae). *Opera Bot.* 130: 1–85.

Pfitzer, E. (1877). Beobachtungen über Bau und Entwicklung der Orchideen. *Verh. naturh.-med. Ver. Heidelb.* 2, 2: 19–32.

——————— (1882). Grundzüge einer vergleichenden Morphologie der Orchideen. Carl Winter's Universitätsbuchhandlung, Heidelberg.

——————— (1888–1889). Orchidaceae. In: Engler, A. & Prantl, K. (eds.), Die natürlichen Pflanzenfamilien nebst ihren Gattungen und wichtigeren Arten, insbesondere den Nutzpflanzen 11(6): 52–220. Verlag von Wilhelm Engelmann, Leipzig.

——————— & Kränzlin, F. (1907). Orchidaceae-Monandrae-Coelogyninae. In: Engler, A., Das Pflanzenreich. Regni vegetabilis conspectus, 4, 50, 2 B 7. Verlag von Wilhelm Engelmann, Leipzig.

van der Pijl, L. & Dodson, C.H. (1966). Orchid flowers – their pollination and evolution. The Fairchild Tropical Garden & University of Miami Press, Coral Gables.

Porembski, S. & Barthlott, W. (1988). Velamen radicum micromorphology and classification of Orchidaceae. *Nord. J. Bot.* 8: 117–137.

Pridgeon, A.M. (1981). Absorbing trichomes in the Pleurothallidinae (Orchidaceae). *Amer. J. Bot.*, 68 (1): 64–71.

Rendle, A.B. (1904). The classification of flowering plants 1. Gymnosperms and monocotyledons. University Press, Cambridge.

Ridley, H.N. (1896). The Orchideae and Apostasiaceae of the Malay Peninsula. *J. Linn. Soc. (botany)* 32: 213–416.

——————— (1907). Materials for a flora of the Malayan Peninsula 1. Singapore.

——————— (1908a). On a collection of plants made by H.C. Robinson and L. Wray from Gunong Tahan, Pahang. *J. Linn. Soc. (botany)* 38: 301–336.

——————— (1908b). A list of the ferns of the Malay Peninsula. *J. Straits Branch Roy. Asiat. Soc.* 50: 1–59.

——————— (1924). The flora of the Malay Peninsula. IV – Monocotyledones. L. Reeve & Co., Ashford.

Roe, F.W. (1965). The geological relationship between Mount Kinabalu and neighbouring regions. *Proc. Roy. Soc. London* 161: 49–56.

Rolfe, R.A. (1904). The genus *Acoridium*. *Orchid Rev.* 12: 219–220.

Rosinski, M. (1992). Untersuchungen zur funktionelle Anatomie der Laubblattstrukturen epiphytischer Coelogyninae und Eriinae (Orchidaceae). Doctoral dissertation, Universität des Saarlandes, Saarbrücken.

Royal Botanic Gardens, Kew (1904). Hand-list of orchids cultivated in the Royal Botanic Gardens. 2nd ed. His Majesty's Stationery Office, London.

Royal Botanic Gardens, Kew (1896). Hand-list of orchids cultivated in the Royal Gardens. Her Majesty's Stationery Office, London.

Sander, D. (1969). Orchids and their cultivation. 7th ed. Blandford Press, London.

Schlechter, R. (1982). The Orchidaceae of German New Guinea. Translated and edited by R.S. Rogers, H.J. Katz, J.T. Simmons & D.F. Blaxell. The Australian Orchid Foundation, Melbourne.

——————— (1986). The Orchidaceae of the Celebes (1911) and the Orchidaceae of the island of Celebes (1925). Translated and edited by H.J. Katz & J.T. Simmons. Australian Orchid Foundation, Melbourne.

Schweinfurth, C.L. (1959). Key to the orchids. In: Withner, C.L. (ed.), The orchids. A scientific survey. *Chronica bot.* 32: 511–528 (i–ix, 1–648).

Seidenfaden, G. (1986). Orchid genera in Thailand XIII. Thirty-three epidendroid genera. *Opera Bot.* 89: 1–216.

———— & Wood, J.J. (1992). The Orchids of Peninsular Malaysia and Singapore. A revision of R.E. Holttum: Orchids of Malaya. Olsen & Olsen, Fredensborg.

Siebert, A. (1901). *Platyclinis glumacea* Bnthm. (*Dendrochilum glumaceum* Lindl.). *Gartenflora* 50: 131–133.

Smith, J.J. (1904). Uebersicht der Gattung *Dendrochilum* Bl. *Recl. Trav. bot. néerl.* 1: 52–80.

———— (1905). Die Orchideen von Java. Flora von Buitenzorg VI. E.J. Brill, Leiden.

———— (1933). Enumeration of the Orchidaceae of Sumatra and neighbouring islands. *Reprium. Spec. nov. Regni veg.* 32: 129–386.

———— (1934). Artificial key to the orchid genera of the Netherlands Indies, together with those of New Guinea, the Malay Peninsula and the Philippines. *Blumea* 1: 194–215.

Solereder, H. & Meyer, F.J. (1930). Systematische Anatomie der Monokotyledonen VI. Scitamineae – Microspermae. Verlag von Gebrüder Borntraeger, Berlin.

Stapf, O. (1894). On the Flora of Mount Kinabalu, in North Borneo. *Trans. Linn. Soc. London, Bot.* 4: 69–263, pl. 11–20.

van Steenis, C.G.G.J. (1965). Plant geography of the mountain flora of Mount Kinabalu. *Proc. Roy. Soc. London* 161: 7–38.

———— (1979). Plant-geography of east Malesia. *Bot. J. Linn. Soc.* 79: 97–178.

Stein, B. (1892). Stein's Orchideenbuch. Beschreibung, Abbildung und Kulturanweisung der empfehlenswetesten Arten. Verlag von Paul Parey (sine loc.).

Thurgood, F.W. (1931). *Platyclinis* notes. *Orchid Rev.* 39: 174–175.

Veitch, H.J. (1889). Manual of orchidaceous plants cultivated under glass in Great Britain V. *Masdevallia, Pleurothallis, Cryptophoranthus, Restrepia, Arpophyllum* and *Platyclinis*. James Veitch & Sons, Chelsea.

Velenovsky, J. (1907). Vergleichende Morphologie der Pflanzen II. Verlagsbuchhandlung von Fr. Rivnác, Prague.

Vermeulen, J.J. (1993). A taxonomic revision of *Bulbophyllum*, sections *Adelopetalum, Lepanthanthe, Macrouris, Pelma, Peltopus* and *Uncifera* (Orchidaceae). *Orchid Monographs* 7: 1–324 & 6 pl.

de Vogel, E.F. (1986). Revisions in Coelogyninae (Orchidaceae) II. The genera *Bracisepalum, Chelonistele, Entomophobia, Geesinkorchis* and *Nabaluia*. *Orchid Monographs* 1: 17–51, 62–82 & 3 pl.

———— (1988). Revisions in Coelogyninae (Orchidaceae) III. The genus *Pholidota*. *Orchid Monographs* 3: 1–118 & 6 pl.

———— (1989). Characters of Coelogyninae – their significance for the taxonomy of the subtribe. In: Vacharotayan, S. (ed.), Proceedings of the Sixth ASEAN Orchid Congress Seminar, Bangkok, Thailand, 10–12 November 1986: 103. Chuan Printing Press, Bangkok.

de Vogel, E.F., Schuiteman, A., Felëus, N. & Vogel, A. (1998). Hortus Botanicus Leiden Catalogue part 1 1998 Orchidaceae. Leiden University.

Wallace, A.R. (1863). On the physical geography of the Malay Archipelago. *J. Roy. Geogr. Soc.* 33: 217–234.

Watson, W. (1903). Orchids: their culture and management. 2nd ed. by H.J. Chapman. L.

Upcott Gill & Charles Scribner's Sons, London & New York.

Webster, P. (1992). The orchid genus book. A study guide for the orchid family. Patricia J. Webster, San Antonio.

Wehmer, C. (1911). Die Pflanzenstoffe botanisch-systematisch bearbeitet. Chemische Bestandteile und Zusammensetzung der einzelnen Pflanzenarten. Rohstoffe und Produkte. Phanerogamen. Verlag von Gustav Fischer, Jena.

Williams, B.S. (1885). The orchid-grower's manual, containing descriptions of the best species and varieties of orchidaceous plants. 6th ed. Victoria & Paradise Nurseries, London.

————— (1894). The orchid-grower's manual, containing descriptions of the best species and varieties of orchidaceous plants in cultivation. 7th ed. by H. Williams. Victoria & Paradise Nurseries, London.

Williams, L.O. (1951). A revision of *Dendrochilum* sect. *Acoridium*. *Philipp. J. Sci.* 80: 281–334.

Wirth, M. & Withner, C.L. (1959). Embryology and development in the Orchidaceae. In: Withner, C.L. (ed.), The orchids. A scientific survey. *Chronica bot.* 32: 155–188 (i–ix, 1–648).

Wood, J.J. (1997). Orchids of Borneo. Vol. 3. Dendrobium, Dendrochilum and others. The Sabah Society, Kota Kinabalu in association with The Royal Botanic Gardens, Kew.

Wood, J.J., Beaman, R.S. & Beaman, J.H. (1993). The plants of Mount Kinabalu 2. Orchids. Royal Botanic Gardens, Kew.

Wood, J.J. & Comber, J.B. (1995). Six new species of *Dendrochilum* from Sumatra. *Lindleyana* 10, 2: 57–67.

Wood, J.J. & Lamb, A. (1994). New orchid records from Borneo. *Amer. Orchid Soc. Bull.* 9: 1010–1015.

Zörnig, H. (1904). Beiträge zur Anatomie der Coelogyninen. *Bot. Jb.* 33: 618–741.

SOURCES OF GOOD PUBLISHED PHOTOGRAPHS OF BORNEAN DENDROCHILUM

Beaman, T.E., Wood, J.J., Beaman, R.S. & J.H. (2001). Orchids of Sarawak. Natural History Publications (Borneo) in association with The Royal Botanic Garden, Kew:
    D. dulitense – Fig. 36.
    D. gibbsiae – Plate 18E
    D. hamatum – Fig. 37.
    D. kingii var. *tenuichilum* – Plate 18F.
    D. oxylobum – Plate 19A.

Cubitt, G. & Payne, J. (1990, reprinted 1992). Wild Malaysia. The wildlife and scenery of Peninsular Malaysia, Sarawak and Sabah. New Holland (Publishers) Ltd., London:
    D. dewindtianum (as *D. sp.*) – p. 182 (top left).

Lamb, A. (1979). The Wild Orchids of Sabah. 2. The Highlands. Nature Malaysiana 4(1):
    D. graminoides (as *D. sp.*) – p. 26 (top left).
    D. lacinilobum (as *D. sp.*) – p. 29 (bottom right).

Moore, W. & Cubitt, G. (1995). This is Malaysia. New Holland (Publishers) Ltd., London:
    D. dewindtianum (as unnamed orchid) – p. 97 (bottom centre).

Sato, T. (1991). Flowers and plants of Mount Kinabalu. Chetsu Co. Ltd., Toyama, Japan:
    D. dewindtianum (as *D. sp.*) – p. 38 (top).
    D. grandiflorum (as *D. perspicabile*) – p. 36 (top).
    D. kamborangense (as *D. sp.*) – p. 37 (bottom).
    D. pterogyne (as *D. grandiflorum*) – p. 35 (bottom).
    D. pterogyne (as *D. sp.*) – p. 37 (top).
    D. stachyodes – p. 36 (bottom).

de Vogel, E., Schuiteman, A., Felëus, N. & Vogel, A. (1998). Hortus Botanicus Leiden Catalogue part 1 1998 Orchidaceae:
    D. crassilabium – p. 154, Plate 87
    D. dewindtianum (as *D. furfuraceum*) – p. 154, Plate 88
    D. kingii var. *tenuichilum* (as *D. kingii*) – p. 155, Plate 89
    D. minimiflorum – p. 155, plate 90

   D. muluense – p. 155, Plate 91
   D. oxylobum – p. 155, Plate 92
   D. pubescens – p. 156, Plate 93

Wood, J.J., Beaman, R.S. & Beaman, J.H. (1993). The Plants of Mount Kinabalu 2. Orchids. Royal Botanic Gardens, Kew:
   D. alpinum – Plate 41A
   D. crassum – Plate 41B & C
   D. dewindtianum – Plate 41D
   D. gibbsiae – Plate 42A
   D. grandiflorum – Plate 42B
   D. haslamii var. *haslamii* – Plate 42C
   D. imbricatum – Plate 42D & E
   D. pterogyne (as *D. alatum*) – Plate 40C & D
   D. stachyodes – Plate 43A & B

Wood, J.J. & Cribb, P.J. (1994). A Checklist of the Orchids of Borneo. Royal Botanic Gardens, Kew:
   D. hologyne – Plate 9C
   D. muluense – Plate 9D
   D. pachyphyllum – Plate 9E

Wood, J.J. (1997). Orchids of Borneo Volume 3. The Sabah Society, Kota Kinabalu in association with Royal Botanic Gardens, Kew:
   D. anomalum – Plate 9B
   D. crassum – Plate 9C & D
   D. cruciforme var. *cruciforme* – Plate 10A
   D. sp. aff. *cruciforme* var. *cruciforme* – back cover
   D. cupulatum – Plate 10B
   D. gibbsiae – Plate 10C & D
   D. grandiflorum – Plate 11A & B
   D. haslamii var. *haslamii* – Plate 11C
   D. hologyne – Plate 11D
   D. imbricatum – Plate 12
   D. kingii var. *kingii* – Plate 13A
   D. lacinilobum – Plate 13B
   D. muluense – Plate 13C & D
   D. ochrolabium – Plate 14A
   D. oxylobum – Plate 14B & C
   D. pachyphyllum – Plates 14D & 15
   D. planiscapum – Plate 16A
   D. pubescens – Plate 16B & C
   D. scriptum – Plates 16D & 17A
   D. stachyodes – Plates 17B & 18A

INDEX TO NUMBERED
BORNEAN COLLECTIONS

Type collections are indicated in **bold**. Species numbers appear in brackets.

Aban: SAN 95230 (69b); SAN 95241 (69b – mixed collection with 78); SAN 95243 (69b).
Amdjah: Bogor cult. no. 108 (74).
Amin et al.: SAN 111793 (32a); SAN 129373 (40).
Anderson, J.A.R.: S.11099 (48); S.30808 (74); S.30838 (9).
Anderson, J.W.: 173 (71a); 196 (67).
Aningguh in *Lamb*: AL856/87 (37).
Argent: 111 (5).
Argent & Coppins: 1126 (16); 1202 (51).
Argent & Jermy: 988 (9).
Ashton: A.205 (56); A.211 (43); A.250 (71b).
Asik: SAN 127822 (69a).
Awa & Lee: **S.47676** (74); S.47827 (39); S.50586 (62); S.50704 (73); S.50731 (3); **S.50754** (12); S.50756 (58); S.50774 (35 – mixed collection with 36); S.50792 (78); S.50812 (39); S.50900 (45); S.50937 (39); S.51102 (43).
Awa et al.: S.50416 (4).
Bacon: 78 (74); 80 (32a); 93 (74); 183 (49).
Bailes & Cribb: 695 (64); 722 (9); 841 (19).
Banyeng & Paie: S.45050 (6a).
Barkman: 1 (60); 2 (22); 3 (22); 4 (23); 6 (23); 7 (5); 8 (28); 9 (28); 10 (10); 11 (19); 12 (19); 19 (22); 20 (22); 49 (44); 51 (9); 63 (25); 64 (64); 142 (23); 193 (29); 194 (13b); **198** (24); 212 (8); 214 (8); 223 (9×8); 227 (56); 228 (16); 235 (52); 259 (73); 260 (40); 261 (21); 262 (69a); 264 (50); 267 (60); 331 (29); 332 (29); 333 (29); 334 (29); 335 (29).
Barkman et al. 16 (24); 17 (15a); 18 (24); 95 (26); 137 (20); 154 (30); 183 (24).
Beaman: 6876 (39); 6950 (69a); 7239 (9); 7309 (49); 8005 (20); 8031 (39); 9888 (44); 9889 (9); 10347 (51); 11106 (71a); 11188 (40); 11196 (44); 11261 (6a); 11448 (60); 11459 (63); 11574 (77); 12068 (77).
R. Beaman: 94032806 (10); 94032810 (25); 94032811 (22×26).
Beccari: **476** (79); **1125** (75); 1683 (77); 2095 (32a); **3036** (6a); 3342 (77).
Blicher et al.: S.59856 (32a).
Bogle: 548 (10).
Bogle & Bogle: 532 (28).
Bogor cult.: 298.11.46 (47); 980.I.669 (79); 995.IX.30 (79); 995.IX.301 (79); 997.II.457 (32a); 3897 (32a).
Boyce: 295 (6c); 295A (6a); **319** (6c).

Brooke: 10525 (6a).

Brooks: 107 (77).

Brunig: S.12015 (6a).

Burtt: B.4943 (32a); B.11325 (43); B.11377 (37); B.11460 (9).

Burtt & Martin: B.4894 (53); B.5241 (61); B.5244 (39); B.5299 (9); B.5319 (44); **B.5328** (55); B.5399 (45); cult. R.B.G. Edinburgh C.5468 (56).

Burtt & Woods: B.2123 (43).

Buwalda: 7728 (6a).

Carr: C.3128, SFN27428 (19); C.3134, SFN26745 (9); C.3172, SFN26668 (5); C.3175 (20); C.3186, SFN26759 (6a); **C.3233, SFN26874** (49); C.3412, SFN27315 (6a); C.3476, SFN27430 & 27430A (10); **C.3477, SFN27431** (26); C.3521, SFN27531 (28); C.3533, SFN27562 (60); C.3534 (60); **C.3541, SFN27597** (22); **C.3545, SFN27624** (23); C.3548, SFN27635 (22); **C.3550, SFN27645** (29); **C.3597** (25); **C.3608, SFN27892** (52); C.3620, SFN27884 (9); C.3622, SFN27908 (64); **C.3623, SFN27891** (69a); C.3653, SFN27998 (73); **C.3663, SFN28020** (40); C.3668, SFN28029 (8); C.3671, SFN28047 (61); **C.3675, SFN28004** (13b); **C.3680, SFN28006** (56); C.3684, SFN28019 (59); C.3695 (9); C.3702 (9); C.3710 (21); C.3717 (60); C.3729 (9); C.3730 (60); C.3742 (64); C.3745 (9); C.3750 (8); C.3751 (64); C.3752 (44); C.3760, SFN27140 (69a); C.5228, SFN27864 (15a); ?SFN27430 (10); SFN27544 (9); SFN27550 (9); SFN27970 (9); SFN28034 (9); SFN28049 (8); SFN36564 (10); SFN36565 (15a); SFN36567 (64).

Chai: S.30098 (9); S.35896 (45).

Chan: 12/86 (5); 17/86 (9).

Chan & Gunsalam: 38/87 (10); 43/87 (10); 53/87 (28).

Chan & Lamb: SAN 87477 (56).

Chew: 1410 (77).

Chew et al.: RSNB 182 (64).

Chow & Leopold: SAN 74511 (40).

Church et al.: 898 (6a); 2204 (37); 2610 (45).

Clemens, J.: 114 (44); 115 (28); **116** (60); 146 (9); 178 (9); **179** (73); **202** (60); **205** (64); 209 (10); 224 (28); 224A (44); 242 (44); **247** (21); **248** (19); **270** (59); 278 (78); **280** (44); **285** (46); 289 (9); 318 (73); **332** (50); **361** (9); 377 (50); **383** (20); 385 (64); 386 (44); **396** (8).

Clemens, J. & M.S.: 04959 (20); 20307 (67–mixed collection with 9); **20313** (71a); 20729 (48); **22587** (68); 26784 (9); 26784A (8); 26841 (9); 26899 (6a); 26930 (9); 27140 (60 – mixed collection with 69a); 27141 (28); 27142 (10); 27143 (64); 27145 (60); 27146 (64); 27147 (44 – mixed collection with 60); 27148 (60); 27149 (10); 27168 (60); 27186 (8); 27256 (8); 27860 (64); 27864 (22); 27866 (10); 27867 (10); 27870 (28); 28949 (49); 29120 (28); 29128 (10); 29235 (9); 29288 (64); 29295 (49); 29361 (49); 29412 (50); 30101 (10); 30102 (10 – mixed collection with 22); 30103 (10); 30142 (28); 30147 (10); 30173 (8 – mixed collection with 60); 30471 (13a – mixed collection with 56); 30573 (9); 30578 (73); 30601 (50); 30640 (8); 30702 (9); 30703 (9); 30797 (9); 30799 (8); 30826 (13a); 31002 (50); 31337 (6a); 31457 (9); 31663 (15a); 31683 (10); 31800 (9); 31830 (15a); 31831 (22); 31835 (10); 32220 (13a – mixed collection with 56); 32244 (19); 32246 (9); 32247 (9); 32322 (22); 32324 (22); 32432 (9); 32548 (19); 32552 (9); 32556 (6a); 32617 (9 – mixed collection with 59); 33130 (10 – mixed collection with 26); 33173 (10 – mixed collection with 26); 33177 (28); 33178 (10); 33180 (10 – mixed collection with 22); 34329 (13a); 35205 (50); 40063 (19); 40134 (13a); 40135 (19); 40196 (78); 40255 (59); 40434 (61); 40467 (59); 40548 (78); 40633 (6a); 40701 (9); 40925 (9); 40931 (61); 50105 (78); 50162 (73); 50177 (9); 50184 (9); 50234 (61); 50235 (61 – mixed collection with 8 & 64); 50236 (69a); 50246 (8); 50256 (8); 50258 (61); 50278 (13a); 50322 (13a); 50347 (78); 50360 (69a); 50370 (9); 50650 (15a); 50652 (20); 50653 (28); 50654 (64); 50661 (10); 50669 (23); 50768 (10); 50774 (13a – mixed collection with 60); 50777 (15a – mixed collection with 20); 50801 (10); 51012 (28); 51079 (22); 51080 (50); 51534 (10); 51626 (44); 51714 (59).

Clements: 3226 (5).

Collenette: 31 (19); 41 (13a); 104 (59); 46/79 (40); 48/79 (78); 49/79 (36); 616 (22); 695 (77); 1042 (73); 21534 (64); 21535 (15a); 2349 (6a).

Comber, H.F.: 4032 (66); 4037 (8).

Comber, J.B.: **102** (3); **108** (70); 122 (8).

Cribb: 89/15 (48); 89/24 (69a); 89/65 (32a); 89/66 (74).

Darnaedi: D.467 (6a).

Dewol & Abas: SAN 89075 (69b).

Dransfield: 6979 (6a).

Edinburgh, Royal Botanic Garden: C.6530 (74).

Elsener: H.114 (77).

Endert: **3221** (58); **3991** (45); **4103** (17); 4129 (44).

Fidilis: SAN 127819 (78)

Fidilis & Sumbing: SAN 96124 (32a)

Ford: 2290 (41).

Fuchs: 21067 (10).

Fuchs & Collenette: 21404 (64).

Gardner: 104 (28).

Geesink: 8989 (62); 9074 (73); **9180** (62); 9190 (73); 9198 (6a).

George et al.: S.42814 (6a).

Gibbs: 4181 (28); **4085** (9): **4087** (9); 4250 (10).

Giles: 659 (34); 964A (34).

Giles & Woolliams: 467 (40).

Gunsalam: 3 (15a); 10 (21).

Hallier: **1312** (32a); **2646** (78); 2647 (6a).

Hani: SH 0045 (69a).

Hansen: 498 (16).

Harrisson: S.405 (43).

Hartley: S.512 (43).

Haviland: 46 (74); **169** (77); **1097** (28); **1142** (10); 1381 (6a); **1814** (19); **2346** (48).

Hewitt: **1** (48); **9** (48); **14** (77); 46 (75).

Holttum: SFN 36567 (64); SFN 36568 (10); SFN 36569 (20); SFN 36570 (28).

Hose: **52** (6b).

Jacobs: 5731 (64).

Jermy: 13200 (69a); 13894 (6a).

Johns: 7341 (79); 7555 (6a).

Jongejan cult. (*de Vogel*) 3976 (6a).

Joseph et al.: SAN 113509 (5).

Jumaat: 3397 (64).

Kato et al.: B.9758 (39); B.10110 (50); B.10627 (6a); B.11071 (45); B.11884 (6a).

Kato & Wiriadinata: B.5834 (6a).

Kidman Cox: 1008 (10).

Kitayama: 11–13 (60); 893 (16); K.2086 (60).

Kokawa & Hotta: 890 (71a).

Kostermans: 6690 (6a).

Krispinus: SAN 119385 (6a).

Lai et al.: 69869 (71a).

Laman et al.: TL 870 (6a).

Lamb: AL 359/85 (32a); AL 425/85 (75); AL 502/85 (32a); AL 506/85 (75); **AL 674/86** (36); AL 687/86 (60); AL 735/87 (20); AL 856/87 (21); AL 876/87 (61); AL 1119/89 (32a); AL 1128/89 (6a); AL 1392/91 (39); LMC 2303, SAN 93489 (56); **MAL 12** (16); SAN 88950 (32a); SAN 89674 (50); SAN 91572 (74); SAN 92253 (40).

Lamb & Chan in *Lamb*: AL 875/87 (9).

Lamb & Comber, J.B. in *Lamb*: AL 345/85 (5).

Lamb & Surat in *Lamb*: AL 1365/91 (34); **AL 1390/91** (66).

Lee: S.39318 (37); S.45322 (32a); S.45514 (32a); S.54013 (77); S.54569 (32a); S.54688 (32a); S.54797 (54); S.54798 (78).

Leeuwenberg & Rudjiman: 13405 (75).

Leiden cult. (*Cantley*): 26836 (38).

Leiden cult. (*Franken & Roos*): 21035 (75); 21085 (75); 21093 (75).

Leiden cult. (*Geesink*): 20313B (74).

Leiden cult. (*Roelfsema*): 950222 (78).

Leiden cult. (*Schuiteman et al.*): 932777 (32b); **932780** (72); **932946** (32b); 932959 (33); 933008 (33); 933028 (33); 933108 (54); 933140 (78); 933273 (9).

Leiden cult. (*Vermeulen*): 26457 (39); 26459 (36); 26484 (8); 26488 (78); 26514 (49); 26521 (36); 26602 (5); 26684 (74).

Leiden cult. (*de Vogel*): 27371 (71a); 27686 (80); 30322 (6a); 911135 (aff. 51); 911214 (56); 911227 (56); 911236 (56); 911245 (56); 911260A (3); 911272 (56); 913346 (34); 913452 (34); 913520 (9); **913960** (7); 913963 (78); 913964 (78); 913965 (78); 914389 (80); 914922 (56).

Leiden cult. (*de Vogel & Cribb*): **913205A** (35).

Leiden cult. (*Vogel et al.*): 980071 (39); 980074 (18); 980087 (18); 980090 (3); 980142 (39); 980299 (9); 980410 (35); 980415 (43).

Leiden cult. (*Wong*): 27853 (53).

Leopold: SAN 71928 (44).

Leopold & Taha: SAN 83505 (4).

Lewis: 343 (44); 345 (45); 346 (69a); 349 (58); 364 (45); 365 (39); **366** (27); **369** (15b).

Lugas: PEK 1048 (59); PEK 1818 (59).

Madani & Ismail: SAN 108904 (32a).

Madani & Sigin: SAN 107709 (32a).

Mantor: SAN 118767 (32a).

Martin: S.37059 (39).

Mason: 92 (76): 986 (77).

McLeod in *Synge*: S.156 (32a).

Meijer: 599 (11); 696 (71a); 715 (71a); 842 (71a); 843 (11); 867 (11); 874 (11); 877 (11); **882** (11); SAN 20326 (9); SAN 20367 (69a); SAN 22067 (28); SAN 24131 (10); SAN 28560 (28); SAN 29165 (10); SAN 48111 (40); SAN 54020 (19).

Mjöberg: 46 (41); 49 (78); 52 (44); 64 (73); 65 (2); 66 (45).

Mogea: 3963 (45).

Mohtar et al.: S.48084 (6a); S.48086 (46); S.48225 (37); S.53913 (32a).

Moulton: **10** (78); 15 (45); **SFN 2762** (31); SFN 6660 (37); SFN 6663 (43); SFN 6670 (46);

(16); 668 (8); 671 (21); 686 (61); 687 (74); 688 (9); 690 (8); 693 (5); 905 (2); 931 (46); 959 (47); 962 (52); 965 (36); 967 (54); **1008** (68); 1011 (2); 1020 (43); 1038 (8); 1042 (46); 1043 (61); **1057** (1).

Vermeulen & Lamb: 324 (78); 325 (4); 337 (32a); 340 (6a); 705 (9); 706 (56).

de Vogel: 115 (79); 195 (6a); 919 (32a); 1013 (16); 1116 (43); 1116A (43); 1245 (9); 1584 (32a); 1585 (32a); 1825 (71b); 1826 (71b); 2043 (9); 8032 (40); 8033 (40); 8036 (52); 8141 (6a); 8150 (75); 8172 (66); 8173 (66); 8339 (2); 8350 (57); **8351** (54); 8375 (50); **8376** (14); 8402 (74); 8426 (51); **8451** (51); 8452 (73); 8569 (3); 8579 (37); 8604 (37); 8632 (52); 8646 (36); 8661 (13a); 8662 (69a); 8663 (69a); 8664 (69a); 8667 (60); 8888 (80); 8953 (6a).

de Vogel & Cribb: 9157 (67).

Warwick: 144A (6a).

Winkler: **1495** (38).

Wong: WKM 297 (6a); WKM 802 (39).

Wood: 573 (69b); 578 (9); 583 (69b); 584 (5); 596 (6a); 605 (28); 608 (10); 618 (73); 621 (9); 646 (2); **657** (57); 694 (50); 703 (37); 733 (5); 749 (6a); 817 (6a); **886** (30); **905** (42).

Yii: S.39541 (48); S.44402 (9); S.44430 (55 – mixed collection with 63); S.44432 (45); S.44616 (39); S.48403 (6a); S.61421 (32a).

Yii & Talib: S.58429 (43).

HERBARIA WHERE NUMBERED BORNEAN COLLECTIONS ARE LOCATED

(Type collections are indicated in **bold**. Herbaria abbreviations appear in brackets. Duplicates of some collections may be deposited in additional herbaria.)

Aban: SAN 95230 (K, L, SAN); SAN 95241 (L, SAN); SAN 95243 (SAN).

Amdjah: Bogor cult. no. 108 (L).

Amin et al.: SAN 111793 (SAN); SAN 129373 (K, SAN).

Anderson, J.A.R.: S.11099 (K, L, SAR); S.30808 (K, L, SAR); S.30838 (A, E, K, L, SAR, SING).

Anderson, J.W.: 173 (K, SING); 196 (SING).

Aningguh in *Lamb*: AL 856/87 (K).

Argent: 111 (E).

Argent & Coppins: 1126 (E, K); 1202 (SAR).

Argent & Jermy: 988 (E, K).

Ashton: A.205 (K, L); A.211 (K); A.250 (K).

Asik: SAN 127822 (K, SAN).

Awa & Lee **S.47676** (K, KEP, SAR); S.47827 (AAU, K, KEP, L, SAR, SING); S.50586 (AAU, K, KEP, L, MEL, SING); S.50704 (K, L, SAR, SING); S.50731 (K, L, SAR); **S.50754** (K, L, SAR, SING); S.50756 (K, L, SAR, SING); S.50774 (K, SAR); S.50792 (AAU, K, KEP, L, MEL, SING); S.50812 (K, L, SAR, SING); S.50900 (K, L, SAR); S.50937 (K, L, NY, SAR); S.51102 (AAU, K, KEP, L, MEL, SING).

Awa et al.: S.50416 (AAU, K, L, MEL, SAR, SING).

Bacon: 78 (E); 80 (E); 93 (E); 183 (E).

Bailes & Cribb: 695 (K); 722 (K); 841 (K).

Banyeng & Paie S.45050 (L, SAR).

Barkman: 1 (K); 2 (K); 3 (K); 4(K); 6 (K); 7 (K); 8 (K); 9 (K); 10 (K); 11 (K); 12 (K); 19 (K); 20 (K); 49; 51; 63; 64; 142; 193; 194; **198**; 212; 214; 223 (K); 227; 228 (K); 235; 259; 260; 261; 262; 264; 267; 331; 332; 333; 334; 335 (first set of all numbered collections deposited at Sabah Parks Herbarium, Kinabalu Park).

Barkman et al.: 16 (K); 17 (K); 18 (K); 95; 137; 154; 183 (first set of all numbered collections deposited at Sabah Parks Herbarium, Kinabalu Park).

Beaman: 6876 (K, L); 6950 (K); 7239 (K); 7309 (K); 8005 (K); 8031 (K); 9888 (K); 9889 (K); 10347 (K); 11106 (K); 11188 (K); 11196 (K); 11261 (K); 11448 (K); 11459 (K); 11574 (K); 12068 (K).

R. Beaman: 94032806 (K); 94032810 (K); 94032811 (K).

Beccari: **476** (FI); **1125** (FI); 1683 (FI); 2095 (FI, K); **3036** (FI, L); 3342 (FI, G, K, W).

Blicher et al.: S.59856 (L, SAR).

Bogle: 548 (AMES).

Bogle & Bogle: 532 (AMES).

Bogor cult.: 928.11.46 (BO); 980.I.669 (BO, K); 995.IX.30 (BO, K); 995.IX.301 (BO, K); 997.II.457 (BO, K); 3897 (BO, K).

Boyce: 295 (BRUN); 295A (K); **319** (K).

Brooke: 10525 (BM, G, L, SING).

Brooks: 107 (BM).

Brunig: S.12015 (SAR).

Burtt: B.4943 (E); B.11325 (E, SAR); B.11377 (E); B.11460 (E).

Burtt & Martin: B.4894 (E, SAR); B.5241 (E, SAR); B.5244 (E, SAR); B.5299 (E, SAR); B.5319 (E, SAR); **B.5328** (E, K, SAR); B.5399 (E); cult. R.B.G. Edinburgh C.5468 (E).

Burtt & Woods: B.2123 (E, K).

Buwalda: 7728 (BO).

Carr: C.3128, SFN 27428 (BM, BO, K, L, SING); C.3134, SFN 26745 (SING); C.3172, SFN 2668 (K, SING); C.3175 (SING); C.3186, SFN 26759 (K, SING); **C.3233, SFN 26874** (AMES, K, SING); C.3412, SFN 27315 (K, SING); C.3476, SFN 27430 (K, SING) & SFN 27430A (K, SING); **C.3477, SFN 27431** (AMES, K, LAE, SING); C.3521, SFN 27531 (K, SING); C.3533, SFN 27562 (K, SING); C.3534 (SING); **C.3541, SFN 27597** (AMES, C, K, SING); **C.3545, SFN 27624** (AMES, K, SING); C.3548, SFN 27635 (K, SING); **C.3550, SFN 27645** (AMES, K, L, SING); **C.3597** (SING); **C.3608, SFN 27892** (AMES, K, LAE, SING); C.3620, SFN 27884 (K, SING); C.3622, SFN 27908 (K, SING); **C.3623, SFN 27891** (AMES, K, SING); C.3653, SFN 27998 (SING); **C.3663, SFN 28020** (AMES, K, SING); C.3668, SFN 28029 (K, SING); C.3671, SFN 28047 (K, SING); **C.3675, SFN 28004** (K, SING); **C.3680, SFN 28006** (AMES, K, SING); C.3684, SFN 28019 (K, SING); C.3695 (SING); C.3702 (SING); C.3710 (SING); C.3717 (SING); C.3729 (SING); C.3730 (SING); C.3742 (SING); C.3745 (SING); C.3750 (SING); C.3751 (SING); C.3760, SFN 27140 (SING); C.5228, SFN 27864 (BO, K, L, SING); ?SFN 27430 (SING); SFN 27544 (SING); SFN 27550 (SING); SFN 27970 (SING); SFN 28034 (SING); SFN 28049 (SING); SFN 36564 (SING); SFN 36565 (SING); SFN 36567 (SING).

Chai: S.30098 (A, BO, K, KEP, L, SAN, SAR, SING); S.35896 (K, KEP, L, MO, SAR, SING).

Chan: 12/86 (SING); 17/86 (K).

Chan & Gunsalam: 38/87 (K); 43/87 (K); 53/87 (K, SING).

Chan & Lamb: SAN 87477 (SAN).

Chew: 1410 (HBG, K, L, SAR, SING).

Chew et al.: RSNB 182 (AMES, K, L, SING).

Chow & Leopold: SAN 74511 (K, SAN).

Church et al.: 898 (L); 2204 (AMES, BO, K); 2610 (AMES, BO, K).

Clemens, J.: 114 (AMES); 115 (AMES, BM, BO, E, K, SING); **116** (AMES); 146 (AMES, BM, SING); 178 (AMES, BM, BO, K, SING); **179** (AMES, BM, BO, K, NY, SING); **202** (AMES, BM, BO, E, K, SING); **205** (AMES, BM, BO, K, SING); 209 (AMES, BM, E, K, SING); 224 (AMES); 224A (AMES); 242 (AMES); **247** (AMES); **270** (AMES); 278 (AMES, BM); **280** (AMES, BM, K, NY, SING); **285** (AMES); 289 (AMES, BM, BO, K, SING); 319 (AMES); **332** (AMES, BO, K); **361** (AMES, BM, BO, K, SING); 377 (AMES, BM, K, SING); **383** (AMES, BM, K, SING); 385 (AMES, BM, E, K, SING); 386 (BM, BO, E, K, SING); **396** (AMES).

Clemens, J. & M.S.: 04959 (BM); 20307 (AMES, K, L, SAR); **20313** (K, L, SAR, SING);

20729 (AMES, BO, K, L, SAR, SING); **22587** (AMES, BO, K, L, SING); 26784 (BM); 26784A (BM); 26841 (BM, HBG, K); 26899 (BM, BO, E, HBG, K, L); 26930 (BM, E); 27140 (BM, E, HBG, K, L, SING); 27141 (BM, E, HBG, K, SING); 27142 (BM, E, K, L, SING); 27143 (BM, BO); 27145 (BM, E, K, L); 27146 (BM, E, K); 27147 (AMES, BO, HBG, K, L, SING); 27148 (BM); 27149 (BM, E, K, L, SING); 27168 (BO); 27186 (BO); 27256 (BM, BO, E, K); 27860 (BM, E, K); 27864 (BM, E, K, L); 27866 (BM, SING); 27867 (BM, E, HBG, K, SING); 27870 (BM, BO, E, HBG, K, SING); 28949 (BM); 29120 (BM, E); 29128 (BM, E, K); 29235 (BM, E, K); 29288 (AMES, BO); 29295 (BM); 29361 (BM, E, HBG, K, L); 29412 (BM, E); 30101 (AMES, BO, E, HBG, K, L); 30102 (BO); 30103 (E, HBG, K, L); 30142 (AMES, E, HBG, K); 30147 (AMES, E, HBG, K, L, SING); 30173 (AMES); 30471 (BM, E); 30573 (AMES, BM, HBG, K); 30578 (BM); 30601 (BM, BO, E, L); 30640 (BM); 30702 (BM, BO); 30703 (AMES); 30797 (L); 30799 (BM); 30826 (AMES, BM, E, K); 31002 (BM, BO); 31337 (BM); 31457 (AMES, BM, BO, E, K); 31663 (AMES, E); 31683 (BM, E, L); 31800 (AMES, BM, E, G, HBG, L); 31830 (BM, K); 31831 (BM, BO); 31835 (AMES, BM, E); 32220 (BM, E); 32244 (AMES, BM, BO, E, L); 32246 (AMES, BM); 32247 (BO, E, G, L); 32322 (AMES, BM, E); 32324 (AMES, BM, E, L); 32432 (AMES, BM, E, K); 32548 (BM, E, HBG, L); 32552 (BM); 32556 (BM, BO, E, L); 32617 (AMES, BM, E); 33130 (BM, E); 33173 (AMES, BM, E, L); 33177 (AMES, BM, E, HBG, K); 33178 (AMES, BM, E, HBG, K, L); 33180 (BM); 34329 (AMES, BM, E, HBG, L); 35205 (BM, BO); 40063 (AMES); 40134 (AMES, BM, E, L); 40135 (BM); 40196 (BM, E, L); 40255 (BM); 40434 (BM, K); 40467 (AMES, BM, K); 40548 (BM); 40633 (AMES, BM, E, K, L); 40701 (BM); 40925 (BM, K, L); 40931 (BM); 50105 (AMES, BM, K); 50162 (BM, K, L); 50177 (BM); 50184 (BM, E, K, L); 50234 (SING); 50235 (AMES, BM); 50236 (AMES, BM, K, L); 50246 (AMES, BM, K, L); 50256 (SING); 50258 (AMES, BM, K); 50278 (AMES, BM, K); 50322 (AMES, BM, K); 50347 (AMES, BM, E, K); 50360 (BM, K); 50370 (BM, K); 50650 (BM, E, K, L); 50652 (AMES, BM, E, K, L); 50653 (BM, E, K); 50654 (BM, E, K, L); 50661 (BM, G, K); 50669 (BM); 50768 (BM); 50774 (BM); 50777 (AMES, BM, E, L); 50801 (BM, K); 51012 (BM, K); 51079 (AMES, BM, E, K, L); 51080 (BM, K); 51534 (AMES, BM, K); 51626 (BM); 51714 (AMES, BM, K, L).

Clements: 3226 (K).

Collenette: 31 (BM); 41 (BM); 104 (BM); 46/79 (E); 48/79 (E); 49/79 (E); 616 (K); 695 (K); 1042 (K); 21534 (AMES, BO, K, L); 21535 (K, KSEPL, L, SAR); 2349 (K).

Comber, H.F.: 4032 (K); 4037 (K).

Comber, J.B.: **102** (K); **108** (K); 122 (K).

Cribb: 89/15 (K); 89/24 (K); 89/65 (K); 89/66 (K).

Darnaedi: D. 467 (BO).

Dewol & Abas: SAN 89075 (SAN).

Dransfield, J.: 6979 (BRUN, K).

Edinburgh, Royal Botanic Garden: C.6530 (E).

Elsener: H.114 (L).

Endert: **3221** (BO); **3991** (L); **4103** (BO, K, L); 4129 (L).

Fidilis: SAN 127819 (K, SAN).

Fidilis & Sumbing: SAN 96124 (SAN).

Ford: 2290 (K).

Fuchs: 21067 (L).

Fuchs & Collenette: 21404 (K, L).

Gardner: 104 (E, L).

Geesink: 8989 (BO, KYO, L); 9074 (BO, KYO, L); **9180** (BO, K, KYO, L); 9190 (BO, KYO, L); 9198 (L).

George et al.: S.42814 (K, L, SAR).

Gibbs: **4085** (BM, K); **4087** (BM, K); 4181 (BM, K); 4250 (BM).

Giles: 659 (K); 964A (K).

Giles & Woolliams: 467 (K).

Gunsalam: 3 (K); 10 (K).

Hallier: **1312** (BO, K, L); **2646** (BO, K, L); 2647 (BO, K).

Hani: SH 0045 (UKMS).

Hansen: 498 (C, K).

Harrisson: S.405 (K).

Hartley: S.512 (K).

Haviland: 46 (SING); **169** (K, SAR); **1097** (BM, K, SAR, SING); **1142** (K); 1381 (K); **1814** (SAR); **2346** (K, SING).

Hewitt: **1** (SING); 9 (SING); **14** (K, SING); 46 (SAR).

Holttum: SFN 36567 (AMES); SFN 36568 (AMES, SING); SFN 36569 (AMES); SFN 36570 (AMES, SING).

Hose: **52** (BM).

Jacobs: 5731 (B, BH, CANB, G, K, L, S, SAN, US).

Jermy: 13200 (K); 13894 (K).

Johns: 7341 (BRUN, K); 7555 (K).

Jongejan cult. (*de Vogel*): 3976 (K, L).

Joseph et al.: SAN 113509 (K, L, SAN, SAR).

Jumaat: 3397 (UKMS).

Kato et al.: B.9758 (L); B.10110 (BO, L); B.10627 (L); B.11071 (L); B.11884 (L).

Kato & Wiriadinata: B.5834 (L).

Kidman Cox: 1008 (L).

Kitayama: 11–13 (Sabah Parks Herbarium, Kinabalu Park); 893 (UKMS); K.2086 (UKMS).

Kokawa & Hotta: 890 (KYO, L).

Kostermans: 6690 (BO, L, SING).

Krispinus: SAN 119385 (K, SAN).

Lai et al.: S.69869 (AAU, K, KEP, SAR).

Laman et al.: TL 870 (AMES, BO, K).

Lamb: AL 359/85 (K); AL 425/85 (K); AL 502/85 (K); AL 506/85 (K); **AL 674/86** (K); AL 687/86 (K); AL 735/87 (K); AL 876/87 (K); AL 1119/89 (K); AL 1128/89 (K); AL 1392/91 (K); LMC 2303, SAN 93489 (K); **MAL 12** (K); SAN 89674 (SAN).

Lamb & Chan in *Lamb*: AL 875/87 (K).

Lamb & Comber, J.B. in *Lamb*: AL 345 /85 (K).

Lamb & Surat in *Lamb*: AL 1365/91 (K); **AL 1390/91** (K).

Lee: S.39318 (K, L, SAR, SING); S.45322 (K, KEP, L, MEL, SAR, SING); S.45514 (AAU, MEL, SAR); S.54013 (AAU, K, L, SAR, SING); S.54569 (K, L, SAR); S.54688 (L, SAR); S.54797 (SAR); S.54798 (K, SAR).

Leeuwenberg & Rudjiman: 13405 (L).

Leiden cult. (*Cantley*): 26836 (K, L).

Leiden cult. (*Franken & Roos*): 21035 (K, L); 21085 (K, L); 21093 (L).

Leiden cult. (*Geesink*): 20313B (L).

Leiden cult. (*Roelfsema*): 950222 (L).

Leiden cult. (*Roelfsema et al.*): 970344 (K, L).

Leiden cult. (*Schuiteman et al.*): **932780** (K, L); 932777 (K, L); **932946** (K, L); 932959 (L); 933008 (L); 933028 (L); 933108 (K, L); 933140 (L); 933273 (L);

Leiden cult. (*Vermeulen*): 26457 (L); 26459 (K, L); 26484 (K, L); 26488 (L); 26514 (K, L); 26521 (K, L); 26602 (L); 26684 (K, L).

Leiden cult. (*de Vogel*): 27371 (L); 27686 (L); 30322 (K, L); 911135 (K, L); 911214 (L); 911227 (L); 911236 (L); 911245 (L); 911260A (K, L); 911272 (L); 913346 (L); 913452 (K, L); 913520 (L); **913960** (K, L); 913963 (K, L); 913964 (K, L); 913965 (L); 914389 (L); 914922 (L).

Leiden cult. (*de Vogel & Cribb*): **913205A** (K, L).

Leiden cult. (*Vogel et al.*): 980071 (K, L); 980074 (K, L); 980087 (K, L); 980090 (K, L); 980142 (L); 980299 (K, L); 980410 (K, L); 980415 (K, L).

Leiden cult. (*Wong*): 27853 (L).

Leopold: SAN 71928 (K, L, SAN, SING).

Leopold & Taha: SAN 83505 (K, L, SAN, SAR).

Lewis: 343 (K, SAR); 345 (K); 346 (K, SAR); 349 (K); 364 (K); 365 (K); **366** (K); **369** (K).

Lugas: PEK 1048 (K, Sabah Parks Herbarium, Kinabalu Park); PEK 1818 (K, Sabah Parks Herbarium, Kinabalu Park).

Madani & Ismail: SAN 108904 (K, SAN).

Madani & Sigin: SAN 107709 (K, L, SAN, SING).

Mantor: SAN 118767 (K, SAN).

Martin: S.37059 (L, MEL, SAR).

Mason: 92 (K); 986 (K).

McLeod in *Synge*: S.156 (K).

Meijer: 599 (BO, L); 696 (BO, L); 715 (BO); 842 (BO, K, L); 843 (BO, K, L); 867 (BO, L); 874 (BO, L); 877 (BO, L); **882** (A, BO, K, L); SAN 20326 (K, L, SAN); SAN 20367 (AMES, K, SAN); SAN 22067 (SAN); SAN 24131 (K, SAN); SAN 28560 (K, SAN); SAN 29165 (K, SAN); SAN 4811 (K, SAN); SAN 54020 (SAN).

Mjöberg: 46 (AMES); 49 (AMES); 52 (AMES); 64 (AMES); 65 (AMES); 66 (AMES).

Mogea: 3963 (BO, L).

Mohtar et al.: S.48084 (AAU, K, L, MEL, SAR, SING); S.48086 (AAU, K, L, MEL, SING); S.48225 (AAU, KEP, SAN, SAR); S.53913 (K, SAR, SING).

Moulton: **10** (SAR); 15 (?BO, L, SAR); **SFN 2762** (AMES, SING); SFN 6660 (K, SING); SFN 6670 (K, SING); SFN 6761 (K, SING); **SFN 6763** (AMES, SING).

Nair: AN 064 (UKMS).

Nais et al.: SNP 4724 (Sabah Parks Herbarium, Kinabalu Park); SNP 5473 (Sabah Parks Herbarium, Kinabalu Park).

Native collector: **68** (E); 90 (E); **99** (E, K); D.162 (E, SAR).

Native collector in *Synge*: S.175 (K, L, SING); **S.558** (K, SING).

Nielsen: **143** (AAU, K); **675** (AAU, K); 806 (AAU, K).

Nooteboom: 1334 (L, SAN).

Nooteboom & Chai: 01890 (L, SAR); 01954 (L); **01995** (K, SAR); 01995A (L); 02097 (L, SAR); 02278 (K, L, SAR).

van Nouhuys: Cult. Bogor no. 16 (BO).

Oxford University Expedition native collector: **2497** (K, SING).

Paie: S.26455 (E, K, L, SAR, SING); S.26585 (L, SAR); S.28076 (E, K, L, SAR, SING); S.31566 (A, BO, K, KEP, L, SAN, SAR, SING); S.36322 (KEP, L, SAR, SING).

Phillipps: SNP 2948 (Sabah Parks Herbarium, Kinabalu Park); SNP 2949 (Sabah Parks Herbarium, Kinabalu Park); SNP 2969 (Sabah Parks Herbarium, Kinabalu Park); SNP

2970 (Sabah Parks Herbarium, Kinabalu Park).

Pickles & Ahmad bin Topin: S.2911 (L, SING).

Price: 145 (K); 217 (K).

Purseglove & Shah: P.4728 (K, SAR, SING); P.4747 (SAR, SING).

Richards: **S.476** (K, L, SING); S.482 (K); S.483 (K, SING); **S.484** (K, L, SING), S.548 (K).

Sabah National Parks: SNP 1587 (Sabah Parks Herbarium, Kinabalu Park); SNP 1603 (Sabah Parks Herbarium, Kinabalu Park); SNP 1833 (Sabah Parks Herbarium, Kinabalu Park); SNP 2369 (Sabah Parks Herbarium, Kinabalu Park); SNP 2397 (Sabah Parks Herbarium, Kinabalu Park); SNP 2411 (Sabah Parks Herbarium, Kinabalu Park); SNP 2705 (Sabah Parks Herbarium, Kinabalu Park); SNP 2712 (Sabah Parks Herbarium, Kinabalu Park); SNP 2890 (Sabah Parks Herbarium, Kinabalu Park); SNP 3038 (Sabah Parks Herbarium, Kinabalu Park); SNP 4705 (Sabah Parks Herbarium, Kinabalu Park); SNP 4709 (Sabah Parks Herbarium, Kinabalu Park); SNP 4717 (Sabah Parks Herbarium, Kinabalu Park); SNP 5526 (Sabah Parks Herbarium, Kinabalu Park); SNP 5572 (Sabah Parks Herbarium, Kinabalu Park); SNP 5872 (Sabah Parks Herbarium, Kinabalu Park).

Said: BRUN 15825 (BRUN, K).

Sands: 3880 (K); 5335 (BRUN, K).

Sands et al.: **5327** (BRUN, K).

Sarawak Museum native collector: 903 (AMES); 1580 (K); 1584 (BM).

Sato: 162 (UKMS); 699 (UKMS); 759 (UKMS); 762 (UKMS); 970 (UKMS); 974 (UKMS); 1046 (UKMS); 1047 (UKMS); 1080 (UKMS); 1081 (UKMS); 1082 (UKMS); 1223 (UKMS); 2158 (UKMS); 2178 (UKMS).

Sato et al.: 040 (UKMS); 0703 (UKMS); 0704 (UKMS); 1436 (UKMS); 1558 (UKMS).

Shackleton: **S.186** (K, SING).

Sidek bin Kiah: S.38 (L, SING).

Sinclair, I. & Argent: 126 (E).

Sinclair, J.: 5604, SFN 38351 (BO, E, K, SING).

Sinclair, J. et al.: 9180 (E, K, L, SAN, SING); 9181 (E, K, L, SING).

Smith: 543 (L).

Smith & Everard: 155 (K).

van Steenis: **1435** (BO, L).

Suhaili et al.: TM 17 (UKMS); TM 18 (UKMS).

Synge: S.100 (K); S.397 (K); **S.406** (K, L, SING); S.415 (K); **S.418** (K, SING); **S.435** (K, SING); S.436 (K); S.446 (K); **S.513** (K, L, SING).

Talip: SAN 70988 (L, SAN).

Tenom Orchid Centre: TOC 863 (L); TOC 1117 (L).

Thomas: 197 (BRUN, K).

Ueda & Darnaedy: B.11544 (L).

Unknown collector (probably *Hose*): 25 (W).

Vermeulen: 530 (L).

Vermeulen & Chan: 413 (L).

Vermeulen & Duistermaat: 544 (L); 545 (L); 663 (K, L); 664 (K, L); 665 (K, L); **666** (L); 667 (K, L); 668 (L); 671 (L); 686 (L); 687 (K, L); 688 (K, L); 690 (L); 693 (K, L); 905 (K, L); 931 (K, L); 959 (K, L); 962 (L); 965 (L); 967 (L); **1008** (K, L); 1011 (K, L); 1020 (K, L); 1038 (L); 1042 (K, L); 1043 (K, L); **1057** (K, L).

Vermeulen & Lamb: 324 (K, L); 325 (L); 337 (K, L); 340 (L); 705 (K, L); 706 (K, L).

de Vogel: 115 (L); 195 (K, L); 919 (L); 1013 (K, L); 1116 (L); 1116A (K, L); 1245 (L); 1584 (K, L); 1585 (L); 1825 (L); 1826 (L); 2043 (K, L); 8032 (L); 8033 (L); 8036 (L); 8141

(L); 8150 (L); 8172 (L); 8173 (L); 8339 (L); 8350 (L); **8351** (K, L); 8375 (L); **8376** (K, L); 8402 (L); 8426 (L); **8451** (K, L); 8452 (L); 8569 (L); 8579 (L); 8604 (L); 8632 (L); 8646 (L); 8661 (K, L); 8662 (L); 8663 (L); 8664 (L); 8667 (L); 8888 (L); 8953 (BRUN, L).

de Vogel & Cribb: 9157 (K).

Warwick: 144A (E).

Winkler: **1495** (HBG)

Wong: WKM 297 (BRUN, K, L, SING); WKM 802 (BRUN, K, L, SING).

Wood: 573 (K); 578 (K); 583 (K); 584 (K); 596 (K); 605 (K); 608 (K); 618 (K); 621 (K); 646 (K); **657** (K); 694 (K); 703 (K); 733 (K); 749 (K); 817 (K); **886** (K, L, UKMS); **905** (K, Tenom Orchid Centre).

Yii: S.39541 (K, L, MEL, SAR); S.44402 (K, L, SAR); S.44430 (K, L, SAR); S.44432 (K, L, SAR, SING); S.44616 (K, L, SAR); S.48403 (K, KEP, L, MEL, SAR, SING); S.61421 (AAU, L, SAR, SING).

Yii & Talib: S.58429 (KEP, SAR).

COLOUR PLATES

Plate 1

A. *Dendrochilum acuiferum* Carr (Photo: T.J. Barkman)

B. *Dendrochilum acuiferum* Carr (Photo: T.J. Barkman)

C. *Dendrochilum angustilobum* Carr (Photo: T.J. Barkman)

D. *Dendrochilum angustilobum* Carr (Photo: T.J. Barkman)

E. *Dendrochilum corrugatum* (Ridl.) J.J. Sm. (Photo: T.J. Barkman)

F. *Dendrochilum alpinum* Carr (Photo: A. Lamb)

G. *Dendrochilum alatum* Ames (Photo: A. Lamb)

Plate 2

A. *Dendrochilum crassifolium* Ames. (Photo: K. Barrett)

B. *Dendrochilum crassum* Ridl. (Photo: K. Barrett)

C. *Dendrochilum cruciforme* J.J. Wood var *cruciforme* (Photo: C.L. Chan)

D. *Dendrochilum cruciforme* J.J. Wood var *longicuspe* J.J. Wood (Photo: T.J. Barkman)

E. *Dendrochilum cupulatum* J.J. Wood (Photo: T.J. Barkman)

Plate 3

A. *Dendrochilum dewindtianum* W.W. Sm. (Photo: G. Cubitt)

B. *Dendrochilum dewindtianum* W.W. Sm., white flowered form (Photo: T.J. Barkman)

C. *Dendrochilum exasperatum* Ames (Photo: T.J. Barkman)

D. *Dendrochilum exasperatum* Ames (Photo: K. Barrett)

Plate 4

A.   *Dendrochilum gibbsiae* Rolfe (Photo: C.L. Chan)

B.   *Dendrochilum gibbsiae* Rolfe (Photo: K. Barrett)

C.   *Dendrochilum gibbsiae* Rolfe, variant from Mount Mulu (Photo: André Schuiteman)

D.   *Dendrochilum sp.* in cultivation at Bogor Botanic Gardens, Java, probably referable to *D. flos-susannae* (Photo: P.J. Cribb)

Plate 5

A. *Dendrochilum glumaceum* Lindl. (Photo: Paul Davies)

B. *Dendrochilum gracile* (Hook. f.) J.J. Sm. var. *gracile* (Photo: André Schuiteman)

C. *Dendrochilum gracile* (Hook. f.) J.J. Sm. var. *bicornutum* J.J. Wood (Photo: C.L. Chan)

D. *Dendrochilum graminoides* Carr (Photo: T.J. Barkman)

E. *Dendrochilum graminoides* Carr (Photo: A. Lamb)

Plate 6

A. *Dendrochilum grandiflorum* (Ridl.) J.J. Sm. (Photo: C.L. Chan)

B. *Dendrochilum grandiflorum* (Ridl.) J.J. Sm. (Photo: K. Barrett)

C. *Dendrochilum grandiflorum* (Ridl.) J.J. Sm. (Photo: K. Barrett)

D. *Dendrochilum hamatum* Schltr. (Photo: P.J. Cribb)

E. *Dendrochilum hamatum* Schltr., *Beccari* 476 (isotype in Florence herbarium) (Photo: P.J. Cribb)

Plate 7

A. *Dendrochilum haslamii* Ames var. *haslamii* (Photo: K. Barrett)

B. *Dendrochilum haslamii* Ames var. *haslamii* (Photo: K. Barrett)

C. *Dendrochilum haslamii* Ames var. *haslamii* (Photo: K. Barrett)

D. *Dendrochilum hologyne* Carr (Photo: J.B. Comber)

Plate 8

A. *Dendrochilum imbricatum* Ames (Photo: C.L. Chan)

B. *Dendrochilum imbricatum* Ames (Photo: K. Barrett)

C. *Dendrochilum imbricatum* Ames (Photo: K. Barrett)

Plate 9

A. *Dendrochilum sp. aff. cruciforme* J.J. Wood var *cruciforme* (Photo: C.L. Chan)

B. *Dendrochilum joclemensii* Ames (Photo: P.J. Cribb)

C. *Dendrochilum joclemensii* Ames (Photo: K. Barrett)

D. *Dendrochilum kamborangense* Ames, pale flowered form (Photo: K. Barrett)

E. *Dendrochilum kamborangense* Ames, dark flowered form (Photo: K. Barrett)

F. *Dendrochilum dewindtianum* W.W. Sm., large flowered form similar to *J. & M.S. Clemens* 50774 (Photo: T. Sato)

Plate 10

A. *Dendrochilum kingii* (Hook. f.) J.J. Sm. var. *tenuichilum* J.J. Wood (Photo: André Schuiteman)

B. *Dendrochilum kingii* (Hook. f.) J.J. Sm. var. *kingii* (Photo: A. Lamb)

C. *Dendrochilum lacteum* Carr (Photo: T.J. Barkman)

D. *Dendrochilum lacinilobum* J.J. Wood & A. Lamb (Photo: A. Lamb)

Plate 11

A. *Dendrochilum lancilabium* Ames (Photo: T.J. Barkman)

B. *Dendrochilum longirachis* Ames (Photo: T.J. Barkman)

C. *Dendrochilum lancilabium* Ames (Photo: K. Barrett)

D. *Dendrochilum minimiflorum* Carr (Photo: André Schuiteman)

E. *Dendrochilum ochrolabium* J.J. Wood (Photo: A. Lamb)

F. *Dendrochilum muluense* J.J. Wood (Photo: C.L. Chan)

G. *Dendrochilum muluense* J.J. Wood (Photo: G. Lewis)

Plate 12

A. *Dendrochilum oxylobum* Schltr. (Photo: A. Lamb)

B. *Dendrochilum oxylobum* Schltr. (Photo: A. Lamb)

C. *Dendrochilum pachyphyllum* J.J. Wood & A. Lamb (Photo: C.L. Chan)

D. *Dendrochilum pallidiflavens* Blume var. *pallidiflavens* (Photo: P. Jongejan)

E. *Dendrochilum pallidiflavens* Blume var. *pallidiflavens, Beccari* 3036 (isotype of *D. bulbophylloides* Schltr. in Florence herbarium) (Photo: P.J. Cribb)

F. *Dendrochilum pandurichilum* J.J.Wood (Photo: André Schuiteman)

G. *Dendrochilum pandurichilum* J.J.Wood (Photo: André Schuiteman)

Plate 13

A. *Dendrochilum planiscapum Carr* (Photo: A. Lamb)

B. *Dendrochilum pseudoscriptum* T.J. Barkman & J.J. Wood, pale flowered form. (Photo: K. Barrett)

C. *Dendrochilum pseudoscriptum* T.J. Barkman & J.J. Wood, dark flowered form (Photo: K. Barrett)

D. *Dendrochilum pseudoscriptum* T.J. Barkman & J.J. Wood (Photo: C.L. Chan)

Plate 14

A. *Dendrochilum pterogyne* Carr (Photo: T.J. Barkman)

B. *Dendrochilum pterogyne* Carr (Photo: K. Barrett)

C. *Dendrochilum pubescens* L.O. Williams (Photo: P. Jongejan)

D. *Dendrochilum pubescens* L.O. Williams (Photo: J.B. Comber)

Plate 15

A. *Dendrochilum rufum* (Rolfe) J.J. Sm. (Photo: P. Jongejan)

B. *Dendrochilum scriptum* Carr (Photo: T.J. Barkman)

C. *Dendrochilum scriptum* Carr (Photo: T.J. Barkman)

D. *Dendrochilum simplex* J.J. Sm. (Photo: T.J. Barkman)

E. *Dendrochilum stachyodes* (Ridl.) J.J. Sm. (Photo: Robert New)

F. *Dendrochilum stachyodes* (Ridl.) J.J. Sm. (Photo: K. Barrett)

Plate 16

A. *Dendrochilum stachyodes* (Ridl.) J.J. Sm. (Photo: T. Sato)

B. *Dendrochilum stachyodes* (Ridl.) J.J. Sm., growing on granite slopes at the edge of upper montane *Leptospermum* scrub, *c.* 3500 m, Mount Kinabalu. (Photo: Robert New)

A

B

Plate 17

A. *Dendrochilum tenompokense* Carr var. *tenompokense* (Photo: C.L. Chan)

B. *Dendrochilum tenompokense* Carr var. *tenompokense* (Photo: C.L. Chan)

C. *Dendrochilum mucronatum* J.J. Sm., variant from Kalimantan Timur, *de Vogel & Cribb* 9157 (Photo: P.J. Cribb)

D. *Dendrochilum transversum* Carr (Photo: T.J. Barkman)

E. *Dendrochilum gibbsiae* × *exasperatum* (Photo: T.J. Barkman)

F. *Dendrochilum trusmadiense* J.J. Wood (Photo: T.J. Barkman)

Plate 18

Dendrochilum pallidiflavens Blume var. *pallidiflavens*. A late nineteenth century specimen in the John Lindley herbarium at Kew collected in Sarawak by Thomas Lobb for the London nurserymen Messrs. Veitch & Sons who sent material to Lindley for determination.

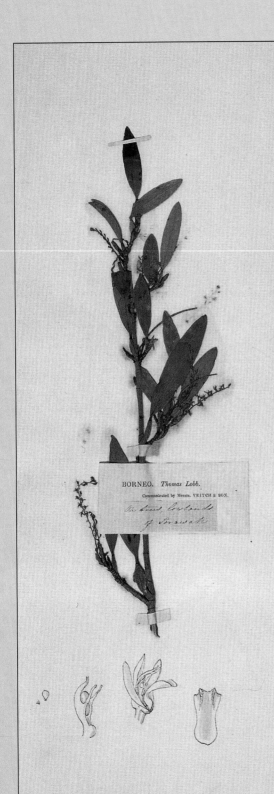

Dendrochilum pallidiflavens Blume
var. *pallidiflavens*
det. J.J.Wood   July 1997

Plate 19

Dendrochilum pallidiflavens Blume var. *pallidiflavens*. *Kostermans* 6690 (L), collected in Kalimantan Timur, showing congested growth-form.

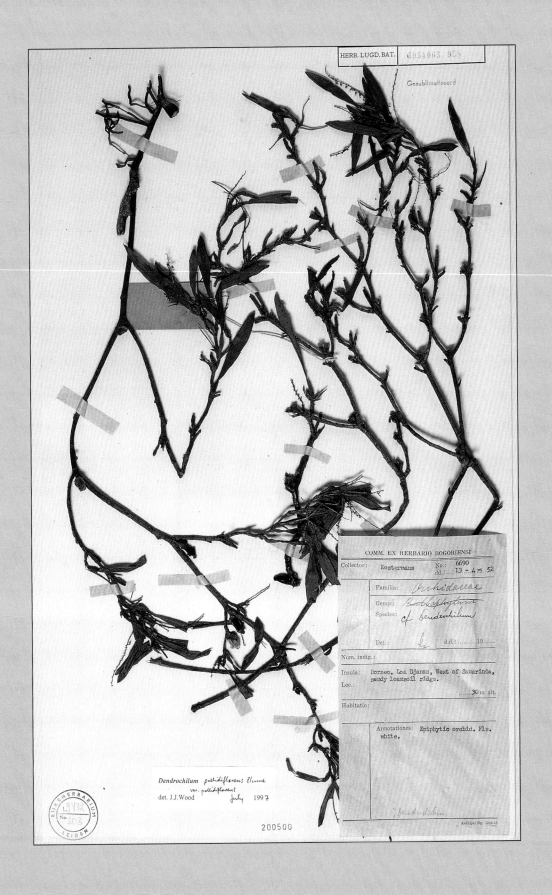

Gesublimatiseerd

COMM. EX HERBARIO BOGORIENSI

Collector:	Kostermans	No.:	6690
		dd.:	13 - 4 - 52

Familia: *Orchidaceae*

Genus: *Bulbophyllum*

Species: cf *Dendrochilum*

Det.: d.d.: 19

Nom. indig.:

Insula: Borneo, Loa Djaran, West of Samarinda,
Loc.: sandy loamsoil ridge.
30 m alt.

Habitatio:

Annotationes: Epiphytic orchid. Fls.
white.

Archipel Dg. 1236-12

Dendrochilum pallidiflavens Blume
var. *pallidiflavens*
det. J.J.Wood   July 1997

200500

Plate 20

Dendrochilum pallidiflavens Blume var. *pallidiflavens*. *Wood* 749 (K), collected in Sabah, showing lax growth-form.

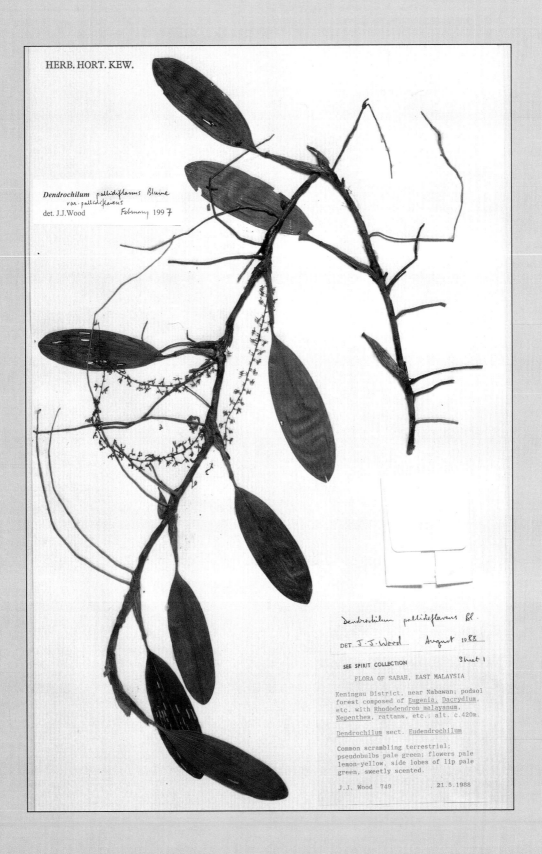

HERB. HORT. KEW.

Dendrochilum pallidiflavens Blume
var. pallidiflavens
det. J.J.Wood   February 1997

Dendrochilum pallidiflavens fl.

DET. J.J.Wood   August 1988

Sheet 1

SEE SPIRIT COLLECTION

FLORA OF SABAH, EAST MALAYSIA

Keningau District, near Nabawan; podsol
forest composed of Eugenia, Dacrydium,
etc. with Rhododendron malayanum,
Nepenthes, rattans, etc.; alt. c.420m.

Dendrochilum sect. Eudendrochilum

Common scrambling terrestrial;
pseudobulbs pale green; flowers pale
lemon-yellow, side lobes of lip pale
green, sweetly scented.

J.J. Wood 749   21.5.1988

Plate 21

Dendrochilum pallidiflavens Blume var. *pallidiflavens. Carr* C. 3186, SFN 26759 (SING), collected in Sabah, showing broad, elliptic mature leaves; previously refered to *D. conopseum*.

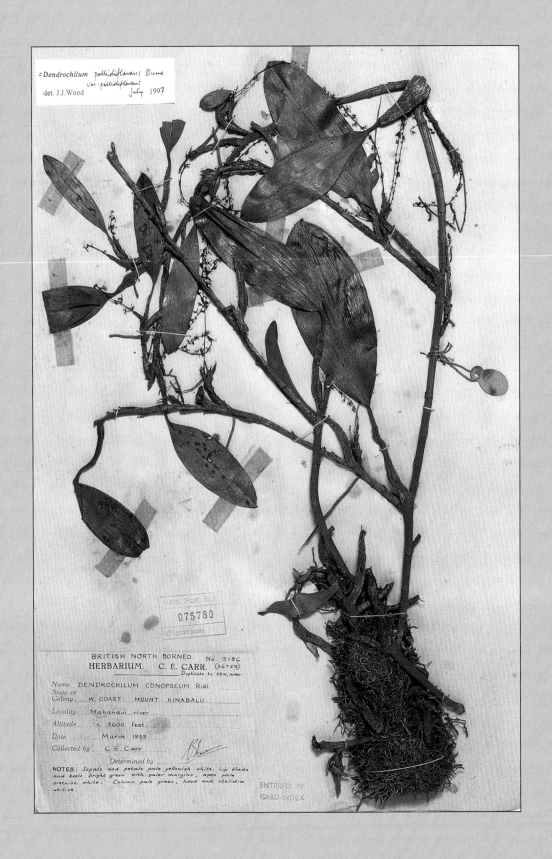

=*Dendrochilum pallidiflavens* Blume
var. *pallidiflavens*
det. J.J.Wood   July 1997

BRITISH NORTH BORNEO.   No. 3186
HERBARIUM.   C. E. CARR.   (26759)
Duplicate to KEW, AMES

Name DENDROCHILUM CONOPSEUM Ridl.
State or
Colony   W. COAST   MOUNT KINABALU
Locality   Mahandui river
Altitude   c. 3600 feet.
Date   March 1933
Collected by   C. E. Carr
          Determined by
NOTES: Sepals and petals pale yellowish white. Lip blade
and keels bright green with paler margins, apex pale
greenish white. Column pale green, hood and stelidia
whitish.

Plate 22

Dendrochilum pallidiflavens Blume var. *pallidiflavens. Native collector* in *Synge* S.
175 (K), collected in Sarawak, showing narrow, elliptic mature leaves;
previously refered to *D. brevilabratum* (Rendle) Pfitzer.

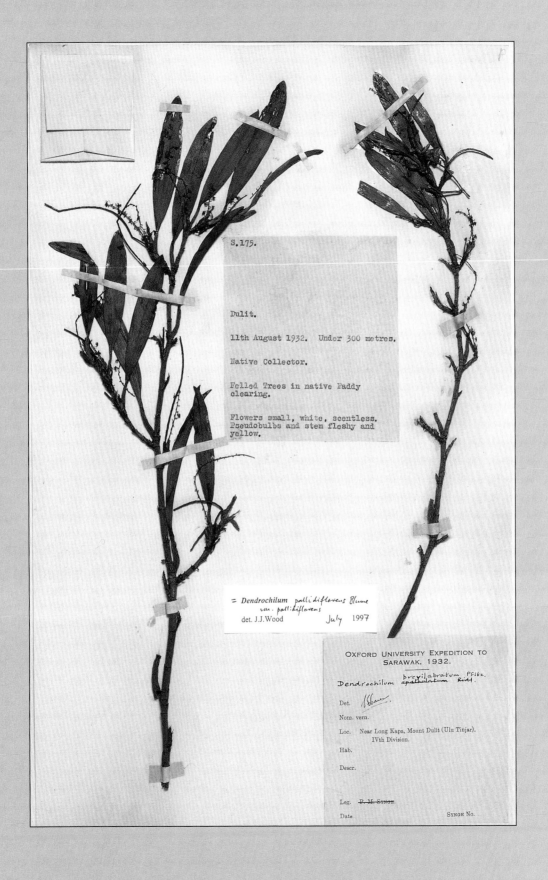

S.175.

Dulit.

11th August 1932.   Under 300 metres.

Native Collector.

Felled Trees in native Paddy clearing.

Flowers small, white, scentless. Pseudobulbs and stem fleshy and yellow.

= *Dendrochilum pallidiflavens* Blume
var. *pallidiflavens*

det. J.J.Wood   July 1997

OXFORD UNIVERSITY EXPEDITION TO
SARAWAK, 1932.

Dendrochilum brevilabratum P.Fib.
epathulatum Ridl.

Det.

Nom. vern.

Loc.   Near Long Kapa, Mount Dulit (Ulu Tinjar).
IVth Division.

Hab.

Descr.

Leg.   P. M. Synge.

Date   SYNGE No.

Plate 23

Dendrochilum pallidiflavens Blume var. *pallidiflavens. Nooteboom* 1334 (L), collected on Mount Trus Madi in Sabah. A broad-leaved variant showing abundant roots produced from nodes.

MALAYSIA, SABAH H. P. Nooteboom 1334
17-3-1969

Fam. Orchideae
Nom. : Dendrochilum

Locality: Ingaran (Between Tambunan
and Mnt Trusmadi) 5°38' N
Alt. 800 m. 116°28' E
Habitat:

Notes: epiphyte, same genus as 1203

expedition sponsored by WOTRO, the Hague

Dendrochilum pallidiflavens Blume
v.m. pallidiflavens
det. J.J.Wood July 1997

200500

Plate 24

Dendrochilum pallidiflavens Blume var. *pallidiflavens*. *J. & M.S. Clemens* 40633 (L), collected in Sabah; previously refered to *D. conopseum* Ridl.

PLANTS OF MOUNT KINABALU
B. N. Borneo
No 40,633   Oct. 9, 1933
J. & M. S. CLEMENS
Dendrochilum
Penibukan. Jungle ridge. Elev.
4,000 ft.
Fl. Cream yellow
Upper Kinabalu

INDEX

Families and genera are printed in CAPITALS. Accepted names are in roman type. Synonyms are in *italics*. Numbers in **bold** refer to where taxa are formally enumerated and figured.

Other titles by *Natural History Publications (Borneo)*

For more information, please contact us at

Natural History Publications (Borneo) Sdn. Bhd.
A913, 9th Floor, Wisma Merdeka
P.O. Box 15566, 88864 Kota Kinabalu, Sabah, Malaysia
Tel: 088-233098 Fax: 088-240768 e-mail: chewlun@tm.net.my

Head Hunting and the Magang Ceremony in Sabah *by* Peter R. Phelan

Mount Kinabalu: Borneo's Magic Mountain—an introduction to the natural history of one of the world's great natural monuments *by* K.M. Wong & C.L. Chan

Enchanted Gardens of Kinabalu: A Borneo Diary *by* Susan M. Phillipps

A Colour Guide to Kinabalu Park *by* Susan K. Jacobson

Kinabalu: The Haunted Mountain of Borneo *by* C.M. Enriquez (Reprint)

Discovering Sabah *by* Wendy Hutton

National Parks of Sarawak *by* Hans P. Hazebroek and Abang Kashim Abg. Morshidi

A Walk through the Lowland Rainforest of Sabah *by* Elaine J.F. Campbell

In Brunei Forests: An Introduction to the Plant Life of Brunei Darussalam *by* K.M. Wong (Revised edition)

The Larger Fungi of Borneo *by* David N. Pegler

Pitcher-plants of Borneo *by* Anthea Phillipps & Anthony Lamb

Nepenthes of Borneo *by* Charles Clarke

Nepenthes of Sumatra and Peninsular Malaysia *by* Charles Clarke

The Plants of Mount Kinabalu 3: Gymnosperms and Non-orchid Monocotyledons *by* John H. Beaman & Reed S. Beaman

The Plants of Mount Kinabalu 4: Dicotyledon Families Acanthaceae–Lythraceae *by* John H. Beaman, Christiane Anderson & Reed S. Beaman

Slipper Orchids of Borneo *by* Phillip Cribb

The Genus Paphiopedilum (Second edition) *by* Phillip Cribb

The Genus Pleione (Second edition) *by* Phillip Cribb

Orchids of Sumatra *by* J.B. Comber

Orchids of Sarawak
by Teofila E. Beaman, Jeffrey J. Wood, Reed S. Beaman and John H. Beaman

Gingers of Peninsular Malaysia and Singapore
by K. Larsen, H. Ibrahim, S.H. Khaw & L.G. Saw

Mosses and Liverworts of Mount Kinabalu
by Jan P. Frahm, Wolfgang Frey, Harald Kürschner & Mario Manzel

Birds of Mount Kinabalu, Borneo *by* Geoffrey W.H. Davison

The Birds of Borneo (Fourth Edition)
by Bertram E. Smythies (Revised by Geoffrey W.H. Davison)

The Birds of Burma (Fourth Edition)
by Bertram E. Smythies (Revised by Bertram E. Smythies)

Proboscis Monkeys of Borneo *by* Elizabeth L. Bennett & Francis Gombek

The Natural History of Orang-utan *by* Elizabeth L. Bennett

The Systematics and Zoogeography of the Amphibia of Borneo
by Robert F. Inger (Reprint)

A Field Guide to the Frogs of Borneo *by* Robert F. Inger & Robert B. Stuebing

A Field Guide to the Snakes of Borneo *by* Robert B. Stuebing & Robert F. Inger

The Natural History of Amphibians and Reptiles in Sabah
by Robert F. Inger & Tan Fui Lian

Marine Food Fishes and Fisheries of Sabah *by* Chin Phui Kong

Layang Layang: A Drop in the Ocean
by Nicolas Pilcher, Steve Oakley & Ghazally Ismail

Phasmids of Borneo *by* Philip E. Bragg

The Dragon of Kinabalu and other Borneo Stories *by* Owen Rutter (Reprint)

Land Below the Wind *by* Agnes N. Keith (Reprint)

Three Came Home *by* Agnes N. Keith (Reprint)

White Man Returns *by* Agnes N. Keith (Reprint)

Forest Life and Adventures in the Malay Archipelago *by* Eric Mjöberg (Reprint)

A Naturalist in Borneo *by* Robert W.C. Shelford (Reprint)

Twenty Years in Borneo *by* Charles Bruce (Reprint)

With the Wild Men of Borneo *by* Elizabeth Mershon (Reprint)

Kadazan Folklore (*Compiled and edited by* Rita Lasimbang)

An Introduction to the Traditional Costumes of Sabah
(*eds.* Rita Lasimbang & Stella Moo-Tan)

Bahasa Malaysia titles

Manual latihan pemuliharaan dan penyelidikan hidupan liar di lapangan
oleh Alan Rabinowitz (*Translated by* Maryati Mohamed)

Etnobotani *oleh* Gary J. Martin (*Translated by* Maryati Mohamed)

Panduan Lapangan Katak-Katak Borneo *oleh* R.F. Inger dan R.B. Stuebing

Other titles available through
Natural History Publications (Borneo)

The Bamboos of Sabah *by* Soejatmi Dransfield

The Morphology, Anatomy, Biology and Classification of Peninsular Malaysian Bamboos *by* K.M. Wong

The Plants of Mount Kinabalu 1: Ferns and Fern Allies
    by B.S. Parris, R.S. Beaman & J.H. Beaman

The Plants of Mount Kinabalu 2: Orchids *by* J.J. Wood, R.S. Beaman & J.H. Beaman

Forests and Trees of Brunei Darussalam (eds. K.M. Wong & A.S. Kamariah)

Rafflesia: Magnificent Flower of Sabah *by* Kamarudin Mat Salleh

The Theory and Application of A Systems Approach to Silvicultural Decision Making
    by Michael Kleine

Orchids of Borneo Vol. 1 *by* C.L. Chan, A. Lamb, P.S. Shim & J.J. Wood

Orchids of Borneo Vol. 2 *by* Jaap J. Vermeulen

Orchids of Borneo Vol. 3 *by* Jeffrey J. Wood

Orchids of Java *by* J.B. Comber

A Checklist of the Orchids of Borneo *by* J.J. Wood & P.J. Cribb

A Field Guide to the Mammals of Borneo *by* Junaidi Payne & Charles M. Francis

Pocket Guide to the Birds of Borneo *Compiled by* Charles M. Francis

The Fresh-water Fishes of North Borneo *by* Robert F. Inger & Chin Phui Kong

The Exploration of Kina Balu *by* John Whitehead (Reprint)

Kinabalu: Summit of Borneo (eds. K.M. Wong & A. Phillipps)

Common Seashore Life of Brunei *by* Marina Wong & Aziah binte Hj. Ahmad

Birds of Pelong Rocks *by* Marina Wong & Hj. Mohammad bin Hj. Ibrahim

Ants of Sabah *by* Arthur Y.C. Chung

Traditional Stone and Wood Monuments of Sabah *by* Peter Phelan

Borneo: the Stealer of Hearts *by* Oscar Cooke (1991 Reprint)

Maliau Basin Scientific Expedition (eds. Maryati Mohamed, Waidi Sinun, Ann Anton, Mohd. Noh Dalimin & Abdul-Hamid Ahmad)

Tabin Scientific Expedition (eds. Maryati Mohamed, Mahedi Andau, Mohd. Nor Dalimin & Titol Peter Malim)

Klias-Binsulok Scientific Expedition
(eds. Maryati Mohamed, Mashitah Yusoff & Sining Unchi)

Traditional Cuisines of Sabah

Cultures, Costumes and Traditions of Sabah, Malaysia: An Introduction

Tamparuli Tamu: A Sabah Market *by* Tina Rimmer